The Chip War

The Battle for the World of Tomorrow

By FRED WARSHOFSKY

Charles Scribner's Sons / New York

Charles Scribner's Sons
Macmillan Publishing Company
866 Third Avenue, New York, NY 10022
Collier Macmillan Canada, Inc.

Library of Congress Cataloging-in-Publication Data
Warshofsky, Fred.
The chip war.
Bibliography: p.
Includes index.
1. Integrated circuits industry.   2. Competition,
International.   I. Title.
HD9696.I582W37   1989   338.4'762138173   88-30537
ISBN 0-684-18927-5

Macmillan books are available at special discounts for bulk purchases
for sales promotions, premiums, fund-raising, or educational use.
For details, contact:

Special Sales Director
Macmillan Publishing Company
866 Third Avenue
New York, NY 10022

10 9 8 7 6 5 4 3 2 1

Designed by Jack Meserole

Printed in the United States of America

To C.H.

CONTENTS

ACKNOWLEDGMENTS

This book is the product of more than a year of personal research in nine countries on three continents. In the course of that research I have interviewed more than a hundred people; executives, scientists, engineers, technocrats, government officials, workers, educators, students, and other journalists. I have been led into areas that at times seemed as complex and labyrinthine as the chips themselves. The diplomatic and often secret maneuvering of trade officials seeking to open or protect markets, the high-tech espionage by nations and corporations, national literacy levels, and even the use of cosmetics by female workers are all factors in the chip war that make the complexities of engineering design and chip manufacture seem simple by comparison.

Sorting this all out would have been impossible without the help of many people who cheerfully shared their own specific knowledge and contacts with me, who guided me through the almost arcane technology of microelectronics and the equally intricate paths of a global industry that influences and affects every aspect of our civilization today.

Arranging for interviews half a world away with busy scientists and executives who spoke little if any English was a logistical challenge worthy of our chip-based information age.

Among the many people who helped me solve these problems was Akio Akagi, a longtime friend who is the science correspondent for NHK, the Japanese national television network. Aki was indispensable. He not only helped set up a number of meetings with key Japanese scientists, government officials, and semiconductor company executives but also served as my interpreter at many meetings and interviews. Even more important were his insights and counsel on Japanese scientific, industrial, and social institutions.

There were, of course, many other people who provided advice, encouragement, information, and assistance during the research and writing of this book. To all of them I say thank you.

The Chip War

1

Infinity in a Grain
of Sand

On a chilly, sunny February afternoon in 1986, in one of the corporate compounds just north of the New York City line in suburban Westchester County, an IBM executive named Sanford L. Kane was preparing for a meeting. Kane was then forty-three years old, a vice-president of the General Technology Division, that part of IBM that develops and produces more than half of all of the semiconductor chips the giant computer company uses. Carefully, Kane packed his attaché case with 35 eight-by-ten inch foils, as they are known in the computer industry, transparencies to be used in overhead projectors. They formed the heart of a presentation he was to deliver in the stratosphere of executive suites, a meeting of the IBM management committee, chaired by the chief executive officer, John Akers.

The news Kane had to deliver was bad. Worse than bad; it was dire. He had started this project at the direction of his boss, the then president of the division, Dr. Paul R. Low. Low was concerned with the ability of the American semiconductor industry to maintain its technological leadership in the face of increasing Japanese competition. Kane had done an extensive research analysis of the situation and was about to present it to IBM's most important executives. The presentation was, barring a sudden and drastic change in the course and direction of the

1

industry, an obituary. And yet, unless Akers and the others on the committee had the foresight to recognize the message at a point in time when IBM's profits were soaring, its stock reaching for all-time highs, it might well be dismissed as alarmist hype on the part of a junior vice-president. Sandy Kane at this point had been with IBM for twenty-three years. An industrial engineer from NYU with a graduate degree in engineering administration from Syracuse University, he had joined the company right out of college. A big, almost hulking man, Kane has a delicacy of movement that is occasionally seen in outsize football players.

"I always liked to use my hands," he told me as we sat in his well-appointed office in Harrison, New York, with its ubiquitous computer screen staring from behind his shoulder. "When I have the time, I'm into woodworking; I build cabinets and furniture. I also like doing other things with my hands, like needlepoint. It's astounding how you can get into the greatest conversations with stewardesses on airplanes when you sit and do your needlepoint."

Kane was introduced to needlepoint by Roosevelt Grier, the ex–New York Giant tackle. "We met at a party given by a mutual friend in Westchester. This goes back fifteen years. He was already into needlepoint in a big way, designing and selling. You have to see this man; he is just large in every dimension. This guy sitting there doing needlepoint is an image that's just hard to forget. We got to talking about it, and I had learned to sew when I was young, but I had just never tried needlepoint. Rosie sent me one of his designs, and I did it and I loved it, and I've done several since. I have this tapestry that's about three feet by five feet, and I do it on airplanes, which is the only time I get to work on it now, and I'm about half-done.

"My mother taught me to sew, and I just wanted to learn. If somebody needs a button sewn on at home, they come to Dad. I was born in Queens and spent most of my life growing up in Forest Hills, where my parents still live. My wife is originally from the Bronx, had just moved to Forest Hills when I met her. I was a sophomore at the time at NYU, and she was a freshman."

After his graduation from NYU in 1962, Kane went to work for IBM in Endicott, a small city in western New York, about an hour's drive from Scranton, Pennsylvania. Here, in one of its oldest laboratories and manufacturing centers, IBM develops and builds its mid-range computers and peripheral devices such as printers. For three years Kane worked in a variety of engineering jobs in Endicott. "I liked what I was doing, but I was born and raised in New York City, and I hated Endicott with a passion. I was concerned that if I did as well on the job for the rest of my career as I was doing in those three years, I could see myself spending the rest of my life in Endicott. Maybe I could have been a successful Endicottian, but I would not have been a very happy person, and my wife would have divorced me."

It took another two years, but still happily married, Kane left Endicott to work on a variety of financial assignments at a number of IBM locations in the Northeast. "I worked in business policy and financial planning, and I was the controller of the laboratory in Poughkeepsie for a couple of years." Then, in 1975, he was introduced to the semiconductor business when he became the controller of the Fishkill site near Poughkeepsie, for four years. Fishkill is one of IBM's major semiconductor manufacturing plants, turning out millions upon millions of the most advanced integrated circuits (ICs) produced anywhere in the world.

Now, twenty-three years into his career at IBM, Kane could only hope that the information contained in the statistics on his foils would be understood and, not incidentally, that the committee would not shoot the messenger for bearing the message. "We looked at the finances; we looked at the marketplace and who's selling what kind of products to whom. This was all done looking at the U.S. industry vis-à-vis the Japanese. We talked about product consumption and how that plays on the game, talked about R & D, talked about capital expenditures, and then, as a separate segment, we also talked about the equipment industry as a subset. All of the data in the presentation are publicly available information."

What it all added up to in Sandy Kane's opinion was nothing less than the ultimate demise of the American semiconductor industry. In its wake, Kane would tell the committee, lay disaster for IBM and, he believed, the United States.

At IBM's Armonk headquarters the six-man management committee, headed by John Akers, the president and chief operating officer (CEO), displayed enormous interest in the presentation. "It was an interesting session," Kane noted in an understated fashion. "The level of attention and interest in the subject was intense. There was a genuine concern on the part of everybody in the room not just in the subject but what its implications to the IBM company were—we talked about that at length—and then what do we do about it. And out of that meeting came a couple of things."

The first was an almost immediate recognition that here was truly a case where what was good for General Bullmoose was indeed good for the United States.

"And we came away with the idea that there's really two concerns that IBM has relative to the health of the U.S. semiconductor industry. And we have been since that time—since the February '86 time—very vocal about those; we haven't been shy about the concerns that we have. The first one is sort of an obvious one. It says that if there's no U.S. semiconductor industry, then we have no choice but to buy our semiconductors, of which we are one of the largest purchasers in the world, principally from Japan. We buy semiconductors today from Japan, and we don't object to buying them from Japan, but we like to be able to have a choice about where we buy, and we prefer to buy domestically. If we can buy the chips we need to use in Europe in Europe, for example, I'm better off that way. To a certain extent, what we need to use in Japan we can buy in Japan. And we like the idea of being able to buy what we need to use in the U.S. in the U.S. If we are precluded from that and have to buy them primarily in Japan, you focus on the fact that almost every single one of the major semiconductor manufacturers is also a vertically integrated systems manufacturer. So every time we send a dollar to Japan to buy a memory chip, or whatever else we buy, they're gonna turn that thing into a system, ship it back across the water, and compete with us here in the United States on our turf. In purely competitive business logic, that doesn't make a hell of a lot of sense.

"Okay, so you would prefer to have an alternative to that, which is to be able to buy what you want in the United States." But Kane's second reason was even more compelling. For in addition to being one of the largest, if not the largest, purchasers of semiconductors, IBM is also the largest manufacturer of semiconductors in the world. "That," he points out, "gives us a constant choice about what do we make and what do we buy. And invariably what we are buying on the outside is commodities, like soy beans or pork bellies. The stuff that makes a difference to us in terms of the leading edge, that gives us the capability of cost performance and price performance inside our box, that we make in our own facilities, in plants like East Fishkill, New York, in Burlington, Vermont, and in Japan and France and elsewhere in the world.

"But in order for us to be able to have the leverage, to be able to ship 3090s, our most advanced computers with the most advanced 1-megabit high-performance memory in them, we needed to produce them in Burlington. We couldn't have depended on any other company, whether it be Japanese or American, to do it for us, because our whole high-end strategy rested upon the performance required. And for that we had to make it inside. So our point is that anything that we want to have where the technology is buying something unique for that box or set of products, then we choose to make that inside. So we are not worried about our supplier, be they Japanese, Americans, or anyone else, holding us up, if you will, from the standpoint of the chip end of the equation.

"Now I move, however, to my second reason why we were concerned. Since we are a large manufacturer and since what we choose to make are in fact those chips that give us leverage, the way you do that technologically is to make sure that you are in fact pushing the state of the art of the technology. The only way you can push the state of the art of the technology is if you're also sure you're using leading-edge equipment. And we have ongoing relationships with most of the key manufacturers of equipment where we have joint-development contracts, or we have the availability of prototypes, or we get to see their machine on the line, and we know what direction they're headed in so that we can adjust our processes accordingly. And we take advantage of that.

"But we are convinced that if there is no U.S. semiconductor industry, it goes without saying there will be no equipment industry. There aren't enough IBMs in the United States, other captive manufacturers like us, to be a strong enough base to support the equipment industry. We're just not enough source of income to these guys, so they have to die away. In fact, my conviction is that once it becomes blatantly obvious that the U.S. semiconductor industry is going to go, the equipment industry will go first. . . . It won't be able to survive.

"What does that mean? That then means we're going to have nowhere to go but Japan to buy our tools. And since most of the

Japanese equipment makers are subsidiaries of our competitors, it is not likely that we will be able to get the sort of ongoing, long-term, close relationship with them that we enjoy here in the U.S. That to us is a much bigger threat, because it speaks right to our technological prominence. If we can't be sure that we can produce the most advanced chips we now make in Fishkill and Burlington and around the world today, then our capability for technological excellence and cost performance in our end products is threatened. That means it isn't any better than anybody else's. And that scares us.[1]

The source of all this concern on the part of Big Blue, of mighty IBM, the largest computer company in the world and one of the most powerful corporations in history, is the chip, so small it rests feather-like on the pad of my index finger. It is a tiny silvery wafer, etched with a minute pattern, less visible and far more complex than the whorls of skin on which it lies. It is a semiconductor, a silicon chip, made of one of the most common substances on earth—sand.

> To see a world in a grain of sand. . .
> Hold infinity in the palm of your hand. . . .

The mystic vision of William Blake has been fulfilled in a manner not even the most visionary dreamer might have imagined. Within those grains of sand that form the silicon chip and the transistors and microscopic circuits embedded in it lie the neurons of the modern world. They form the processing core and memory cells of every modern computer and electronic device, a maze of tiny transistors, diodes, capacitors, and other microscopic electronic components sealed in a silicon chip no larger than a newborn's thumbnail. Linked together in circuits on wafers of crystal, these chips store information, add and subtract, read and write text, play music, enable television cameras to see and compact disks to play noise-free music, run factory robots, and perform hundreds of other tasks.

In their countless billions they are the brain cells driving the modern information age, essential to the nation's economy and absolutely vital to national defense. They are the cornerstones of the $60 billion computer and electronics industry in this country, an industry that did not exist forty years ago. Chips not only run computers, medical systems, and equipment, advise bankers, stockbrokers, and salesmen, run power plants, telephone and telecommunications systems, televisions, VCRs, and Walkmans, but virtually all radar, fire-control systems, missiles, ships, fighter planes, bombers, and command-and-control systems that are needed to fight a war. And more uses are being found for them every day. And, in keeping with their commonplace origin, sand, their essential function is extraordinarily simple.

Every semiconductor is basically a tiny platform on which are mounted thousands, and in some cases millions, of tiny transistors, capacitors, diodes, resistors, and wires that connect them. The critical operation in the chip takes place in the transistor. Like spring-loaded gates, the transistors are either open, allowing electrons to flow, or closed, in which case the signal is blocked. Open or closed, on or off, these two states, represented in binary code as 1 or 0, are the "words" of computer thought. Possessed of no moving parts, generating almost no heat, consuming almost no power, it is as if individual grains of sand had been granted the power of thought and memory.

Without the right chips, Star Wars, the strategic defense initiative, is not only impossible but unthinkable. Without the ability to produce and sell chips, the American economic future is in doubt. At stake is a worldwide electronic-equipment market that reached $770 billion in 1988.[2] And since the chip is the brain cell of computers, telecommunications, and electronics, the winner of the Chip War will also dominate the anticipated $1 trillion plus "computronics" market and, with it, the world economy.

Advances in biotechnology, new energy sources, superconductivity, and other emerging and hitherto undreamed of technologies are dependent on advances in semiconductor technology. For the chip is far more important than the devices it makes possible. "For two generations, analysts have foreseen a new industrial revolution based on the processing of information," Michael Burros, James Millstein, and John Zysman, a trio of scholars at the Institute of International Studies at the University of California, Berkeley, wrote in a policy paper on the semiconductor industry prepared for the U.S. Congress in 1982.

Although the computer has been the symbol of this transformation, the semiconductor has in great measure been responsible for it. Innovation in and diffusion of semiconductor technology has helped to create markets in data processing, automated production, and robotics, and it has fundamentally altered communications, instrumentation, transportation, consumer goods and military systems. The videogame may be no more than a new diversion, but the automated factory portends fundamental changes in the organization of work. The semiconductor industry is therefore strategically vital to the future growth of knowledge-intensive industrial development within the U.S. economy.[3]

Not surprisingly, the greatest challenger to American leadership in the semiconductor industry, the Japanese, voiced almost the identical idea in the June 1982 *Japan Economic Journal*:

Semiconductors are very likely to determine the level of a country's computer, telecommunications, robotics, aerospace and other high technology industries in the future. The reason is that microchips now constitutes the "core" components of highly-sophisticated products. Some people call semiconductors 'the crude oil of the 1980s.'

The very ubiquity of the chip confers an overarching importance on all aspects of its design, development, and production. The ability to design and produce chips creates with it a technological ripple effect, a synergy that stimulates R & D in other fields. The chip thus becomes the base upon which the entire economy not only rests but draws sustenance and vigor.

The ability to make these chips cheaper, faster, and better than anyone else will determine the winner of the Chip War and the ultimate controller of the world's economy. With so much at stake, the United States has suddenly awakened to the fact that it is in an economic and technological war. The prize is control of an industry and technology so central to modern civilization that the winner will unquestionably be the strongest economic power in the world. Ironically, this is a war in which we find ourselves struggling against virtually everyone—our allies, our enemies, and even ourselves. From Japanese dumping of chips to gain new markets to Soviet industrial espionage to steal American technology to U.S. government policies that consistently strip competitive advantages from American producers, the Chip War is pursued relentlessly by friend and foe.

"The Japanese," says MIT's Charles H. Ferguson, "are moving toward control of semiconductors and computers." Ferguson, a one-time IBM executive turned academic crusader, conducts a tireless campaign to alert politicians, business leaders, and the media to what he considers a deadly threat to a seminal American industry. He writes and lectures and testifies to congressional committees, industry conferences, magazines and newspapers, about the Japanese threat to American dominance in the semiconductor and computer industries. "This will be a watershed in geopolitical relations as well as in Japanese industrial power," he warns. "For the first time the U.S. will no longer dominate the critical technologies needed for military power and future industrial development."[4]

Jack Clifford, director of the Department of Commerce's Office of Microelectronics and Instrumentation, has been a keen observer of the semiconductor industry for twenty-five years. In his cluttered office in the cavernous Commerce Department building near the Washington Monument, he told me of some other worrisome considerations. "Semiconductors play such a significant role in all other advancing technologies, in avionics, in computers in telecommunications, it is a key product in areas where we still hold technological leadership. The United States is still a good instrument manufacturer, we still produce a lot of good medical electronic equipment, computers, telecommunications, we are still a strong competitor in many of those areas. But if we lose the semiconductor and especially the integrated-circuit production base here, then we are going to lose it in all those other areas."[5]

The loss of American dominance has been as sudden as it has been shocking. "We had been ahead for so long that when everything began to fall apart no one in the Government was ready to listen," said Andrew Grove, the chief executive officer of Intel Corporation, one of the nation's largest chip makers.[6] The United States, which as recently as five years ago dominated the world market in semiconductors, is faced with the unpleasant reality that it is no longer able to meet the needs of its own marketplace and may soon be forced out of the business altogether. The impact on our economy has already been severe, plunging the American computer and electronics industries into a steep depression. Losses by American chip makers were in excess of $2 billion in 1985 and 1986, and more than 25,000 people were laid off.[7] The situation brightened considerably in 1987 and 1988, partly due to trade sanctions against Japan, the lower dollar, and a renewed surge in the sale of personal computers (PCs). But the sudden upturn, which has heartened industry leaders, carries the seeds of caution as well as hope for the future. The semiconductor industry is notoriously cyclical, lurching from boom to bust with dramatic frequency. In the depressed years of 1985 and 1986, memory chips were dumped at prices as low as $1.25 for a 256K chip. In 1988 those same chips are in such demand, prices have soared to twelve dollars each.[8] The most advanced memory chips now in wide use, the 1-megabit DRAM (dynamic random access memory) which sold for sixteen dollars at the end of 1987, had by March of 1988 jumped to sixty dollars a chip.[9] Part of the shortage has been caused by cutbacks in production by the Japanese, in response to trade agreements with the United States, and by increased demand by U.S. computer makers.

The shortage of memory chips are but one example of the increasingly complex future facing the semiconductor industry, wherein the rules of the game have been forever changed. New competitors from the Pacific Rim, the so-called NICs, or newly industrialized countries, such as South Korea, Taiwan, Singapore, Malaysia, and Hong Kong, are now producing chips and will in the next few years become major players in this critical game. In Europe, what had been a largely moribund industry is gaining new vigor and energy and is harnessing the capital and cooperation needed to become yet another factor in the Chip War.

But it is the Japanese who have become and continue to be the major threats to American dominance in the semiconductor industry. For a decade now, the Japanese have been accused of dumping chips, of predatory pricing, of stealing American chip designs and technology. But the Japanese also work extraordinarily hard and apply enormous skills and a highly educated, highly motivated work force to designing, producing, and selling their chips. And in the process they have become a major threat to American dominance not only in semiconductors but in computers, semiconductor manufacturing equipment, medical electronics, and a host of other fields. In some, such as consumer electronics, they have actually eliminated American competition entirely. The sad fact is that there is no longer a viable American consumer-electronics industry. We produce no VCRs, no Walkmans, no compact-disk players. There is only one American maker of color TVs, and it is losing vast sums of money and is trying desperately to sell out.

But consumer electronics and other civilian-sector markets are essential to continued growth and development of the semiconductor industry. U.S. military needs and purchases once set the pace of developments in electronics, but that is no longer the case. Military requirements now lag behind civilian capabilities, and according to the National Research Council (NRC), "Even programs designed to accelerate the pace such as the Very High Speed Integrated Circuit (VHSIC) program, barely keep up with the civilian world."[10]

In fact, the Japanese are leading the Americans in many areas, producing chips that are often not just cheaper but better. Although Americans invented the transistor and the IC and at one time, barely fifteen years ago, produced much of the entire world's supply, U.S. semiconductor technology has suffered dramatic reverses to the Japanese. In thirteen different categories of semiconductor manufacture, concluded the Defense Science Board, the United States leads in only two, the Japanese in nine, and we are holding our own in only one area and are actually ahead in just one field.[11]

"Japanese firms are reputed to be leading most U.S. merchant semiconductor companies in developing reliable, low-cost, 1-megabit dynamic random access memory DRAM chips and in the early development of 4-megabit designs," reported the National Academy of Science's (NAS) Panel on the Impact of National Security Controls on International Technology Transfer. The report continued:

And Japanese companies now are the only source of the highest-quality fused quartz glass required for mass producing state-of-the-art chips of all types. Japan also rivals U.S. capability in semiconductor production equipment technology. The erosion of traditional U.S. semiconductor technology has occurred almost entirely within the last 5 to 10 years.[12]

Even quantitatively, the once numerical superiority of U.S. firms has been sharply reduced. Six of the world's ten largest open-market producers of semiconductors are Japanese.[13] Notes MIT's Charles Ferguson:

Less widely appreciated, but probably equally important, is the concomitant and similar decline of U.S. capital equipment, materials and services technology. Over the past decade, the Japanese equipment industry's world market share has more than doubled to over 30 percent, primarily at the expense of U.S. firms. Moreover, Japanese suppliers have reached parity or even superiority in major technologies, including packaging, automated assembly equipment, various ultrapure materials, some categories of fabrication equipment and specialized procedures such as mask making. For example, Hoya and Shin-Etsu now hold 90 percent of the world market for mask-quality glass and quartz; IBM's new East Fishkill facility is being built by Shimizu; and Japanese firms supply nearly half of Intel's masks.[14]

Under the onslaught of Japanese and other Asian chip makers in Korea and Taiwan, and the Europeans, the U.S. semiconductor industry has lost 20 percent of the world market in the last four years. In 1980, for example, U.S. high tech produced a trade surplus of $27 billion. But in 1986 high tech became part of the American trade deficit—$2 billion of the staggering $170 billion import burden.[15] Just how severe and rapid the decline can be seen in one of the most important and certainly the largest-volume segment of semiconductor production, advanced memory chips. In 1975 the American share of the world market was virtually 100 percent. Today we have managed to hang on to only 42 percent of the total memory-chip market. What is even more disturbing is that in the new generation of chips, the 1-megabit DRAMs, for example, the United States has only 5 percent of the market, a market that will grow significantly in the future as more and more systems are designed to take advantage of the greater memory capacity.[16]

America's high-tech industries, which were to be the dynamic engine that would propel us into the Information Age and replace our creaking smokestack economy with trade surpluses, seems destined to follow steel, automobiles, television sets, and other once robust industries into decline.

Most of the efforts currently being made to win the Chip War, which seem largely to center around protectionism, government bailouts in the form of Defense Department contracts, and research funds and joint ventures with the Japanese, are not likely to be successful. Even one of the most hopeful and highly hyped efforts, SEMATECH, the Semiconductor Manufacturing Technology Institute, a joint industry–government-funded consortium that will develop new manufacturing technologies, methods, and equipment to increase the U.S. company's ability to compete with the Japanese and the rest of the world, took more than eight months after its formal incorporation to find a president and is facing other delays in starting up.

Even when fully geared up and fulfilling the most optimistic of hopes held for it, SEMATECH will not by itself restore the American semiconductor industry to its once dominant position in the world. For the problems that plague the semiconductor industry are in many ways the problems that have afflicted American industry in general for many years.

Once the envy of the entire world, American industry is now second-rate in many areas. Even the once self-perceived best-in-the-world reputation of many of our products may have been arrogant self-delusion. "U.S. companies overall never have known what top quality is or produced it," Professor Martin Starr, director of the Columbia University Graduate School of Business Center for Operations, told me in his cramped office on the Columbia campus. "Other than computers, aircraft, and a few other products, the United States rarely produced world-class quality. And the reason for that is as soon as U.S. products reached a level of acceptability, that was considered good enough. By contrast, the Japanese are always asking for something better."[17]

The simple fact is that the United States has lost its competitive edge. The process of decline has been steady and long-term. "We used statistical quality control to produce an acceptable product," says Professor Starr, an engineer turned teacher whose focus is clearly on the plant floor where the products are produced. "But we never had the drive to produce fewer and fewer defects. We always assumed that fewer defects inherently cost more. We stopped even thinking about quality. The consumer became an adversary. Consumer complaints were treated as enemy propaganda. Companies did not stand up for their products, and they in essence said, 'Screw you,' to the consumer."

The problem is not new. "By 1970," says Professor Starr," we had picked up data that indicated really basic reversals in productivity that were long-term, that had started about 1963, and they kept varying year by year in terms of our export-import relationships. All kinds of supporting data were appearing, and we were able to trace these productivity declines right down to the plant floor. There was definitely something wrong, but we were having trouble determining just what it was. There seemed to be two components: One was that things were changing, things that had gone from right to wrong, and the other was things that were wrong in the first place. They were beginning to show up for reasons that had to do with aging of equipment, decisions on the part of management to really accent marketing in the sixties, decisions to move from marketing into finance, to change who was really running the strategic aspects of the business.

"In addition, we didn't watch our competitors at all, that is, our foreign competitors, and we almost never do. Take the Zenith Radio case. In 1962, Zenith Radio faced a decision as to whether they wanted to automate their circuit lines or to continue with hand wiring. They were looking at RCA and GE and assuming they were the competition. Zenith opted for a marketing slogan. They said, 'We build quality in; we do hand wiring.' But they didn't see that Sony was down there in Japan developing printed-circuit technology. Zenith's management thought it would take the Japanese ten years to bring it to an appropriate level of reliability. It was, in fact, just a couple of years before printed circuit boards were on the production line. As a result, Sony simply took over the transistor radio business not only from Zenith but from GE and RCA as well."

American chip makers began hearing words of warning long before the Japanese began to dominate the memory-chip market. On April 14, 1981, in a speech to the Electronic Industries Association, William J. Weisz, vice-chairman and chief operating officer of Motorola, must have seemed like a high-tech Cassandra when he said, "Unfortunately, some in America were complacent; not sensitive to what was going on—not listening to customers who reported how well the Japanese competitors were doing. The signs were ignored. Many preferred to believe that it was just a lot of propaganda. I don't think anyone in our industry is complacent anymore."

Perhaps not, but in the seven years since Weisz's warning, the competition has become even more fierce. American chip makers, hitherto indifferent, are now keeping an eye on the Koreans, the Taiwanese, Singaporeans, and the other NICs that have made significant moves into the memory markets. Nor can the Europeans be ignored. Major companies such as Philips and Siemens have always laid claim to a minor but meaningful portion of the market, and recently two of the major European firms, the Italian chip maker SGS and the French electronics firm of Thomson have merged. All are determined to do battle in the Chip War.

And all seem determined to follow the Japanese example of working harder and waiting longer for profits, an approach we have largely disdained. "We were doing very well as long as the game was being played on our court with our rules, our balls and at our levels of proficiency," notes Columbia's Professor Starr. "But here comes somebody who says, 'I play the same game, with the same balls, but I have this new racquet and this new technique. And I've been practicing a lot, and I'm a perfectionist. And whereas those guys practice three hours a day and then go to the clubhouse for a drink, we practice twelve hours a day, and we do it seven days a week. And we learn everything there is to know about the game.'"

Americans do not. We haven't yet as a nation decided that we, too, want to pay the price to be world-class. Despite an astonishing energy and creativity in a number of areas, such as science, agriculture, the arts, military weaponry, computers, finance, etc., we seem to be spread awfully thin. There is in this country a huge gap between the thinkers and achievers, the entrepreneurs and professionals, and the majority of workers. There are vast pockets of poverty and illiteracy as well as goof-offs and don't-give-a-damn work habits that have in too many cases replaced the "good day's pay for a good day's work" ethic of an America that seems long past. Indeed, it is evident that the problems besetting America's semiconductor industry are a microcosm of American industry in general. And what is even more troubling is that America's industrial woes represent a complex of deep-seated problems that afflict our entire society.

"We've made education easier and easier, and our system is in decline," says Professor Starr. "The Japanese have done the reverse. Their expectations at every level are higher; they spend many more hours per day and days per year in school. Maybe their system does stultify innovative capability, as some Americans would like to believe, but I don't know that is so."

The simple truth is that we stand in grave danger of losing a seminal technology that carries with it profound geopolitical and economic consequences for the American future. That cannot be allowed to happen, for what is at stake in the Chip War is more than the possible loss of yet another industry to the Japanese. The very future of America as a great nation may be the ultimate prize.

The dynamism that led to the creation of the semiconductor industry still exists. The creativity of America's engineers and scientists remains as a major strength of the industry. How that creativity and drive are applied to the structural problems of the industry and the competition from abroad will determine the American future.

That future was once so bright that it exerted an irresistible attraction for the young engineers who created an industry from the impetus of a single technological breakthrough—the invention of the transistor. One need only look at the stunning rise and decline of Silicon Valley as the center of the semiconductor universe and see within its brief, incandescent course in brilliant start-ups and failed companies, in swaggering corporate giants suddenly overwhelmed by even larger companies, a historical parallel with almost every civilization from Sumer to Rome, from Islam to the British Empire. Indeed, we may also be witnessing in microcosm a course that might unfortunately be that of the American civilization.

For the Chip War is ultimately a worldwide struggle for dominance of an industry and technology that may well determine the geopolitical and economic leadership of the twenty-first century.

2

Silicon Valley Days

Perhaps more than anyone else in the world, Americans are devoted to hyperbole. We do not simply have baseball stars; we have superstars. People no longer give 100 percent, but 110 percent. Boxes of detergent are not merely large; they are giant. Olives are not big; they are colossal, and so on. But if ever there was a fitting use of hyperbole, it is in describing the incredible pace of semiconductor technology since the invention of the transistor in 1947. Alex Osborne, a onetime boy wonder whose computer company flashed cometlike across the Silicon Valley horizon, observed in 1979 that if transportation technology had progressed from the stagecoach to the Concorde as swiftly as electronics technology since the transistor, the Concorde would be able to "carry half a million passengers at twenty million miles per hour." Equally dramatic have been the precipitous reductions in the cost per unit of performance. According to Osborne, "a ticket for a Concorde flight would have to cost less than a penny if it were to compare with the rate at which microelectronics has gotten cheaper."[1]

Those are stunning achievements for the transmuted grains of sand that came into being a scant forty years ago with the invention of the transistor at Bell Laboratories in 1947. It came at a time when the computer had seemingly been pushed by wartime research to its outer limits.

To fully understand the dimensions of the invention, one needs a vantage point closer to World War II than to the Chip War. The computer then was an arcane, even forbidding device. It had grown from ideas and dreams that began with the ancient Greeks of somehow constructing a calculating machine. As early as 1642, the French philosopher Blaise Pascal built a mechanical adding machine that was the model for early twentieth-century versions. The idea of the modern computer, however, can be traced directly to the theories of a nineteenth-century English mathematician and engineer named Charles Babbage. Described by contemporaries as an "irascible sort," Babbage was the protégée of Ada, the countess of Lovelace and daughter of Lord Byron. Through this connection, the countess has often been hailed as the world's first computer programmer. The Defense Department paid homage to this myth by dubbing its recently developed universal computer language "Ada." Unfortunately for the romantically inclined, the esteemed countess's contributions were limited to encouraging Mr. Babbage and slipping him occasional sums of money. With this assistance, Babbage, who had already found a place for himself among the British scientific community by inventing the cow catcher, was able to pursue his less acceptable ideas for a machine that could perform mathematical computations. In 1822 he designed an all-purpose calculating machine he dubbed the Difference Engine. It was to be constructed of an extraordinarily complex arrangement of gears, cranks, levers, pulleys, wheels, and cogs, all to be driven by the then half-century-old steam engine. Punched cards, similar to those used to create the patterns in the Jacquard weaving loom, would carry the numbers to be calculated into the engine. The Difference Engine was followed by the Analytical Machine. In it, Babbage combined arithmetic functions with logic and could therefore not only perform computations but make decisions based on the results. Unfortunately, the Difference Engine and Babbage's even more grandiose concept, the Analytical Engine, were far too complicated and the tolerances demanded between parts, far beyond the technology of the time. But Babbage's theory was sound; it needed only a confluence of pressing need and burgeoning technological base to bring it into being.

Part of the base was mathematical, a new form of logic invented by a contemporary of Babbage's named George Boole. With it, Boole reduced all logical relationships to simple expressions such as And, Or, and Not. So-called Boolean algebra enabled mathematical functions to be expressed in binary numbers, 1 or 0. Thus, a simple switch or binary state of on or off could be used to represent any number. The electronic digits the computer would need were thus created.

A string of calculating machines and primitive computers based on mechanical and finally electrical energies began to emerge during the late nineteenth and early twentieth centuries. One of the most useful was invented by a statistician named Hans Hollerith of the U.S. Census Bureau. His punched-card machine tabulated the 1890 census in one-third the time of the previous hand-computed census.

But it remained for that most violent of all man's progenitors of invention, war, to bring the modern computer into being. Like radar, sonar, the jet engine, and a host of other dazzling technological breakthroughs of World War II, the midwife was the U.S. Department of Defense (DOD). The War Department, as it was then known, needed ballistics trajectory tables for every combination of gun, shell, and projectile the artillery fired. It was an impossible problem. To attempt a solution, the army's Ballistics Research Laboratory embraced an idea proposed by John Mauchly, then an assistant professor of physics, and J. Presper Eckert, at the time an engineering graduate student at the University of Pennsylvania's Moore School of Engineering. On June 5, 1943, the government signed a $400,000 contract with the University of Pennsylvania to develop the electronic numerical integrator and computer—ENIAC. Mauchly and Eckert had available something undreamed of in Babbage's day—the vacuum tube. An outgrowth of Edison's light bulb, the vacuum tube contained a metal plate or a zigzag wire instead of a filament inside the glass bulb and had the remarkable ability to receive and amplify radio signals. The vacuum tube could also act as an electrical switch, either on or off. Unlike the mechanical switches used until that time as electrical relays, the vacuum tube had no moving parts and could shift states from off to on or the reverse far faster, in millionths of a second.

One or two small computers using a few hundred vacuum tubes had already been built. Mauchly and Eckert proposed to build a machine out of 18,000 vacuum tubes. The size of a small house, it also contained 70,000 resistors, 10,000 capacitors, 6,000 switches, and miles of wires connecting the various components together. ENIAC was 100 feet long, 10 feet high, and 3 feet deep. It weighed thirty tons and gulped power like a fat man in an eating contest. When first fired off, it drew so much electricity that the lights of Philadelphia dimmed, or so the legend has it. But it was the 18,000 vacuum tubes, which could switch states so much faster than mechanical relays, that were both the key and the *bête noir* of the ENIAC. For the cucumber-shaped vacuum tubes generated vast amounts of heat. Despite dozens of fans and blowers stirring the air around the monstrous machine, the temperature in the ENIAC room would swiftly soar to 120 degrees Fahrenheit. The great heat would burn and blow out tubes by the hundreds. To keep the ENIAC functioning, teams of soldiers carrying peach baskets filled with replacement tubes darted in and out replacing blown vacuum tubes. Others were charged with the task of literally debugging the giant computer, chasing, swatting, and removing the thousands of moths fluttering around the myriad lights produced by the vacuum tubes.

With it all—the heat, the need to rewire all the miles of connections each time the ENIAC was asked to solve a different problem—it worked. This astonishing collection of glass, metal plates, and electrical wires solved mathematical equations in hours rather than the days previously required. But it was clear that only the gigantic resources of a government or giant corporation could keep the calculating monster running. Tube changing alone required a small army of men constantly on the run. Not even firms like Hollerith's Automatic Tabulating Company, which had become the International Business Machines Corporation after being acquired by Thomas Watson in 1924, was interested.[2]

But the vacuum tube was another story. Even before its debut in the computer, the vacuum tube had become the essential component of a new telephone switching network envisioned by physicist Mervin J. Kelly, then the director of research for the American Telephone and Telegraph Company's Bell Laboratories. In 1936 he left the parklike campus at Murray Hill, New Jersey, for a scouting trip to MIT in Cambridge, Massachusetts. There he met and hired William Shockley, a young physicist from Palo Alto, California. Shockley was finishing up his Ph.D. in what was then the rather esoteric field of solid-state physics.

When he arrived at Murray Hill, Shockley was shoved off into a training program in the vacuum-tube department. The vacuum tube was being designed to replace the mechanical switches the telephone exchanges of the time depended on. Kelly gave Shockley a pep talk on the need to do telephone switching electronically. The same problems of fragility, heat, and power consumption that were to prove so troublesome in the ENIAC were dogging the use of the vacuum tube in telephone circuits.

After a year of working with vacuum tubes, "I expressed a strong leaning to resume research in the field of my Ph.D. on the behavior of electrons in crystals," Shockley later said. "This research introduced me to Walter Brattain, later to become, along with John Bardeen, one of the coinventors of the point-contact transistor. With Brattain's help, I experimented with some very crude models that were total failures."

Brattain had been with Bell Labs since 1929 and had been exploring the electrical properties of semiconductor materials since 1931. World War II put a stop to the research. Shockley went on to work in the Pentagon and did not return to Bell Labs until 1945. John Bardeen, the third member of the team, came to Bell Labs at about the same time from the Naval Ordinance Laboratory. A postdoctoral fellow at Harvard, he had known Shockley at MIT during the thirties and, based on that student acquaintance, followed him to Bell Labs. Here, in the hectic aftermath of the war, space was at a premium, and Bardeen was squeezed into an office already occupied by Walter Brattain and another physicist named Gerald Pearson. Bardeen was fascinated by Brattain's experimental work on copper oxide, one of what was then an exotic group of semiconducting compounds. Semiconductor materials had been discovered in 1874 when a German physicist noted the peculiar flow of electrical current in certain minerals. Eventually these minerals were called "semiconductors" because they conduct current better than insulators like glass but not as well as good electrical conductors like copper. Furthermore, when certain semiconductors were held in contact with a point wire, they could detect the vibrations that radio waves produced in receiving antennas. This phenomenon became the basis for the "cat's whisker" detectors of the early crystal radio sets.

With Shockley and Bardeen as theoreticians and Brattain as the experimentalist, the group picked up on the war-postponed efforts to create a solid-state device that would amplify electronic signals. Shockley would explain the concept of amplification in this way: "If you take a bale of hay and tie it to the tail of a mule and then strike a match and set the bale of hay on fire, and if you then compare the energy expended shortly thereafter by the mule with the energy expended by yourself in the striking of the match, you will understand the concept of amplification."

Often Bardeen and Brattain would play a round of golf together when it seemed they were getting nowhere in the lab. The problem seemed to be centered on the materials being used to build the transistor. Brittle crystals such as germanium were formed from ordinary sand but were hard to shape, and their electrical conductivity seemed to change with no apparent reason in different samples of the same material.

Nor were the Bell Labs scientists the only ones working on semiconductor materials. Research teams at MIT, GE, Purdue, and Penn were all in the race. One of the prime targets was another crystal material called silicon. After oxygen, silicon is the most abundant element in the earth's crust. Most often found in combination with oxygen in the form of sand and quartz rocks, it is mined and refined for industrial uses. The Bell Labs researchers were finding silicon, even in its purified form, too erratic to provide the reliability needed for switching. In trying to obtain the purest silicon possible, scientists would melt the crystal and then pull ingots and study their behavior when exposed to electrical currents. They soon discovered there were two types of silicon, one dubbed "n" and the other "p," based on the positive or negative direction in which they allowed the current to flow.

Then came the discovery of a new type of silicon. It looked like every other dull black ingot but turned out to be very different. All other silicon had been either n type or p type. This new specimen was not homogenous; it was p in one section and n in another. Even more intriguing was a mysterious event that took place at the interface between the p and n sections. Here the black opaque crystal unexpectedly converted light into electricity. Connecting the silicon to a voltmeter and then shining a flashlight on the interface produced a measurable voltage, just as if it had been connected to a battery. And the wonders continued, for it was then discovered that the interface acted as a rectifier; that is, it converted alternating current into direct current by permitting the electrons to flow through in only one direction.

Brattain and Bardeen kept experimenting with silicon, but it was very brittle, crumbled easily, and was filled with impurities. Finally, they switched from silicon to germanium because the impurities in germanium were easier to control. Then, in December 1947, a pair of metal cat's whiskers, 2,000ths of an inch apart, were stuck into a slab of germanium. The crude transistor (the word was coined from the device's ability to *transfer resistance*) then demonstrated one of its remarkable properties that was to revolutionize communications. "This circuit," Brattain recorded in his laboratory notebook on December 24, the day after the first demonstration, "was actually spoken over, and by switching the device in and out, a distinct gain in speech level could be heard and seen on the scope presentation with no noticeable change in quality. . . . It was determined that the power gain was on the order of a factor of eighteen or greater." That power gain was the amplification they had sought for so long.

The three researchers were elated. In a car pool the next day, Walter Brattain told his fellow riders that he had "just taken part in the most important experiment I ever expect to do." Later, in his office, Brattain was horrified by his announcement and the next day swore his fellow riders to secrecy. When finally the news was announced seven months later, its significance was almost universally overlooked. The *New York Times* carried the story in its "News of Radio" column as a modest four-inch announcement that the Bell Telephone Laboratories had invented an electronic device to replace the vacuum tube. Despite being buried on the entertainment page, the transistor soon attracted the attention of the military. There were rumblings in the Pentagon of classifying the discovery, but executives of Bell Labs argued success-fully that it was in the public interest to release detailed information about the transistor. Bell executives were also keeping a wary eye on the Justice Department, which had begun to investigate AT&T's "monopolistic" manufacturing practices. And so a critical decision was made. The transistor would be given wide publicity in the technical world and be licensed to whoever was willing to pay the freight. At the same time, Bell engineers and scientists swiftly set about converting it from an experimental device into a commercial product. Covered with a protective tin cap with three slender legs poking out to connect it to the outside world, transistors were embedded in circuit boards and wired together. Taking advantage of its small size and amplification abilities, its first use was in the development of small, lightweight hearing aids. And, as a gesture to Alexander Graham Bell, the inventor of the telephone and a teacher of the deaf, hearing-aid manufacturers were granted a royalty-free license to the transistor.

Those charged a royalty, however, soon got their money's worth. In April 1952, Bell held a symposium for its licensees. "There was nothing new about licensing our patents to anyone who wanted them," recalled Jack Morton, vice-president of Bell Labs Electronic Components Division, sixteen years later. "But it was a departure for us to tell our licensees everything we knew. We realized that if this thing was as big as we thought, we couldn't keep it to ourselves and we couldn't make all the technical contributions. It was to our interest to spread it around. If you cast your bread on the water, sometimes it comes back angel food cake."[3]

As a result of these so-called cookbook sessions, there was more than enough cake to go around. By 1955 there were more than twenty U.S. companies producing transistors, most funded by $25 million in Defense Department contracts. Bell charged a license fee of $25,000 against royalties for its patents, but in 1954 the Justice Department filed an antitrust suit against AT&T. Rather than fight, Ma Bell signed a consent decree releasing all existing patents to U.S. companies free of all royalties. Only foreign companies had to pay the $25,000 license fee. In Europe, the Dutch electronics company Philips began building transistors, followed soon therafter by Siemens, the German electronics giant. On the other side of the world, the license was purchased by a small Japanese company called Sony. Sony used the transistor to make pocket-sized radios, which soon came to be called "transistor radios," after their semiconductor amplifiers.

Not only the transistor but the people who developed it were passing from the exclusive control of Bell Labs. Walter Brattain returned to his alma mater, Whitman College in Walla Walla, Washington, where he taught undergraduate physics until his death in 1987. John Bardeen also left Murray Hill to teach physics at the University of Illinois. William Shockley, however, wanted to make money. The place for that he believed was his old hometown of Palo Alto, California. Here, in an effort to reverse a perceived brain drain to the East, the assistant provost of Stanford University, an engineering professor named Frederick Terman, had set aside some of the vast tracts of land owned by the university for the creation of an industrial park. Among the companies located there was the Spinco Division of Beckman Instruments. Arnold O. Beckman, the founder of the company, had been Shockley's chemistry professor at Cal Tech. He happily put up the money for a new company called Shockley Semiconductor Laboratories.[4]

With his towering reputation, Shockley attracted several very bright young physicists and engineers away from East Coast firms. Among them was a preacher's son from Grinnell, Iowa, named Robert Noyce. Today Noyce is one of the most respected scientists and executives in the semiconductor industry. On July 27, 1988, he left his longtime post as vice-chairman of Intel, one of the major American merchant chip makers, to become the president of SEMATECH. A joint industry-government research consortium, SEMATECH is supposed to develop the new manufacturing technologies American chip makers need to challenge the Japanese. At the age of sixty, Noyce declared himself too old for the job, but took it on only after a fruitless search among industry leaders failed to turn up anyone else either willing or able enough to fill it. I believe he was moved by a deep sense of obligation to the industry and the nation. It's the sort of thing Bob Noyce does.

Among the technocratic royalty of Silicon Valley, Bob Noyce might be considered the philosopher king. He remains eminently approachable, and when I called from New York to arrange an interview, he not only answered his own phone but took the time to give me detailed directions to his office. At Intel he works in a vast labyrinth of office partitions, a maze that echoes the complicated geometries of one of his company's microprocessors.

The one concession to his status is that his office has a window, providing him with a gas jockey's eye view of the Intel parking lot and the freeway just beyond. On the shoulder-high portable walls of his thrift-shop modern office hang two awards of staggering implications: the National Medal of Science for his contributions to the semiconductor industry (he received a second Medal of Science from President Reagan a week after our meeting) and the Farraday Medal of the Institute of Electrical and Electronic Engineers (IEEE).

He is as informal as a pair of worn loafers and as quick to respond to the monumental problems that plague his industry as the electrons that whiz through the integrated circuits he invented. The plastic-topped table at which we sat inside the maze of cubicles is perhaps just another manifestation of his low profile, an "Aw shucks" sort of farm-boy philosophy that characterizes this headquarters building of Intel and reflects his small-town Iowa upbringing. From his open collar shirt to his open manner, Bob Noyce is a remarkable combination of brilliant intellect and down-home warmth. His personality is a plus, his intellectual brilliance a national treasure.

At the time the transistor was invented, Noyce was a student at Grinnell College in Iowa. "My physics professor out there had been a roommate of John Bardeen's," he recalled for me. "He was all excited; we were studying vacuum tubes and that sort of thing, and the idea of being able to get amplification without a vacuum was a revolutionary idea. I got interested in that field at that time, went on to graduate school, and did work in solid-state physics, which was the closest they could get to anything having to do with transistors, because there was no course work at that time."

After getting out of graduate school, young Bob Noyce had offers from all the major electronics companies, Bell Labs, GE, RCA, etc. "I went to work for Philco," He said, "because it was a small organization and I felt they needed me. I worked there for three years and ran into Shockley at one of the solid-state physics conferences. When he came out here to organize Shockley Labs, he whistled and I came. I worked for him for a year and a half, during which time he got the Nobel Prize, along with John Bardeen and Walter Brattain."

The Nobel Prize did little to soften Shockley's prickly personality. "He was very attractive to bright young people," Fred Terman once said of Shockley, "but hard as hell to work with."

"The relationships there were not very good," Noyce recalls. "I think the main problem I had with Shockley was that if you had done a piece of work, then he would call up Bell Labs and check on it to see if that was correct or not. He didn't trust you was the way we interpreted it. He was insecure enough himself so that he had to rely on other authority rather than on his own resources.

"On the other hand, scientists like to check their ideas with somebody else to see if they will fly, so that could have been a more charitable interpretation. Anyhow, as a result, there was a house revolution, and a group of people left finally and formed Fairchild Semiconductor.

"After they had their first meeting with what looked like a

reasonable chance of getting financial backing, they asked me if I would join them, and I did. It was with a great deal of personal agony, I might say."

The backing had come from an East Coast company, Fairchild Camera and Instrument. The Shockley eight had become the Fairchild eight, who would go on to provide the remarkable nucleus of Silicon Valley. Among them was a Swiss physicist named Jean Hoerni, who devised the so-called planar process, which finally made the high-volume production of silicon transistors possible. The planar process laid down a thin insulating coating over the silicon. It also made possible the development of the IC.

"There was a lot of emphasis then in the military on getting small electronics, and the interconnect problem was really severe," added Bob Noyce. Interconnections were indeed the key to electronic performance. Designers were creating communications and computer circuits that made lavish use of thousands of transistors and equal numbers of other electronic components, diodes, resistors, and capacitors. But they all had to be wired together by hand, a tedious, expensive, and unreliable process. "And if you looked at failure rates," Noyce pointed out, "it was always the solder joints, not the transistors. The earlier failure rates had been the result of surface phenomena, leakage currents around junctions, and so forth. The planar transistor took care of that because it coated everything with this nice, adherent, solid, insulating silicon dioxide."[5] The planar process replaced the wire connectors that had been used to link the transistors with metallic lines that were diffused directly into the silicon.

The idea of siting dozens, even hundreds, of transistors on a single piece of silicon was being explored by many companies, including a Dallas-based electronics firm called Texas Instruments (TI). In 1951, TI had been known as Geophysical Service Inc., a small company servicing the oil giants of Texas, when it took a $25,000 flyer to obtain a license to produce transistors. Two years later, it changed its name to Texas Instruments, developed silicon as a replacement for germanium, and applied itself to overcoming what had become known as the "tyranny of numbers," the difficulty of connecting all those transistors together.

In 1958 that task was given to a newly hired six-foot-six-inch engineer from Kansas named Jack Kilby. Although the transistor was the key to most electronic circuitry, other elements, such as resistors, capacitors, and diodes, were also essential. Kilby wondered what would happen if those components could, like the transistor, be made from silicon. If so, could they all be built on the same single block of silicon? The more Kilby thought about the idea, the more he liked it. After describing it in his lab notebook—"the following circuit elements could be made on a single slice: resistors, capacitor, distributed capacitor, transistor"—he proceeded to build the first primitive integrated circuit, of germanium, which at the time was still easier to work with than silicon. It was composed of five electrical components, a transistor, a capacitor, and three resistors. Tiny gold wires connected the five components together. It was to function as a phase-shift oscillator, a device that would convert direct current to alternating current and back again. Hooked up to an oscilloscope, the crude germanium IC, looking more like a toothpick that had fallen into a glue pot than an electronic component, functioned exactly as it was supposed to, tracing a straight, narrow green line across the face of the scope and then converting it into a spiraling sine wave, an undulating curve as the current shifted from direct to alternating.[6]

For all its demonstrated virtues, Kilby's version of the integrated circuit was not a production reality. Although the components had been integrated on a single chip, they still had to be interconnected by hand. That problem of interconnection was solved by Bob Noyce in January 1959, barely four months after Kilby had built his first germanium IC.

"I didn't see it as being comparable with what Kilby was doing. I was always interested in how you did it in a practical way. What Jack was working on looked totally impractical to me. They were different methods of trying to approach the same objective." In his notebook, Noyce wrote, "It would be desirable to make multiple devices on a single piece of silicon in order to be able to make interconnections between devices as part of the manufacturing process and thus reduce size, weight, etc., as well as cost per active element."

Noyce's lab notebook entry followed Kilby's by six months. "Jack's work preceded mine," he freely admits. "On the other hand, what he did was never going to fly in terms of a practical way of making integrated circuits.

"My idea just seemed to be the easy way to do it. What we were doing was making thousands of transistors on a wafer and then cutting them up into little tiny pieces and wiring them all back together. It just seemed so easy to say okay, here's all the transistors; you know exactly where they are, and you can print the circuit board on top of them rather than take them off the wafer and put them on the circuit board. It makes it much smaller and the most reliable interconnect you can get. And it solved a lot of problems.

"It seemed obvious once you decided that would save a lot of work. It was an idea whose time had come. If I hadn't done it at that time, somebody else would have within the next few years—there's no question about that."

Noyce and Kilby are acknowledged today as coinventors of the integrated circuit. Since their discovery, in 1958, that more than one transistor could, like angels on pinheads, be fit on a chip, the chase has been on to pack more of them into less space. The closer the individual circuits are on the chip, the shorter the distance electrons have to travel. A chip with a thousand transistors packed onto its surface does more work and is much faster than one with only ten transistors. Moreover, it costs no more to manufacture than the one with only ten transistors.

Gordon Moore, the chairman of Intel and one of the Shockley eight, who became director of research at Fairchild, made the astonishing prediction in 1964 that the number of individual circuits on a chip would double every year. Now known as Moore's law, that prediction has held firm, and today the most densely packed chips contain about a million transistors, and still the race continues.[7]

Much of it was to be centered in the Santa Clara Valley and the countryside around Stanford University. For in addition to encouraging Shockley to come to Palo Alto, Stanford's Fred Terman had induced other high-technology companies to set up shop in what was to become the Stanford Industrial Park. Fledgling companies such as Hewlett-Packard and Varian Associates opened in the new industrial park. An idea loop linked the new start-up companies with the university, and in a remarkable example of intellectual synergy, Stanford faculty and students attracted industry, which in turn attracted research money, which in turn attracted industry, and so on.[8]

Real estate speculators also moved in, buying up the apricot and prune orchards of the once exclusively agricultural valley and dotting the dun-brown hills with forty-eight-foot-long slabs of concrete that were yanked up by cranes to become the walls of long, shedlike buildings that seemed more sweatshop than think tank. Nonetheless, inside the sun-splashed buildings young men dreamed of packing more and more transistors onto smaller and smaller bits of silicon, and hundreds of women hunched over microscopes and sliced silicon wafers with diamond cutters, picked at them with tweezers, and hand wired and soldered them together to become the brain cells of the most dazzling technology of the twentieth century.

The new industry grew rapidly, and although the ICs were developed at Fairchild and TI without any government R & D funding, 95 percent of the initial sales were to the Defense Department and NASA. The attraction of the California Valley, with its remarkably easy life-style, pleasant climate, and big money, was proving irresistible. Fairchild was growing dramatically and, like Shockley Labs before it, kept reaching back East for new talent. Among the company's needs was a skilled production manager, for as the IC became more complicated, it grew increasingly difficult to manufacture. Fairchild ran a want ad in the *New York Times* and dispatched two of its people to interview candidates. Among the respondents was a young engineer who had been working at a GE plant in upstate New York. His name was Charlie Sporck.

It was 1959, and the tall, gangling, young engineer had reached a point where "I decided that GE was not where I wanted to be, and so I answered the ad. I went to New York City for an interview with these two characters who were drunk at the time. One guy was the operations manager, the other the personnel manager. We hit it off, and they offered me a drink and a job. I thought it was standard on the West Coast that that's what you do. The job paid $13,000, almost double what I was getting at GE. I accepted on the spot, came home, sold the house, packed up, and went to California."

Out in California, Sporck got his first glimpse of a management style quite different from that practiced in the East. "When I showed up at Mountain View, they didn't remember me. I had to go through quite a bit of shouting and yelling before they finally owned up to the fact that they had made me a job offer. They put me on the payroll and brought me into this office where they introduced me to the one guy that would be working for me, the general foreman. And they didn't introduce me to a second guy in this office who it turned out had also been hired for the same job. So they had two production managers, and it was typical of the industry at that time. Fairchild had just been started; they were growing rapidly. The founders knew nothing about manufacturing; they were tremendous scientists, but they know nothing about manufacturing, and things just sort of happened. The second production manager disappeared after a couple of months, and obviously they made a choice, the right choice."[9]

For almost the next ten years Fairchild was the place to be. NASA came into being to redeem President John F. Kennedy's pledge to place a man on the moon by the end of the decade. They chose Fairchild to build the ICs used first for the Gemini and later Apollo spacecrafts' on-board computers. Orders and money poured in, sales soared from a few thousand dollars to $130 million a year, and the number of employees jumped to 12,000, including some of the brightest young talents in the country. But they and the original founders were beginning to chafe under the detached yet autocratic management of the parent company back East.

Bob Noyce had been running Fairchild almost from the time he had come over from Shockley. And for the most part he was allowed to run things his own way—loose, confident, and approachable by everyone from the production line to the research laboratory. The hierarchy that characterized the old smokestack industries of the East Coast and the Midwest was flattened out. There were no reserved parking spaces at Fairchild Semiconductor, no paneled offices, no chauffeured limousines, no layers of middle management. Young engineers and scientists were hired directly out of graduate school and thrown immediately into the middle of major projects. It was sink or swim. Noyce and his director of research, Gordon Moore, would provide guidance if asked, but the responsibility and authority to go with it went with whoever was assigned to the project.

At the same time, the example that had been set with the first mass exodus from Shockley Laboratories was being repeated at Fairchild. The original Shockley eight were splintering off. Like yeast budding, they formed new companies in the growth-rich environment of the Valley. In 1962, Jean Hoerni and two other original Shockley scientists left and formed a company called Amelco. Three years later, Hoerni left Amelco to form still another semiconductor company, this one named

Signetics. The pattern was firmly established as scientists and engineers came to the conclusion that brainpower was all that was needed to create a new company in the semiconductor industry. Soon concrete slabs were being tilted up all over the valley, and technosilly names such as Silconix, Intersil, Qualidyne, etc., were created as fast as venture capitalists could throw their money down and draw incorporation papers up.

Overnight, a young scientist or engineer would leave a high-paying job to become the chairman of his own company and an instant millionaire. In a few short years there were more than twenty companies making chips and competing with Fairchild. Most of the new companies were being formed by the so-called Fairchildren. Indeed, by 1971, twenty-one of the twenty-three semiconductor companies in Silicon Valley were Fairchild offspring.[10] In 1987 a genealogy chart produced by the Semiconductor Equipment Manufacturing Industry Association showed more than a hundred companies whose lineage could be traced to Fairchild Semiconductor. But the late sixties saw an almost biblical genesis of new companies, a string of entrepreneurial begats unequaled in the history of capitalism.

In 1967 it was Charlie Sporck's turn. Sporck by now had become the general manager of Fairchild Semiconductor, and he watched the swelling tide of departures with increasing interest. "There was another reason people left," he explained. "Fairchild Camera and Instrument, the parent company, was headquartered in New York, and these guys had an East Coast mentality, the classic business philosophy in which by and large the people who really got rewarded were the guys on top and it never seemed to seep down into the working levels. We tried very hard to get major stock options distributed down to the guys who were truly important to the corporation. And we were unsuccessful, basically because the view back there was they were not important.

"So these guys looked at the opportunities to go out and start companies and to make significantly more money. After a while, it got to be kind of tedious to sit back at the ranch and try and talk these guys out of leaving, recognizing that it was kind of attractive to start your own company. Many of us started to think that maybe ten thousand Frenchmen can't be wrong, and then the major guys started to leave. Jean Hoerni founded Amelco, and then Signetics. Then there were a number of lower-level guys who left and founded a bunch of companies.

"I decided maybe it would be better to run a company that I was the president of instead of general manager. And there was the possibility of making a significant amount of money. I did it somewhat differently. I looked at all the companies in the industry. I looked for one that was in trouble. And I found National Semiconductor. It was a $7 million company in 1967, and they were in terrible shape, on the brink of bankruptcy, and I made a very good deal with them. Then we just grew a lot. The first year I was there we did $11 million, then $23 million, then $60 million something, then $120 million, and last year we did almost $1.9 billion.

"Basically, we changed the direction of the company. I would say at the very beginning it was more a price and uniqueness-of-product approach. And it happened very rapidly. You have to go back to that period of 1968–71, when you had the big TTL [a type of chip that used a so-called transistor-transistor logic] wars, which were really between ourselves and Texas Instruments. Of course, TI had a very large percentage of the market; we had an invisible percentage. Because we were so small, it was easier for us to maneuver, and we grew very rapidly. But I'd have to say at that time it was price aggressiveness in certain areas that accounted for our rapid improvement."

The manner in which some of the moves were made has since become the stuff of legend. One industry watcher tells of Charlie Sporck's departure in this manner. "Charlie emptied Fairchild of talent. He went over and took his best manufacturing guy, a fellow named Bob Mullen, who now runs some company in San Diego. He took Pierre Lamont, who is now with Capital Management, who was his second-best manufacturing guy. Then they took the next level, then the foremen. They virtually drove buses up to the building and filled them up with ladies. In our business, women are the manufacturing people. Tests show they have more attention to detail, more capability to do repetitive functions, more digital dexterity.

"Fairchild didn't do anything about it. Management of Fairchild just smiled." The smiling must have stopped, however, when Bob Noyce left.

"Everything at Fairchild was sort of a mishmash," he explains. "It was East Coast authority over a West Coast company. It's old industry versus new industry, not a regional conflict. It's sort of General Motors management versus Hewlett-Packard style. Anyhow, we basically decided to start over again and maintain our independence. So Gordon

Moore and I started Intel, and we recruited Andy Grove almost immediately. He had been working with Gordon over in the laboratory. Gordon was the director of research for Fairchild at that time."

Jerry Sanders, the thirty-year-old worldwide marketing director, was also at Fairchild then, and he, too, decided to leave. "Jerry, Bob Noyce, and Charlie Sporck were the three legs of the tripod on which Fairchild rested," said Elliot Sopkin, now the head of public relations for Advanced Micro Devices (AMD). "You had Bob Noyce inventing them, Charlie Sporck making them, and Jerry Sanders peddling them. And they remain the three biggest names in our industry. Then, in August of '68, Fairchild brought in an entire team of upper management from Motorola. Headed by a guy named Les Hogan. Charlie Sporck and Bob Noyce had already left.

"Jerry was the last of the three legs, and he was going to stick it out. Then, depending on who you listen to, he left Fairchild. Les fired him, says Jerry. Jerry at the time was thirty years old, and he headed a worldwide marketing staff of seven hundred people. They had just given him a raise from $30,000 to $45,000, and they then gave him a year's salary to leave. In typical Sanders fashion, he went out and rented a beach house in Malibu, and he and Linda, his ex-wife, and one child spent the summer of '69 there. While he was in Malibu, he considered acting. Various people approached him about starting his own company.

"He had lunch with Charlie Sporck, and Charlie said start a company. And so he did, with eight other guys, in 1969, and we were out with our first product ten months later. Bob Noyce a year earlier had raised $8 million in five minutes on the phone. Jerry raised $7 million, but he crawled from one end of Wall Street to the other to do it. Who was going to invest in a guy who was an engineer and a peddler? He had no management experience, no CEO experience, and no financial experience."[11]

Hollywood handsome, with the zeal of an evangelist and the flair of a circus impresario, Jerry Sanders soon made AMD one of the major merchant producers in Silicon Valley. Typical of his approach was a 1973 in-house advertisement showing a white-haired, white-bearded Sanders standing on the seat of a new Mercedes convertible. Behind him is a typical Silicon Valley building, poured concrete pillars and glass walls. "That's Jerry Sanders, president of Advanced Micro Devices," reads the caption. "That's Jerry Sanders's new $20,000 car. That's Advanced Micro Devices' new $4 1/2 million LSI complex. Help pay for one (or both)—get an order for Advanced Micro Devices."

AMD, National Semiconductor, Intel, and dozens of other companies seemed to spring like Athena fully grown from the brow of the Zeus of Silicon Valley, Fairchild. And what Olympian feats they performed! Innovation and ingenuity were transformed into magical bits of silicon chips that would remember prodigious amounts, chips that would tell time, compute arcane three-body problems in space, serve as the brains for new computers with remarkable power and astonishingly large memories packed in smaller and smaller devices. Indeed, it was the chips developed in Silicon Valley that enabled the complex problems of traveling from the Earth to the moon to be managed by a box only two feet long, one foot high, and six inches thick—the on-board computers on the Apollo spacecraft.[12]

During this period, the structure of the U.S. semiconductor industry became firmly established. The industry had sorted itself out into two distinctly different segments. Most of the smaller companies were the so-called merchant producers manufacturing ICs by the millions and selling them like commodities, much like bushels of wheat or ears of corn. The other group that produced just about as many chips, but for their own purposes, were "captive" producers such as IBM, Hewlett-Packard, and AT&T. These companies developed in-house semiconductor research, design, and manufacturing capabilities to fit their own special needs. For the most part, these were custom chips geared to the computers, telephones, and other equipment they produced. Such chips could provide distinctive qualities to the equipment, enabling the systems makers to differentiate their products from the competition.

Quantities were usually small, and the captive producers did not seek outside sales. Nor did the merchants look to compete with the captive producers for this market. Charles Ferguson explains in his monumental 1985 study "American Microelectronics in Decline: Evidence, Analysis, and Alternatives":

Such circuits also tended to require technologies distinct from those used in commodity circuits produced by merchant firms, or to require extensive coordination between the technology of the circuit and that of the final system of which it would be a part. Frequently the merchant producers were therefore unwilling to manufacture them; their design talent was a scarce resource and the integrated circuit technology rendered production and testing of many short runs of complex components much less remunerative than commodity mass production. . . .

Thus by default as well as by strategic choice the systems firms (and later other major consumers of special purpose circuits such as aerospace firms) developed technology and expertise appropriate to custom functionality circuits and their systems integration requirements. Yet since the merchants adequately supplied general purpose circuits, captive producers tended to confine themselves to their parent firm's special needs and, consequently, refrained from entering the open market in competition with merchants.[13]

It was an ideal arrangement for both sides, and in the continuing glow of economic demand, companies and ideas flourished. New technologies didn't just evolve; they flashed into being. More transistors were crowded onto chips; more functions were performed. At the same time, prices dropped rapidly and dramatically. In 1963 the average cost of an IC was fifty dollars. In 1965 the same chip cost nine dollars. Moreover, sales were moving from the military to the civilian sector. In 1962 virtually 100 percent of IC sales were to the U.S. government.[14] The critical factor was the inherent ability of the IC to add value far in excess of its cost to the system into which it was put. Or, to put it another way, the IC provided a synergy unequaled in all of industrial history. And the first device to profit logarithmically was the computer.

The infant computer industry quickly saw the advantages of the discrete semiconductor over the bulky, blowout-burdened vacuum tube. In 1958, IBM began replacing vacuum tubes with transistors in their computers. Still, the cost of computers remained fairly high. In 1960, for example, a large mainframe computer contained over 100,000 diodes and more than 25,000 transistors. Each had to be wired together into large, complex circuits at commensurate cost. The ICs reduced the amount of hand wiring dramatically and permitted an entire logic circuit to be etched onto a single chip of silicon. Packing fifty or more transistors onto a single chip placed them much closer together, thereby increasing speed, reducing power consumption, and improving reliability. Using ICs, computer companies such as IBM, RCA, Burroughs, and Sperry Rand developed ever more efficient and less costly computers, thus increasing the market for these new thinking machines.

All computers were now based on two types of ICs: memory chips, which did nothing more than store information, and logic chips, which would make decisions such as to take the information stored in memory and move it to a place in the computer where a combination of memory and logic chips were wired together to form the CPU, or central processing unit, of the computer. The CPU was the place within the computer where the information was processed.[15]

Intel and most of the other companies in Silicon Valley were producing memory chips by the millions. "We started with memory as the first thing," says Noyce, "because that was the application where you could see no upper limit in how far you could go and still be useful. Once you got to a given level of complexity, the idea of doing the whole microprocessor [literally a computer on a chip], which was Ted Hoff's idea, popped up and it seemed obvious."

Dr. Marcian E. (Ted) Hoff, Jr., had joined Intel as director of R & D soon after its founding. He was thus the obvious person to work with Busicom, a Japanese manufacturer that wanted Intel to produce a series of complex specialized chips to drive their newly designed desktop calculators. Hoff thought their chip design was far too complex, even more complex than the combination of chips that drove the PDP-8, the large computer he used for his own research. It struck him that it should be possible to do what the Japanese wanted by combining on one chip all of the functions performed in the computer's CPU. To it, Hoff attached two other chips, one containing a read-write memory, a sort of electronic scratch pad; the other was an ROM, or read only memory, into which the program that would drive the CPU could be installed.

The CPU thus became the microprocessor, containing all of its functions on a single chip. In January 1971, Intel produced its first working microprocessor. The microprocessor was literally a computer on a chip, a universal brain capable of running an almost infinite variety of machines—everything from hand-held calculators to cash registers, microwave ovens to digital watches.[16]

The microprocessor created the second computer revolution, making possible several new classes of computers—the minicomputer, the engineering work station, and most revolutionary of all, the PC, or personal computer. The PC literally fit on a desk. The most powerful computers in use just before the microprocessor was invented relied on a CPU that was literally as big as a desk.

The microprocessor also benefited from one of the most important factors operating in the semiconductor industry—the learning curve. The first microprocessors sold for $360 each. Within a few short years, the 8080, the redesign of the original Intel microprocessor, had become the workhorse of the computer and electronics industries. Like the DC3 or the Model T Ford, it made its way all over the world. It can today be purchased in quantities for as little as $2.50 each.

TI, Motorola, and several other companies followed with their own versions of the microprocessor, and a new computer revolution was launched.

The newest of the line, Intel's 80386, is a 32-bit microprocessor. In the computer, each bit, or electronic on or off signal, is strung together in necklaces or strings. A 32-bit microprocessor, for example, can manipulate strings containing thirty-two binary digits, "1"s and "0"s at a time. (The original microprocessor by contrast was a 4-bit device). It's as if traffic had suddenly been funneled from a four-lane highway to thirty-two lanes, greatly increasing flow.

It took a new generation of engineers to develop the new microprocessor, a team of engineers, in fact, headed by Gene Hill. A bearded young man from Oregon, Hill graduated from Oregon State in 1969 and went to work for Rockwell International and then a company called American Microsystems. For the next seven years he tested chips designed by others for use in calculators, printers, and computers. Then, in 1976, he came to Intel and was put to work in the microcontroller design group.

In 1982, the group completed the design of a microprocessor called the 80286. Until that time most PCs had been built with 8-bit microprocessors. The power of a chip is measured in electronic words or bits it can either store or manipulate at one time. An 8-bit microprocessor, for example, utilizes information in the form of eight-digit words. A string of binary numbers such as 01010101 is 8 bits long. The 80286 was a 16-bit microprocessor; it doubled the amount of instructions that could be accepted at one time, thereby more than doubling the speed at which the computer could work.

Even before the 80286 had been shipped to customers, Intel

decided to design the next-generation microprocessor, this one a 32-bit chip that would run at three to five times the speed of the 80286. This was the 80386, a 32-bit microprocessor that has given rise to the newest generation of PCs, scientific computers, and engineering work stations. And the power stored in its 32-bit muscle provides it, according to its spec sheet, with 64 terabytes of virtual memory per task. "How much memory is that?" asked *New York Times* computer columnist Peter H. Lewis.

Is it bigger than a breadbox? Smaller than an elephant? To find out, *Online Today,* the magazine for CompuServe owners, did some calculations, which follow:

"Eight bits make a byte, which is roughly the equivalent of one letter or numeric character. About a thousand bytes (1,024 to be exact) make up a kilobyte, or K. Until recently, 64K was the base memory of a personal computer. Now 640K is becoming standard. Beyond 640K one reaches a megabyte or 1024K, or about a million bytes.

"Take a million of those megabytes, and you have a terabyte. Put another way, a terabyte is one million million bytes.

"If one byte represented a second, a terabyte would be 31,700 years, roughly the span that modern mankind has existed.

"If a byte were a dollar bill, it would take 15,742 standard 50-foot boxcars, making a train 167 miles long, to hold a terabyte's worth of dollar bills.

Got that? Good. Now, multiply all that by 64.[17]

But it is not enough simply to decide to design a new chip. Intel sent out a number of bright young engineers they call chip architects to talk to their customers.

"Architects sit between the chip designers and the customers, and they act as a customer's agent negotiating features on a chip," Hill explained to me as we sat in another Intel rabbit warren of a building, some two miles from the site of my interview with Bob Noyce. "There are really two customer types, the software writers and the hardware board builders. One of the proposals, which really carried the day, was to 'design in' the ability not only to run PC software but engineering software, which is in another operating system called UNIX and other computer languages such as 'C' and Pascal."

Other inputs were needed as well. "The hardware people looked at how much work can be processed by a certain-size die at a certain speed. So they needed to set the boundaries, saying it is economical to build a chip this big that you can run at a certain clock rate. It will run existing DOS [disk operating system] programs at three times their normal rate of speed. But with programs written to take advantage of 32-bit architecture, you will get tremendously more performance, maybe five times, and an ability to tap into more applications with that.

"What happens is after you have the document which says what software you're going to run and how you're going to run it, that gets turned into an architectural spec. That's what you start from, what the end customer will see. Then you take that and break it down into different pieces of chip, an arithmetic unit, registers, microprogram storage, and various other pieces the chip will need. So now you are getting down from the rooms of the house, which is the external spec, to saying we are going to need two-by-fours this far apart, electrical outlets here and here, and getting into what the individual pieces are."

To me, the most astonishing aspect of the project was that, in essence, Gene Hill and his group were developing a chip for a computer that did not then exist. The design was based on their idea of what would be required as a result of the performance of yet another chip designed for a next-generation computer that also did not then exist. "It's a lot of forward thinking," admits Hill, "and that's why it did take a long time and called for a lot of discussion."

Even chips that do not exist cannot be designed in a vacuum. What sort of materials would be available to build the chip? Would silicon still be the material of choice? What about the ability of the manufacturing tools? Could they etch lines of the proper size? Could the masks be made fine enough? In effect, the designers were forced to guess at the manufacturing technology that would be available when their chip was ready to be produced. "We went to our own silicon process people, who were working on next-generation process technology, to see if what we wanted to do would be compatible with what they were aiming at," explains Hill. "So in essence you've got an awful lot of forecasting on which you are placing a very large bet."

Development cost $100 million over a four-year time period. That includes developing the wafer-process technology, the languages, and all the chips that ultimately went to test. They would be saying we can make a line this wide, and we would say well, it's going to take N number of lines in that direction to make a chip, and the chip has to be a certain size, or it will cost a certain amount.

"It's impossible to build these chips without some significant computer-aided design [CAD] tools, primarily in the checking area. We have tools just to make sure everything is wired properly. And tools to cross-check the wiring on the chip with the wiring on the documents, the documents to the test programs we run, all to make certain there's consistency in everything we do. In terms of the number of wires, if you compare it to a builder's drawings of how a house is wired for 110, we have drawings just like that. But the scale is closer to a city of five thousand houses. It would identify every wire and every light switch, every electrical outlet, in a city of five thousand. That's the kind of detail we have to deal with.

"So you see a process that's trying to predict a lot of events, where the market is going to be, where silicon technology is going to be, and where we can get the chip design to be in relation to all that. If you can call all those correctly, you've got a winner."

At this point he gave me a chip about the size of my pinky nail. On it were a few lines etched into the surface that could only be seen if the chip was tilted toward the light in a certain way. Each line was about 1/16 inch wide and perhaps 3/16 inch long.

"Four of these 32-bit Intel microprocessors are equal to the computing power of the IBM 3090 downstairs," Hill noted admiringly, referring to IBM's most powerful general-purpose computer. "If you had four engineers each using a 386 instead of a 3090, they would probably get more done, and it would cost a lot less."

The 3090 sells for more than $10 million. IBM's PCs, based on the 80386 chip, sell for less than $5,000 each.

We left the conference room and walked downstairs to the lobby of the building. There Hill steered me to one of the walls on which hung what looked like a huge lithograph. It was fifteen feet high by about ten feet wide and at first glance seemed to be a piece of modern art, a giant Mondrian with stark geometries of blues, reds, and purples leaping at you from behind the Plexiglas frame. "This is a top-view drawing of the 386, and it might take a microscope," Hill said with a hint of pride, "but I can identify every transistor on the 386 from this. The 386 has about ten levels. After you take off the glass on top that protects it, there are two layers of metal for interconnecting, colored blue and red here for identification, plus ways to connect those two levels. There are two full layers of metal plus another layer of polysilicon, which is another layer of interconnect, which also becomes one of the terminals in the transistor. Beneath that are diffusions, and where the diffusions and the poly intersect, you get a nice square electrical transistor, which is where all the neat switching happens. It's a CMOS [complementary metal oxide semiconductor] process, so there are two kinds of transistors: One likes to pull things to ground, and one pulls it up to the positive power supply. Then there are layers underneath where you can have each of these two types of transistors. This is a composite plotting of almost all the layers. There are about 250,000 transistors in the 386.

"There are moments when you step back and say, 'Wow, that's really neat.' Occasionally, on a project like the 386, you say, 'Maybe I shouldn't do anymore. This is as good as it will ever get.' But then you start chewing away on the next one. So yes, Virginia, there will be a 486. It will probably have a million transistors, and it will be strictly compatible with everything that's gone before, just higher performance. That's speed and applications. What are computers going to need in the 1990s? The obvious trend is graphics, engineering-station stuff. We are moving things onto our own boxes instead of main frames.

"Our charter is very evolutionary. It's largely in the mainstream. The impact can be transmitted to so many people so rapidly because there's software that exists. Every once in a while you do need revolutionary advances. What we found in going from 8086 to 286 to 386—the volumes we can ship out to the marketplace, the amount of power we can get into the users' hands—is incredible. And these people know what to do with it and can immediately put it to work."[18]

The microprocessor provided yet another great stimulus to the American electronics industry, propelling Silicon Valley into a Golden Age. "From 1960 to 1977," according to the U.S. Department of Commerce, "U.S. domestic shipments [of semiconductors] rose from $571 million to more than $4.8 billion; an average annual growth of more than 13 percent, in spite of rapidly declining unit prices." Exports of semiconductors also soared. "From 1967 to 1977," reported the Department of Commerce, "U.S. exports of all semiconductors (including parts) increased from $152 million to $1.5 billion, or at a compound annual rate of nearly 26 percent."[19] As sales soared, Silicon Valley became a magnet for engineers, scientists, salesmen, and money.

By 1976, every major industrial nation had established a domestic semiconductor industry with varying degrees of success. A 1978 survey by the Semiconductor Industry Association (SIA) identified thirty-six captive manufacturers and sixty-three merchant producers in the United States. Japan at that time had thirty semiconductor companies, while Europe had fifty-five firms producing semiconductors. The United States was consuming half of the world's production and producing more than 60 percent of the entire world's supply. That position was not, however, to remain unchallenged. "The growth rate of semiconductor output has been increasing more rapidly in Japan, Europe and in less-developed countries than in the United States," reported the Department of Commerce in a 1979 report on the semiconductor industry. "This is a significant development since Japan is the second largest producer of semiconductors, and both Japan and the European countries have been developing their production capabilities to support their rapidly growing electronic equipment industries."[20]

But from the standpoint of innovation and creativity, the United States was the unchallenged leader by so wide a margin as to be almost unbelievable. Summarizing the most important technological developments in the semiconductor field from the invention of the transistor in 1947 to the development of the 16K random-access memory (RAM) chip in 1976, the Commerce Department listed thirty-eight innovations and inventions. Only two of those thirty-eight had been developed by countries other than the United States.

Such an astonishing burst of innovation led inevitably to a dynamic entrepreneurial style unequaled in business history. "Though every firm in every industry was once a newly formed start-up," notes Charles Ferguson, "no prior industry ever exhibited the rate of new venture formation which characterized the semiconductor and related industries by the late 1960s."[21] Start-up companies blossomed in the fertile soil of Silicon Valley. Between 1966 and 1972, the rate of start-ups was rapid, with six new "merchant" companies entering the semiconductor business in the United States every year.[22] People with new ideas were encouraged by venture-capital firms (which quickly became known as vulture capital) to start new companies.

"People were constantly coming up with permutations of the same idea and starting a new company to bring it out," recalls a young engineer named Bob Flowers, who went out to Silicon Valley in 1978.

We had originally met in New York in the cavernous Javits Convention Center, in April 1987, at an IEEE conference at which semiconductor manufacturers and other electronics makers showcase their products. "All you needed was a good idea about how to do something slightly different with the same set of standard tools available to everybody else. Because the market was so new and appeared infinite, all these forecasting companies and stock analysts were painting a picture of infinite demand, and so it was very easy to get funding for a slightly different idea," he elaborated at breakfast the next day.

"Companies were starting up left and right with what in my estimation were only minor differences from established products. But because there was so much available money, companies started up easily."[23]

What appeared to be an almost unending source of new ideas on which to found new companies was not always beneficial to the semiconductor industry or the economy as a whole. For some companies it was often a disaster. In the 1970s, three of Intel's top microprocessor designers left to form a company called Zilog. The defection, according to Intel president Andrew Grove, set back the company's microprocessor development by a year. "What is more," he added, "some start-ups merely duplicate what is being done elsewhere."[24]

As the semiconductor industry has matured, the defection of engineers and managers has created a twofold problem. It saps the original company of some of its best talent, and the small start-up companies they create are too small and financially weak to compete in the industry now that large economies of scale and huge capital investments are required. And so the entrepreneurial spirit that is the great strength of Silicon Valley is suddenly conceived by many experts to be its Achilles heel.

"Most U.S. merchant producers were established less than twenty years ago as small firms whose primary business was semiconductor manufacture," explains Charles Ferguson. "Market leadership, employee loyalties, and supplier-customer relationships were short-lived; semiconductor and equipment producers rose and fell rapidly.

"But with the advent of very large scale integration (VLSI), microelectronics came of age. Capital-intensive, automated mass production became essential, as did a wide technology base and large-scale R & D. Semiconductors became ever more critical to a widening spectrum of products, including consumer goods, computers, machine tools, weapons systems, and other products. These developments offered a potentially large comparative advantage to the Japanese semiconductor industry as a consequence of technical diversification, large resources, and vertically integrated structure. This structural advantage also offered the possibility of large rewards in downstream industries based upon leadership in microelectronics."[25]

During the 1960s and 1970s, new ideas and technological innovations were rapidly diffused through the industry as a result of a system known as second sourcing. The large sytems manufacturers that were the primary customers for the new semiconductor products were extremely reluctant to rely on new, often inadequately financed start-up companies as their sole sources of supply. They insisted that a second company be licensed to produce the new product, thus providing them with a backup or second source of supply.

It began to look like a perpetual-motion machine. Second sourcing and technology licensing spread ideas around like peanut butter on white bread, attracting more and more people to new start-ups and to fill the ever expanding needs of existing companies. People rapidly became the most volatile, most important and scarce element in the American semiconductor industry.

Charles Ferguson says:

Semiconductor firms have never had any compunction about raiding other companies, and recognized that they would be raided in turn. By the early 1970s stealing employees was an art form, and innumerable professional and executive search firms had sprouted in the Valley. Many firms began offering employees thousand dollar rewards for referring new hires.[26]

The San Jose *Mercury News,* the largest newspaper in Silicon Valley, was running forty pages of want ads for engineers each day during this period. Bob Flowers answered one of those ads in 1978, after graduating from Gannon College in Pennsylvania with a degree in electrical engineering. His experience is typical of most people in the industry. "I went into a training program at Intel. I did everything from expediting the shipment of a product to setting up price negotiations with customers, market forecasting, and competitive analysis. In 1981, Intel moved me to Chandler, Arizona. I didn't like Arizona, and beyond that, I didn't think I was moving fast enough at Intel. In those days, if you didn't like your job, you just drove into the next driveway and got hired for 35 percent more."

Salaries, stock options, bonuses, and perks grew to astounding proportions, leading demographers to dub Silicon Valley workers "gold collar" workers. In 1988, the average pay for two-thirds of all Valley workers was in excess of $35,000 a year. Nationally, only one-third of the working population in America earned as much. For the top executives of Silicon Valley companies, pay and bonuses reached seven figures.[27]

But the semiconductor industry is among the most economically volatile in all manufacturing. Recessions sweep across the industry in worldwide swaths with almost biblical intensity, but the good years and the lean seem to follow two- and three-year cycles rather than seven. In its annual survey of executive salaries, for example, the San Jose *Mercury News* noted that 1986 was a poor year for the semiconductor industry. "Reflecting the woes in the semiconductor and semiconductor equipment industries, executives accounted for half of the 20 biggest pay cuts last year," they reported on May 11, 1987.[28] Even so, executives like Jerry Sanders, the chairman and CEO of AMD, received $447,118 in 1986 cash compensation, and that was *after* a pay cut of 62.4 percent.

Perks and bonuses in the Valley can be equally dazzling. The *Mercury News* reported:

Besides the $461,204 he earned in salary and bonuses last year at Xidex Corp. of Sunnyvale, executive vice-president Bert L. Zaccaria received $120,570 to cover the interest due on a loan the company gave him in March 1984.

Such loans—$1.2 million in Zaccaria's case—are common in Silicon Valley. Usually the money is loaned so top officers can exercise their stock options and pay any resulting federal income tax. Frequently companies will forgive a loan if the executive stays with the company for a prescribed period of time. Zaccaria, for example, doesn't have to repay his loan, according to company records.

The spectacularly profitable early years of the sixties and the mostly good years of the seventies established the trend toward high salaries, perks, and other forms of compensation that sent housing costs in the Valley skyrocketing. And so mortgage allowances become more bait for the Valley. "I wouldn't have moved here without it," the *Mercury News* quotes Frederick Hayes-Roth, a vice-president at Teknowledge Inc. of Palo Alto.

Hayes-Roth moved here five years ago from Los Angeles, where he lived in a house that cost him $250,000. A comparable home in Palo Alto would have cost $675,000, he said. He settled for a less expensive home, but still got $43,994 last year from Teknowledge to cover the difference between his Silicon Valley mortgage payment and what he paid in Los Angeles.

The almost desperate search for talent is not always fruitful and can sometimes attract parasites rather than bees to the honeypot. Consider the case reported a few years ago by the Hedda Hopper of Silicon Valley, Don Hoefler, whose newsletter, *Microelectronics News,* was required reading until his death in 1987. It was Hoefler who coined the name Silicon Valley in 1971 while writing for *Electronic News,* a weekly published by Fairchild Press, no relation to Fairchild Semiconductor. Fairchild, the publisher, is in fact much better known for its flagship paper *Women's Wear Daily.*

A few years ago, Hoefler reported a noted Silicon Valley company offering a top job to a man who claimed to hold, seemingly, more patents than Thomas A. Edison. He also claimed Ph.D.s from both Cambridge and Oxford and to have written several scientific books on the semiconductor process.

The job, wrote Hoefler, "included the title of corporate scientist and a salary of $95,000, a $2,500 bonus, a new Corvette, and a $12,000 advance for suits made of imported English wool."

The new scientist then demanded $1,000 to wine and dine a delegation of German scientists and their secretaries, who he claimed could deliver $20 million in sales.

"Sixteen rooms were reserved at the Pasatiempo Inn in Santa Cruz," noted Hoefler, "each furnished with a bottle of schnapps and fresh flowers. The innkeepers waited through delay after delay for three days, meanwhile flying the German flag each day."

A luxurious bus with a built-in bar and a waitress was sent to meet the Germans at San Francisco Airport. But the Germans never showed up.

"After two weeks . . . management began to suspect there was some skullduggery afoot and began checking out . . . references. Nothing was verifiable." Except apparently for the checks the new employee began bouncing all over the county. "It turns out he was still on probation for a similar scam in Southern California."[29]

In addition to charlatans, the Valley has attracted a unique mix of scientific, engineering, and entrepreneurial talent the *London Economist* has called "an engine of growth that is the envy and obsession of the rest of the world." It has also developed its own Valley style of management and industrial structure that initially generated tremendous innovation and creativity. New products and enhancements of existing chips not only led to the development of new companies and increasing profits but also ironically carried the seeds of self-destruction. Many of the chips had become commodity items. Memory chips, for example, were being produced in such vast quantities they became known as jelly beans. R & D costs were also rising as companies plowed back profits at an ever greater rate to produce the next generation of chips.

New developments such as the so-called metal oxide semiconductor, or MOS technology, allowed ever greater numbers of transistors to be jammed onto chips while consuming less and less power. The increasing density was accompanied by a dramatic drop in price. From the time the first ICs were introduced, the cost per electronic function has fallen by a factor of 100,000. The cost per memory bit, which was about $50 in the mid-1960s had dropped to less than $.0005 in the 16K RAM.[30] Aiding this dramatic reduction was the so-called learning curve whereby for each doubling of cumulative output the cost per electronic function has declined an average of 28 percent. By anticipating the effect of the learning curve, most, if not all, American semiconductor manufacturers engaged in forward pricing, introducing their new products at a price far lower than the actual cost of R & D and manufacture. At some point in the cycle, the actual cost would drop far below that forward price, enabling the chip maker to first expand the total market and then increase the firm's share of the market. This became an essential tactic as the time between chip generations shortened and the cost of R & D began to soar.[31]

The successive generations of chips were named according to their ability to pack more and more transistors into smaller spaces. Chip generations progressed from SSI, or small-scale integration, the first generation, wherein fewer than a hundred components were fitted on a chip, to MSI, for medium-scale integration, then LSI, to the present

VLSI, where a minimum of 250,000 transistors are packed onto a chip and the feature size, that is, width of the lines and transistors, is less than 1 micron. A human hair measures approximately 150 microns in diameter. In the not too distant future, certainly by the early 1990s, ultra-large-scale integration, or ULSI, will be the norm, with more than one million electronic components on a chip. At this point, the physics of light will begin to play a major role, for feature size will shrink below 0.25 micron. The wavelength of red light is by contrast only 0.7 micron. In other words, the beams of light now used to photograph the circuits on each layer of a chip will be too thick to visualize them. Despite the problems, engineers envision the possibility of a chip jammed with as many as one *billion* transistors by the turn of the century.[32]

Each chip generation has produced a corresponding improvement in chip capability. Memories went from 256 bits of RAM in 1968 to 1K RAM with 1,024 bits in 1970. The 1K RAM ushered in the era of LSI. By 1979 the 16,000 bit, or 16K RAM, had become the standard, only to be supplanted a year later by the 64K, 64,000-bit chip. The 64K chip marked yet another generational change, moving the semiconductor industry into the VLSI generation. The new VLSI generation produced a dramatic change in the ability of American chip makers to compete in world markets. Much of American chip production had long since moved offshore, where labor costs were much lower than in the United States.

Many of the nations that had proved so attractive to the American garment industry offered the same compelling advantages to the semiconductor companies—a large pool of young, unmarried, illiterate women who could be paid twenty-five cents to a dollar an hour, often at piece rates. TI set up plants in El Salvador and several Asian nations. National Semiconductor, Intel, Motorola, and many other companies established plants all over Asia, in Indonesia and the Philippines, as well as Mexico and Central America. Dependence on low-wage offshore labor increased until, by the late 1970s, 80 percent of the chips produced by American merchant companies were assembled outside the United States.[33]

But labor was less a factor in VLSI production than the development of automated state-of-the-art manufacturing technology. And so the shift to VLSI technologies accelerated the loss of American dominance to the Japanese. "VLSI fabrication is becoming one of the world's most expensive, complex and capital-intensive technologies," points out Charles Ferguson. "It is no longer a garment industry analogue in which competitive advantage can be obtained by paying Salvadoran or Indonesian women thirty cents an hour."[34]

The stakes became much higher; the cost of start-up, prohibitive. Indeed, just to remain a major player in the game required the expenditure of enormous outlays of capital. "Minimum efficient scale and initial cost requirements," notes Ferguson, "have escalated at an astonishing rate; a world-class plant now costs $200 million, a tenfold increase since 1975."

The relatively small American merchant producers were hard-pressed to come up with the cash. Construction costs forced a jump in capital expenditures from 6 percent of revenues in 1974 to 20 percent in 1984. By contrast, the Japanese producers, with their deeper pockets, increased their outlays for semiconductor plants from 6 percent to more than 28 percent over the same period; this despite the fact that their revenues rose more rapidly than did that of the Americans.

The Japanese proved to be fierce competitors, often selling chips below cost, dumping, according to their American competitors, in order to increase their share of the market. Within a few years they were able to virtually wrest the entire memory segment of the market, which accounts for more than 50 percent of all semiconductor volume, away from the Americans.

During the economic slowdown of 1985 and 1986, the Japanese continued their investments in new plants, whereas the Americans for the most part cut back drastically on plant construction. Thus, when worldwide demand picked up, the Japanese were able to supply the market with high-quality advanced memory chips, and using the American techniques of forward pricing, at low prices.[35]

Japanese success in mastering the enormously complicated techniques of VLSI manufacturing pays off in higher yields. And profits in the semiconductor industry are sharply affected by yields. The production of an IC is one of the most complex and demanding of any manufacturing process devised by man. The densely packed VLSI chips might be compared to high-rise apartment houses. Each floor, or layer of a chip, contains the equivalent of specialized rooms, bedrooms, kitchens, etc. Just as the rooms and floors of a high rise are connected by water and heating pipes, elevators and electrical cables, so, too, are the layers of the chip and the tens of thousands of individual transistors and components on each layer, joined by miles of microscopic metal lines called interconnects. Today some VLSI chips are composed of as many as twelve distinct layers of silicon, and as designers dream of reducing the size of individual elements and cramming even more of them ever closer together, layers may be stacked as much as twenty deep on a chip half the size and more brittle than a single corn flake in your cereal bowl.

The actual chip-making process, from the time raw silicon is grown from a crystal to the final packing of the chip in its protective package, may involve as many as 150 separate manufacturing steps. Moreover, it is enormously costly and technically very demanding. In the low-profit, highly competitive commodity markets where memory chips are sold, the costs have become too high and the rewards too few for most American companies.

And so it has become one of the great ironies of the Chip War that the creative surge that has led to the formation of so many small merchant companies has also put the bulk of American merchant chip makers at a distinct disadvantage vis-à-vis the giant Japanese consortia against whom they must compete in the global marketplace. With far fewer capital resources, American companies spend less money on R & D than do the Japanese. Nor are they able to replace aging manufacturing equipment with state-of-the art machines. The need to produce a profit in every quarter to satisfy stockholders and venture capitalists does not permit the kind of long-range projects the Japanese and the larger European companies are able to undertake. SEMATECH, the newly formed government-industry consortium designed to improve American semiconductor manufacturing skills and develop new manufacturing technologies, may help restore some competitiveness to U.S. firms.

However, by remaining small and disdaining vertical integration, American companies have lost the ability to weather economic downturns. Given the high stakes of the Chip War, to compete successfully, companies must not only sell chips but a broad spectrum of chip-based products as well—computers, telecommunications devices, medical instruments, audio and video equipment, and so on. Such companies can afford the rocketing expenditures needed to keep up with a constantly improving technology. With the swift diffusion of new ideas, success is measured more by the ability to translate them into mass production, and that in turn requires the investment of huge amounts of money to develop and perfect fabrication techniques, so-called process technology, which defines the tools and individual manufacturing steps of semiconductor production.

Without the profits from other products that use chips, such as computers, VCRs, television sets, and so on, to subsidize process technology, the stand-alone merchant chip maker is at a distinct economic disadvantage.

How much longer, then, will these constraints continue to allow American firms to innovate with the same verve and frequency they have in the past? Today the million-bit chip, the 1-megabit RAM, is being produced in huge volumes by the Japanese and American captive chip makers such as IBM. Only two American merchant companies, TI and a very small Idaho firm called Micron Technologies, are producing 1-megabit memory chips. Companies such as Motorola and Intel are buying 1-megabit chips from Toshiba and Mitsubishi and reselling them under their own labels. But the manufacturing capability that is created by production of those chips remains for the most part in Japanese hands.

And the pace of development is speeding up rapidly. At the annual International Solid State Circuit Conference held by the IEEE in February 1987 in New York, five Japanese companies and IBM announced a 4-megabit chip. IBM, however, was the only company capable of actually producing 4-megabit chips in volume at this writing. To fully comprehend the implications of 4 megabits, consider the fact that it can store 400 pages of double-spaced typewritten text. Moreover, it can "read," or make available to the computer for immediate use, those 400 pages in one-quarter of a second. The cells in which memory is stored on the chip are so small that 48,000 of them fit inside a space equal in circumference to the period at the end of this sentence.

The innovation that has long characterized the industry is no longer limited to American firms. At the same conference, Japanese researchers from the Nippon Telegraph and Telephone's (NTT) Atsugi Electrical Communications Laboratories have leaped two generations ahead with the announcement of a design for a 16-megabit chip. Furthermore, in what may be an ominous portent of the direction in which events are moving, of the 116 papers delivered at the conference, 40 percent were by the Japanese. It marked the first time there were more Japanese than American papers read at the annual conference.

In the unbelievably complicated labyrinths of the 16-megabit memory chip lies yet another of the historic ironies in American-Japanese confrontation. Consider that the transistor, originally developed in the parklike New Jersey headquarters of Bell Labs, had to await the development of crystals used in World War II radars for the final breakthrough. Now, some 9,000 miles away, the 16-megabit chip is developed at the research laboratories of NTT, in the rolling parkland outside the Japanese town of Atsugi. I remember Atsugi well. From here, almost forty-five years earlier, kamikaze pilots flew their deadly but hopeless missions against the American fleet. Then, after the surrender, Atsugi was converted to a U.S. naval air station, where I was stationed for a time after the Korean War.

From Atsugi we flew electronic eavesdropping missions along the Kamchatka Peninsula to within 200 miles of Vladivostok. I imagine those missions are still flown today, albeit now with the aid of chips perhaps developed not very far from the runways of USNAS Atsugi, the former kamikaze base.

3

The $25,000 Transistor:
Japan Takes the First Byte

Atsugi lies approximately thirty miles southwest of Tokyo. In Japan's feudal age, when Tokyo was known as Edo, Atsugi was a village surrounded by lush rice fields and small farms. Its main importance was to serve as a rest stop for the hundreds of pilgrims journeying from Edo to worship at the shrines of Ohyama (Big Mountain). In more recent times, in the years prior to World War II, the road connecting Atsugi and Tokyo became a major military highway. The Japanese navy maintained a naval air station at Atsugi, and it was a major kamikaze base. Along the road, infantry and mechanized army units were posted. And in Tokyo itself, in the western end of the city known as Aoyama (Blue Mountain), where the road officially begins, a large barracks housed an entire army regiment.

Today the old military highway has been replaced by a new national roadway, officially labeled with the number 246. It, too, begins in Aoyama, which has become a neighborhood of fashionable foreign shops such as Brooks Brothers, boutiques such as Benniton, a university, and the offices of such major Japanese companies as Honda Motors and TI Japan. Beneath the 246 roadbed run the fiberoptic cables that form the largest-capacity telecommunications lines in Japan, providing the very wide frequency bands needed for diversified voice, data, and video communications. Twenty years ago, NTT, then owned by the government but since privatized, built a special telephone-exchange office in Aoyama as well as the microwave relay center that sends the programs of NHK, the nation's largest television network, to the rest of the country. All of Japan's major banks have established their computer centers along 246. Many software houses, small companies for the most part, with ten or twenty young hackers, work in small offices of one and two rooms in the new buildings that have sprung up along 246. As a result, 246 has become known as Communications Road.[1]

As it was for the pilgrims from Edo and in World War II, Atsugi remains a critical nexus along the road, for here NTT and several other major Japanese electronics firms have established research centers. As the train pulled into Atsugi, I looked in vain for the sleepy Japanese town where I had once been stationed. In its stead was a modern city with new office buildings and apartment blocks, bright signs, fashionable shops, and restaurants. It was only as the taxi wound up into the hills and we passed the rice paddies I had remembered ringing Atsugi that anything looked vaguely familiar. Then, under the low clouds, I saw Ohyama, once worshiped with almost as much reverence as Fuji. Now the mountains are simply a dramatic backdrop to the research park where NTT maintains a semiconductor research center. The entrance is guarded by blue-uniformed private guards who threw a pair of very smart salutes when we pulled up. For an instant I was transported thirty-odd years back, when, as a young ensign, I passed through the gates of NAS Atsugi. Now I had to sign in, was given a badge, and walked up the road into the enormous building that is NTT's semiconductor research center.

Its mission is to develop design technology for VLSI and ULSI chips and the manufacturing technology for silicon devices. In a sense, the laboratory functions very much like Bell Laboratories, although in this case exclusively for semiconductor research. But unlike Bell Labs, whose research is generally applied to products made by AT&T, NTT-developed technology is usually transferred to other Japanese companies. The NTT lab gains a patent to maintain proprietary knowledge, and the patents are then transferred through NTECH, a subsidiary of NTT, to commercial manufacturers. Or NTT often simply supplies a manufacturer with the patent and asks them to make the device for NTT.

With that information as background, I was escorted into a large conference room that overlooked a pond filled with dozens of large, brightly colored carp. The pond, I later learned, was actually a filtration pool for the state-of-the-art manufacturing facility NTT uses as a research laboratory for developing and evaluating new process technology. Then Tsuneo Mano entered. We exchanged name cards in what is a ritual in Japan and sat down. Mano is a small, slender man, and as soon as he started to talk, I realized he was what the Japanese call *shyu sai*, a superior talent. Tsuneo Mano was born in 1947, the year the transistor was invented, and his talent began to show itself early in life. He grew up in the mountainous Nagano prefecture in central Japan. Here he learned to ski and showed an early affinity for tinkering. At the age of ten he built a vacuum-tube radio and then read of the early use of transistors in radios. By the time he entered high school, he was sophisticated enough to make the six-hour, 200-kilometer train journey to Tokyo by himself. Here he ransacked the stalls and stores of Akihabara, the bustling electronics bazaar in downtown Tokyo. With the transistors and other components he found in Akihabara, he returned home to build radios, audio amplifiers, and even an electronic lie detector.

At the age of eighteen he entered the Tokyo Institute of Technology, Japan's MIT. Six years later he was awarded a master's degree in applied physics and went to work for NTT Labs. "I worked at first on CCDs, charged coupled devices [a type of chip used in video cameras], for three or four years. Then I shifted to high-density MOS memories, especially dynamic memory."

That experience provided the background for the design and experimental fabrication of the 16-megabit DRAM memory chip that so startled the large group of semiconductor researchers from all over the world who had assembled in New York's Hilton Hotel in February 1987 for the annual International Solid-State Circuits Conference. His paper came after the announcement that IBM had developed and begun production on a 4-megabit chip.

"The 16-megabit chip was developed as a test device for verifying our technology, not for commercial DRAM production," Mano patiently explained to me. "We used it to test both our design technology and our manufacturing technology. The test device was produced by the Division of Process Technology of this institute. There were only seventy or eighty chips produced for experimental purposes to verify the technology. So I really couldn't understand all the newspaper furor over it."

Indeed, everyone who heard about the project seized on the staggering size of the memory, 16 megabits, capable of storing the contents of a book three times longer than this, but the fact is that it was originally meant merely to verify advanced circuit design. "What we wanted to do," says Mano, "was to develop a technology to accurately test all of the memory cells in so dense a chip and to test the low-voltage capability of DRAMs. On the other hand, there is a group here specializing in process technology. They wanted to test their ability to build a three-dimensional structure. The two groups met and decided to test and verify their most advanced technology. So 4, 6, or even 10 megabits was not meaningful—any number they decided will do—and they finally settled on 16 megabits."

The key element in Mano's design of the 16-megabit DRAM was something he calls ECC (error checking and correcting) technology. "This," he explains, "is essential for 16-megabit and higher-density chips. The 1-megabit and the 4-megabit devices that have just gone into commercial production do not require it. But anything higher will need it. It's a circuit designed into the IC. The other technology we verified was a very high density three-dimensional memory cell structure. It is very small, 5 square microns per bit. Most memory cells occupy an area of 40 or 50 square microns. Low-voltage memory circuits were another aspect we wanted to test; for example, our 16-megabit chip has a voltage of only 3.3. The IBM 4-megabit chip uses 5 volts."

Higher voltage means more heat, which means circuits must be placed farther apart to dissipate that heat. The greater the distance separating circuits, the slower the chip.

The 16-megabit DRAM, with its *40 million* parts—transistors, capacitors, and resistors—seems to be pushing very hard against the barriers of silicon technology, but Mano thinks the crystal, which has been the mainstay of the semiconductor industry, has a lot more life in it than was ever thought possible. "I think," he declares emphatically, "that the limits of silicon technology will not be reached until around 64 megabit or even 100 megabits. For the moment I think that is the upper limit. We have succeeded in building the 16 megabit, and as a result I think we can go higher. The ECC technology that I developed is essential in building very high density memories. It will be essential in getting effective yields in mass production. We think this is a very big problem for MOS memories and so implemented a new testing circuit for test devices."

In the highly competitive memory-chip market, not only are high yields essential, but also high-speed production. And so the time it takes to test high-density DRAMs becomes a very important factor. Mano's goal was to shorten testing time by a factor of 100. In fact, he reduced it 200 times, bringing it down from 20 nanoseconds to under 5 nanoseconds.[2] Measuring time in such microfragments—after all, a nanosecond is a mere *billionth* of a second—may seem hairsplitting of a rather extreme nature, but in the electronic world of the chip, nanoseconds are intolerably long stretches of time.

In 1987 this was an astonishing achievement, especially given the abysmal state of Japanese industry and technology in 1947, the year of Mano's birth and the year the transistor was invented. The Japanese economy had been shattered by World War II; its once mighty industrial plant lay in ruins. The task of rebuilding had largely fallen to a Japanese government bureau known by the acronym MITI, the Ministry of International Trade and Industry. MITI was in fact a phoenix, hastily assembled with personnel from the wartime Ministry of Munitions.

Immediately after the *gyokuon hoso,* the broadcast by Emperor Hirohito announcing the surrender on August 15, 1945, Japanese bureaucrats, fearful that the American conquerors would arrest anyone connected with war production, ordered the dissolution of the old Ministry of Munitions and its personnel reassigned to the Ministries of Commerce and Industry and Agriculture and Forestry. Those two bureaus had been merged during the war to create the Ministry of Munitions to give the military virtually total control over all industrial production. The old MCI, or Ministry of Commerce and Industry, subsequently took on its present name of MITI.[3]

Among the physical assets MITI inherited were a number of laboratories that had been engaged in wartime research and were now assigned to develop industrial technology. For electronics, the most important was the Electrotechnical Laboratory (ETL). In 1947, ETL was located on the western fringes of Tokyo, and it was here that physicist Dr. Makoto Kikuchi came to work immediately after graduation from Tokyo University.

Today Dr. Kikuchi is the director of research for the Sony

Corporation and one of the most famous scientists in Japan. We met at Sony's headquarters building in Tokyo. Like most Japanese electronics plants I have visited, it is guarded by a sort of gate house where one registers and receives a badge. A huge staircase hangs suspended like a steel Tinkertoy in the air of the lobby. At the head of the stairs is a long desk behind which sit three pretty young women. All rise and bow as each visitor arrives. One of the young women led me past an enormous reception area dotted with low chairs and Sony television sets, through a pair of tall doors that had to be opened with a key. The same key opened an elevator door. We went up, the doors slid noiselessly open, and another young woman escorted me down a long hallway to a conference room. Practically skipping from the other end of the hall came Dr. Kikuchi. An ebullient man, well traveled in the United States and Europe and far less formal than most Japanese, Kikuchi wore a vest I suspect may have been of his own design, patched with pockets that were stuffed with pens and pencils. He greeted me with bubbling warmth and asked whether I would take coffee or tea.

Like so many of Japan's intellectual leaders, Kikuchi was a graduate of what was then Tokyo Imperial University. But, he explained, most of his university career had been spent not in Tokyo but in Suwa City, a town in the mountainous Nagano prefecture. In his autobiography, Kikuchi wrote:

The tatami straw matting of the Osachi Village Elementary School sewing room became our seats for studying differential and integral calculus. We would trudge down the mountain path to the school and trudge back up again to our lodgings. Our boarding house was a hot spring spa called Kaminoyu, literally the hot spring of the gods. That was a strange joke. A spa may sound quite grand, but in fact, 25 of us were jammed into a single room measuring some 80 meters square. A washboard straddling two tangerine crates served as our communal desk. Once we had an outbreak of lice and the dorm committee boiled water in an old steel drum to sterilize everyone's shirts.

That August the war ended. When autumn came, we pulled together our belongings and descended from our mountain retreat to Tokyo."[4]

Three years later, Kikuchi graduated and went to work at ETL. Six months after that, he read the announcement of the invention of the transistor at Bell Labs. Amazingly, he then proceeded to build the first Japanese transistor. What was most remarkable about the feat was that he had no scientific description of the original Bell Labs work. His only information was an account in *Newsweek* magazine.

"I was one of the first in Japan who even knew about the announcement. I started by trying to trace the amplification in a crystal. At the time, we didn't even have germanium crystals here, but we did have silicon. Those silicon were so impure no amplification was possible with them, but we didn't know that. My boss gave me a small piece of silicon and told me to try. I tried for weeks, and after a month I still had not succeeded. Every day I would look through a microscope at the wires stuck in the crystal. And every day I would look at the two ampere meters, input and output. My expectation was that I would see parallel movement, but there was never any change in the output meter. Then one morning I got parallel movement on both input and the output meters. I called my boss and showed him. He pointed out that the two wires in the crystal were contacting each other. That was pushing the ampere needles, not amplification.

"We finally succeeded when the director of my laboratory, the Electrotechnical Laboratory of MITI, where I spent twenty-six years, brought back a diode from the U.S. It contained a very small amount of germanium. I broke it and carefully extracted a small piece. Then I redid the experiment with the germanium, and I got amplification. It took a long time for us. And even after we built a transistor, we did not understand the principle behind it. I asked Professor Kubo, a famous physicist at Tokyo University, why we can have amplification this way, and he couldn't explain it. Only a few people at Bell Labs, RCA Labs in Princeton, and at GE knew the principle.

"Despite the seminars Bell Labs held to explain the transistor, under the terms of the occupation we could not get the scientific literature. Then, in the early fifties, we began to learn about transistor science. The occupation turned out to be a blessing. We had no army, no defense forces, and so we began to concentrate on the field of electronics. I went to Montreaux, Switzerland, two years ago when I was invited to deliver a lecture. I watched the trains going up and down the mountains, and I found that the Swiss were masters of gear systems of mechanical engineering. That is why the Swiss delayed in getting into electronics; they were too satisfied with the state of their mechanical engineering. After World War II, the Japanese had no science to be satisfied with."

To Dr. Kikuchi, the transistor offered a tantalizing opportunity. The question was how best to seize it. The same idea occurred to two other Japanese, Masaru Ibuka and Akio Morita. In 1946, in a garage in Tokyo, they started what was to become the Sony Corporation, one of the world's electronic giants. By 1953 they had 125 employees. That was the year they licensed the rights to the newly invented Bell Labs transistor for $25,000. It ranks among the greatest bargains in history. According to Morita, however, MITI was adamantly opposed to the purchase.

Morita confessed in his autobiography *Made in Japan:*

The transistor was so new and foreign currency was so scarce in Japan, which was just then beginning to accelerate its recovery from the war, that the bureaucrats at MITI could not see the use for such a device and were not eager to grant permission. Besides, MITI thought that a small company such as Totsuko (as we were known then) could not possibly undertake the enormous task of dealing with brand new technologies. In fact they were adamant against it at first. Ibuka was eloquent on the possible uses of this little known device, but it took him six months to convince the bureaucrats.[5]

Ibuka was in fact possessed by the idea of the transistor. After he first heard about it, he went to his friend Makoto Kikuchi. "He was and is still like a boy, filled with curiosity," says Kikuchi admiringly of Ibuka. "When he first heard of the transistor, he asked me what will it mean for me and what will it mean for Sony? Could we, for example, make the world's smallest radio with it? Let's try. That was 1953.

"That same year he was invited by Western Electric to visit them in the United States. He told the Western Electric engineers that he had recently learned about the transistor and wanted to make the world's smallest radio. He was laughed at. At that time, the transistor was still in its infancy, not high in reliability and not very good in its frequency characteristics. There were many problems in getting to the target he had. But he was determined, and he came back and formed a group of engineers to study the problem. And this is a very important point; the target was very clear: Mr. Ibuka wanted to have the world's first and smallest pocket radio.

"So the Sony engineers tried to solve the problem. The biggest problem was the high-frequency characteristics of the transistor were not high enough. The only transistor available at that time was the alloy-junction transistor. The ground-junction transistor, which was better suited, was still in the laboratory. To solve the problem, Sony engineers developed the ground-junction transistor, which was the first in the world and received signals as high as 17 megahertz. Even Bell Labs was very pessimistic about being able to do it.

"Reona Esaki, one of the Sony engineers working on the problem, found very unusual characteristics between the base regions and the middle regions while examining a faulty transistor. This led to the development of the tunnel diode, for which he received the Nobel Prize some years later. But finally, in 1954, Sony succeeded in making not a pocketable but still the world's smallest radio. But the next year Sony did produce a pocketable radio."

For Kikuchi, the tiny transistor came along at what was the exact perfect moment in Japanese history. "The announcement of the birth of the transistor three years after the end of the war had a very important meaning for us," he maintains. "If it had come out earlier, Japan was still too poor and not ready to start learning. If it had come out much later, our catch-up process would have been much slower. Transistor physics was an entirely new field, really the dawn of modern electronics. And electronics is the technology Japan started to learn immediately after the war ended as one of the most important bases upon which to rebuild her industry."

It proved to be among the most brilliant national choices ever made. "As it is widely recognized now," says Dr. Kikuchi with a smile, "electronics is not merely 'one of' the fields of technology, but it is a special field of extremely influential and versatile technology, just like the humidity, diffusing into almost everything."[6]

Although the principles of transistor physics were still to be learned, the small size of the new electronic devices was a perfect fit with the thousand-year Japanese tradition of hand-crafted miniaturization. "Miniaturization and compactness have always appealed to the Japanese," says Morita. "Our boxes have been made to nest; our fans fold; our art rolls into neat scrolls; screens that can artistically depict an entire city can be folded and tucked neatly away or set up to delight, entertain and educate, or merely to divide a room."

That, however, can scarcely account for the astonishing success of the Japanese in the exquisitely sophisticated technology that is semiconductor manufacturing. Rather, the Japanese success can be traced to a variety of different factors, all springing from Japan's history, culture, and traditions, much of it developing in isolation from the rest of the world. Although the first contact with the West was in the sixteenth century, Japan had been effectively isolated by the Tokugawa shoguns for 250 years, until the middle of the nineteenth century. Then, in 1853, Commodore Matthew Perry sailed a squadron of U.S. Navy warships into Japanese waters.

The British had by then become the colonial masters of India. China's closed doors had been battered open a hundred years before by the maritime powers of Europe. Only Japan had resisted the insistent demands of the Europeans to open its shores to trade. But with the arrival of Commodore Perry, even the feudal daimyos realized their swords and samurai were no match for the modern cannon of the U.S. Navy. Treaties were drawn between the United States and Japan, and then with the other Western nations. The same unequal system the European powers had forced upon China, with its extraterritorial legal privileges, limits on tariffs that left the economy helpless to the onslaught of Western manufactured goods, was applied to Japan.

But somehow things didn't work quite the same way as it had in the rest of Asia. The Japanese did not readily accept the strange foreign goods the Western nations tried to sell them, a situation that seemingly persists to this day. The Tokugawa government, although outwardly acquiescent, created a bureaucratic morass that enmeshed the Yankee and European traders. And then, as if the gods of Japan themselves seemed determined to lend a hand, a silk blight in Europe led to a heavy demand for Japanese silk that helped offset the trade imbalance.

However, the feudal Tokugawa government, which had ruled Japan for two and a half centuries, was tottering. The Japanese people were opposed to opening their country to foreigners, and the security and stability the Tokugawas had provided for the nation for so long seemed no longer within their competence. Finally, the leaders of a number of outlying provinces revolted against the shogunate and on January 3, 1868, announced the reestablishment of direct imperial rule. The fifteen-year-old emperor was moved from Kyoto, the ancient capital of Japan, eastward to Edo, the city of the shogun, where all administrative power had been wielded. As befit its new status as both the imperial and administrative capital of the nation, Edo's name was changed to Tokyo, or "Eastern capital."

The new "year period" as the Japanese mark imperial reigns, was known as Meiji, and the great transformation from the age of shoguns became known as the Meiji Restoration.

Harvard's Dr. Edwin O. Reischauer in his classic study *The Japanese* writes:

The leadership of the "imperial" forces was ostensibly provided by imperial princes, court nobles and a few feudal lords, but actually the initiative for policies and their execution fell largely to a group of able, young reformers of largely middle or lower samurai rank and mostly from Satsuma and Choshu. These men realized that, although the cry of "expel the barbarians" had been useful in their seizure of power, it would be disastrous for Japan to attempt to carry out such an unrealistic policy. Instead they made clear at once their acceptance of the Western treaties negotiated by the Tokugawa. But they still faced the herculean task of replacing the old feudal system by a more effective centralized rule and starting Japan on technological modernization that would give it security from the powerful and predatory nations of the West. The new regime had fallen heir to no more than the broken down, bankrupt shogunal domain, in a country long divided into autonomous feudal units and still limited to a pre-industrial economy.[7]

One of their first steps was to concentrate economic power in the hands of a few businessmen. In doing so, they abolished the centuries-old social structure of *shi-no-ko-sho,* a four-tier hierarchy of warrior-farmer-artisan-merchant. The samurai, the highest class, were not permitted to farm or engage in trade. Samurai were further distinguished from everyone else by the Sword Hunt of 1589, when the shogun Toyotomi Hideyoshi confiscated all weapons from the peasantry. Farmers were bound to the land, artisans to their jobs, and merchants were simply despised.

In 1869, when the Meiji Restoration officially abolished the four-class system, the samurai constituted 6.4 percent of the population of slightly more than 30 million. The samurai lost their stipends and special privileges, their social status, and finally even the distinctive badges of their rank and station—their swords. In 1876 a law was passed prohibiting the samurai from wearing swords. Almost overnight, Japan was transformed from a feudal aristocracy to a meritocracy.

Bushido, the way of the samurai, was to be replaced by *shonin-do,* the way of the merchant. A handful of budding industrialists were selected, sold government-owned enterprises at fire-sale prices, and were given government subsidies and even the raw materials to start new enterprises. And in what was to become a standard practice of Japanese industry, Western technology was targeted as an import as essential as any raw material.

Professor Reischauer writes:

To carry out all these innovations, the government needed a great deal of Western technical knowledge. It dispatched students abroad to acquire new skills and hired Western experts at great expense to come to Japan. In these efforts, the Japanese were carefully selective, utilizing the specific national model they felt was best in each field. Since they paid for foreign assistance themselves, they appreciated it more and used it better than have many countries which in more recent times have received aid gratis.

The primary recipients of the newly acquired Western technologies, the newly established industrialists, were then left alone, even by the imperial tax collectors, to pursue success as rapidly and ruthlessly as they wished. These early entrepreneurs started companies in a number of different industries, linking all under the banner of a single bank and trading company that sold the output of all the companies both in Japan and abroad.

The bank and the trading company became the central core of a network of companies that became known as *zaibatsu,* or "financial empires." Vast economic power was concentrated in the hands of a few families, which created central holding companies that held the stocks of major affiliates, which in turn owned stock in dozens of lesser companies. The Mitsui family alone controlled an empire of commercial, financial, and industrial enterprises that employed almost 2 million workers in Japan and overseas.

To fully comprehend the astonishing power of these zaibatsu, consider this assessment of Mitsubishi's market position in 1944 by Eleanor Hadley, an economist attached to the American military government in Japan. Acknowledging the fact that Japan was on its last legs and its industry a shambles, she declared Mitsubishi to be, in relative terms, equal to "U.S. Steel, General Motors, Standard Oil of New Jersey, Alcoa, Douglas Aircraft, Dupont, Sun Shipbuilding, Allis-Chalmers, Westinghouse, AT&T, RCA, IBM, U.S. Rubber, Sea Island Sugar, Dole Pineapple, U.S. Lines, Grace Lines, National City Bank,

Metropolitan Life Insurance, Woolworth Stores and the Statler Hotels."[8]

Writes Rodney Clark in *The Japanese Company:*

Groups like Mitsubishi, Yasuda, Sumitomo or Mitsui, having started advantageously, continued to benefit from government influence in, for example, the allocation of import licenses for technology. They also gained immeasurably from their privileged access to capital. Each *zaibatsu* had a bank, which acted as a money pump. Deposits from the public were channelled towards the other member companies of the group, by loans or by the underwriting of share and debenture issues. The ability to raise capital easily allowed the *zaibatsu* to take the lead in the development of heavy, capital-intensive industries like engineering and chemicals between the two World Wars.[9]

At the end of World War II, the postsurrender instructions given Gen. Douglas MacArthur, the supreme commander of the Allied powers, recommended the dissolution of "large Japanese industrial and banking combines or other large concentrations of private business control," notes historian Michael Schaller in his book *The American Occupation of Japan.* "This concern reflected the fact that some ten *zaibatsu* families controlled nearly three fourths of Japan's industrial, commercial and financial resources."[10]

As a consequence, General MacArthur in effect had the Japanese Diet pass a series of laws designed to break up the *zaibatsu.* Interlocking directorates and cartels were declared illegal, and many of the more labyrinthine pieces of the zaibatsu were scissored off and became stand-alone companies. But when the Korean War broke out, the American detemination to restore the Japanese economy as quickly as possible allowed the zaibatsu to retain key company groupings. The financial institutions that were at the heart of the zaibatsu were left untouched, and after the occupation ended, many of the old zaibatsu began to coalesce once again.

This time, however, the relationships were looser, economic power was less highly concentrated, and many of the companies were linked more by a desire to work with one another than actual financial ties.

The elements that made the zaibatsu so powerful and successful, easy access to low-interest financing and a firmly established global trading network, are two of the major factors in thrusting Japanese semiconductor makers into their present dominant position. "The *zaibatsu* structure itself provides important advantages for the Japanese electronics firm that can draw on its resources," states *U.S. Japanese Competition in the Semiconductor Industry,* a study by the University of California, Berkeley. "First, for example, the *zaibatsu* members provide an important internal market for the firm's products (as, for example, when the Dai-Ichi Bank replaced its IBM banking system with a Fujitsu product). Second, each *zaibatsu* usually includes a large trading company which is frequently used by Japanese firms to perform overseas sales, distribution, and financing. The trading company thus provides increased access to international semiconductor markets." This later advantage becomes increasingly important as the semiconductor industry has become a truly global business.

All of the major Japanese semiconductor manufacturers are subsidiary companies of the major zaibatsu, and two, NEC, which is part of the Sumitomo group, and Mitsubishi, are direct lineal descendants of original zaibatsu. The vertical integration of zaibatsu, wherein one company in the group manufactures the parts while another uses them to build final products, is also a major asset in the manufacture of semiconductors. Japanese semiconductor manufacturers, for example, are all part of highly diversified, vertically integrated companies that in themselves use many of the semiconductors they produce in their production of consumer electronics, computers, and telecommunications products.[11]

Although some of the largest chip producers in the world such as IBM and AT&T might be considered vertically integrated companies, for the most part they do not sell their chips to other companies. Virtually all of their production is destined for their own use. Of the American merchant semiconductor companies, not one can be considered vertically intergrated to anything like the degree of the Japanese chip producers.

Cross-fertilization of ideas within the zaibatsu, across company lines, leads to technological innovation in chips as well as computers and communications products. In addition, the profits from other companies within the zaibatsu can and are used to support R & D and the purchase of new technology in the hugely expensive semiconductor industry. A new production line, or fab facility, as it is called, to manufacture ICs can cost upwards of $200 million.

Gene Gregory, professor of international economics at Tokyo's

Sofia University, in his book *Japanese Electronics Technology: Enterprise and Innovation,* points out:

Quite clearly the fact that Nippon Electric is Japan's leading manufacturer of telecommunications equipment and its third largest computer maker is especially relevant to its leadership in microelectronics. Not only has cash flow generated in the sale of telecommunications equipment tended to sustain a high rate of investment in both computer and semiconductor R & D, plant and equipment and market development. By giving the management of technology highest corporate priority, NEC has assured that innovation and production in each of these major product areas is strengthened by innovation and production in each of the other areas. And, since all the technologies must be optimized on a global scale, NEC has organized for worldwide marketing and manufacturing in all three—communications equipment, computers and semiconductors—to obtain the full synergistic effect of their linkage.[12]

Nippon Electric Company's (NEC) entry into the semiconductor business in 1954 was a logical extension of the zaibatsu philosophy. NEC was formed in 1899 as a joint venture between a Japanese group and an American company, Western Electric (WE), the manufacturing subsidiary of AT&T. Using WE's technology as well as parts, NEC-IWE, as it was first known, began manufacturing telephones in Japan. After some initial difficulties in making the new magneto telephones, Kunihiko Iwadare and Takeshiro Maeda, the Japanese founders of the company, decided to concentrate on engineering as the basis for improved productivity. Moreover, in what might be considered the philosophy that seemingly motivates the entire Japanese electronics industry, Maeda coined the slogan "Better Products, Better Services." Successful at home, NEC was soon selling its telephones overseas in Korea and China, markets that had been opened by Japan's victory in the Russo-Japanese War that ended in 1905. As a result, Japan controlled part of Manchuria, a large chunk of northeastern China and Korea. Treaties in 1905 with Korea and in 1908 with China effectively gave Japan control over those nations' communications systems. But five years later, Japan tumbled into a depression, and NEC's telephone sales plummeted. By 1915, sales had dropped by 60 percent, and NEC turned once again to American companies for help. To offset the loss of telephone sales, NEC began importing a number of American-manufactured electrical appliances, including toasters, ovens, washing machines, and the electric fan, a product never before seen in Japan.

Through depressions and World War I, NEC-IWE prospered, and in 1919, when WE proposed another joint venture, this one to manufacture electrical cables, NEC's president, Iwadare, demurred. He preferred to concentrate on the production of telephone equipment, the company's strength. He proposed instead a joint venture with Sumitomo Densen Seizosho, the electrical-cable manufacturing unit of the Sumitoma zaibatsu. After long, complicated negotiations and considerable stock swapping, the joint venture was abandoned, and the WE cable patents as well as all of NEC-IWE's cable-manufacturing equipment were transferred to Sumitomo. In addition, NEC became a de facto member of the Sumitomo zaibatsu.

Japanese industrialization received a great boost from the military in the 1930s. Japan invaded Manchuria and then China and began an expansionist policy that lead inevitably to Pearl Harbor. After World War II, NEC's operations, which had shifted to almost full military production, were at a virtual standstill. Many of its plants had been totally destroyed or badly damaged by U.S. bombing. Production capacity was also sharply reduced by a severe shortage of materials, personnel, and facilities. The company, along with the rest of Japan, began the painful process of rebuilding. A combination of help from the United States and the Korean War, which sharply increased U.S. military procurement of Japanese products, helped put Japanese industry back on its feet.[13]

One of the first priorities was to restore the shattered telecommunications system. NTT, the government-owned telecommunications system, instituted a series of five-year plans designed to install new telephones in homes across Japan and to introduce nationwide direct dialing. Radio and the newly developing television industries also clamored for new equipment, giving a tremendous impetus to all Japanese communications and electronics companies. The telephone and electronics equipment in the 1950s was for the most part still based largely on the vacuum tube. But NEC, having restored its technology relationship with WE and GE in 1950, acquired licenses to the newly invented transistor and began research in the area.

At about that same time, Dr. Atsuyoshi Ouchi, then a young electrical engineer at NEC, was stricken with tuberculosis and granted an indefinite sick leave. Ouchi had joined the company immediately after his graduation from Tokyo Imperial University in 1942. By 1953, he was recuperating and increasingly bored with sitting home. In trying to keep up with his field, he followed with great interest developments in transistor physics and then learned that these fabled and still mysterious devices were actually available in Akihabara.

"At the time," the now sixty-seven-year-old vice-chairman of NEC told me at our meeting in the NEC headquarters building in Tokyo, "nobody knew how a transistor worked. I bought some—one transistor then cost 3,000 yen [then about $15]—and took them home to study. In those days not many Japanese understood anything about transistors. They didn't know they could be used in radios or would work as amplifiers. In order to see just how they would work, I built a radio with those transistors. I then wrote up the results of my tinkering for magazines, a series of articles on how transistors worked. They were not very technical reports and not even about their engineering aspects, just accounts of how I used them to build a radio. But in those years even descriptions of those features intrigued readers."[14]

NEC's research was proceeding independently of Dr. Ouchi's, and in 1958 they built their first transistor manufacturing plant. Then, in 1966, they decided to plunge into the manufacture of ICs. Koji Kobayashi, the current chairman of NEC, was then its president. He thought the development of ICs would involve systems engineers and circuit engineers who had been working independently. He put together a group of about 250 engineers in a separate division for the research, development, and production of ICs. As its head he chose Atsuyoshi Ouchi.

"I was manager of medical electronics equipment at that time. Mr. Kobayashi summoned me to his office and told me I was to head up the new IC division. 'You will be the manager of this new division; it will mean a promotion to you, so do a good job.' Of course, he said, I was free to make recommendations and he, Kobayashi, would listen."

Ouchi said, "If that's the case, I do not want the job. I want to stay with the medical electronics division. I was nearsighted at the time; I thought medical electronics would make a great contribution to the welfare of human beings, and of course it has. I had a hint that ICs would become a big business in the future, but I never dreamed how pervasive they would become."

Kobayashi refused to accept Ouchi's refusal. "You have been moonlighting, making money on transistors"—he chuckled, referring to Ouchi's 1954 magazine articles—"and now it is time to make money for the company on integrated circuits."

NEC has since gone on to become the world's leading producer of ICs. Ironically, however, the company that was first to use the transistor in Japan and has been at the forefront of the consumer-products revolution, Sony, was surprisingly slow to move into IC manufacture. Sony was in fact the reluctant bride in an arranged marriage with TI that the Japanese government had bitterly resisted for years. Until recently, all Japanese IC makers needed three different licenses to produce ICs. There was the original AT&T patent on transistor technology and the many patents on ICs held by TI and Fairchild, based on the Kilby and Noyce inventions. Both companies agreed early on to pool their patents and not to sue each other over possible conflicts. Thus, any manufacturer in the world could obtain a license from either TI or Fairchild, and that license embodied the right to use all of the technology covered by the patents held by both companies. Any manufacturer, that is, except in Japan.

"I don't know the rationale," Tom Pugel, an associate professor of economics and international business at NYU's Graduate School of Business Administration, told me. Pugel has studied the history of Japan's semiconductor industry closely for many years. A Harvard-trained economist, he is a slender, bearded, friendly, relaxed man in his early forties. We met at his office in the business school overlooking the tiny graveyard and steeple of Trinity Church and in turn overlooked by the towers and temples of Wall Street. The walls and bookcases were covered with pictures, scrolls, and other mementos of Japan that once might have seemed incongruous on Wall Street. Pugel had only recently returned to the school's Center for Japanese-U.S. Business and Economic Studies, following a year as a visiting professor at Aoyama Gakuin University in Tokyo. "Japanese firms needed licenses from both Fairchild and from TI," he went on to explain. "Fairchild was fairly generous. They licensed NEC and used them to sublicense the rights to anybody else in Japan. But they still needed TI, whom I perceive as one of the smartest companies in the semiconductor industry. And TI said, 'Sure, we'd love to give you the license, but in turn, we'd like to get a production facility, in effect a subsidiary in Japan.' At that time, because of the Japanese laws, that was almost impossible. The Japanese hemmed and hawed and, and for some strange reason, TI's patents had never been approved in Japan. They were pending for many years. Consequently, Japanese firms could produce IC's for use in Japan without violating TI's patents, but they could not export them.

"Then Japan began to liberalize foreign direct investment restrictions in about 1968. That was under pressure from the OECD [Office for Economic Cooperation and Development], whose rules said that foreign direct investment should move freely between OECD countries. Japan joined the OECD in 1964, but the Japanese did it in a very deliberate way. They first liberalized the industries that were least vulnerable, but the automobile situation kind of forced the hand of MITI around 1970. The weaker Japanese auto companies started to link up with American companies. They told MITI, 'You can try to stop us, but this is what we are going to do.'

"At that point MITI went to fairly broad liberalization in 1972 but kept a few sectors out, including computers and semiconductors. Nonetheless, by 1976, both computers and semiconductors were liberalized."[15]

To get the rest of the story, some months later I visited TI Japan's Tokyo headquarters, a modest twelve-story building in Aoyama. Symbolically, it sits astride route 246, the Communications Road. I met with Hideo Yoshizaki, the chairman of the board, who ironically was a senior bureaucrat at MITI while TI was struggling to set up shop there. He is less formal than most upper-executive-class Japanese, perhaps because of the many years he served in Europe and Washington. Indeed, about halfway through the interview, on a sweltering early September afternoon, he invited me to take my coat off. After the interview, to the shock and surprise of a Japanese friend who was waiting for me in a coffee shop down the street, Yoshizaki, still in his shirt-sleeves, escorted me to meet him.

"TI's proposed entry into Japan was strongly resisted by many in MITI, who looked for ways of keeping them out, including the use of laws that would have permitted the abrogation of their patents if it were in the public interest," he recalled for me in fluent, colloquial English. "I was against it. At the time, NEC, Toshiba, Mitsubishi, and at least three others were producing semiconductors in Japan." And finally, in 1968, TI struck a deal with the Japanese government, was allowed in, and began operations in 1969. Presumably it was a *quid pro quo,* for the deal that put them in Japan also gave all Japanese companies the right to the single-license Fairchild TI patents that held for everyone else throughout the world. TI, however, did not win a wholly owned subsidiary. They were forced into a fifty-fifty joint venture with Sony, which allowed them to produce ICs in Japan up to 10 percent of the Japanese market. Then TI licensed all the Japanese firms, and they were able to export.

By 1972, Sony, the reluctant bride, had found the marriage of convenience to be convenient enough but irksome in the extreme. The agreement between the two companies broke down, leaving TI a wholly owned American subsidiary in Japan.

They also wisely chose to make TI Japan as Japanese as possible and after the divorce hired Yoshizaki. Whether they knew of his support at MITI or not, he became chairman of the board of TI Japan. "I left MITI in '72. Towards the end of my career there, one of the senior guys called me in. I was on my way to Paris for a meeting on the oil crisis. He suggested I consider joining TI when I retired from MITI. It's the way MITI people retire; they go to a private company. An intermediary set it up with TI, and they thought having an ex-MITI official in their company would be useful."

TI Japan has done remarkably well and has annual sales of more than 100 billion yen (approximately $75 million). That makes them about one-third the size of the major Japanese chip producers. Like virtually all semiconductor makers, they lost money during the great chip slump of 1985 and 1986 but are now well in the black again.

TI has four plants in Japan, one devoted to electronic components and the other three to chips. Its biggest customers are NEC, Toshiba, and Fujitsu, accounting for 50 percent of TI Japan's revenues. Most sales are in Japan, but they also sell to the rest of Asia, the United States, and Europe. "We operate under Japanese law and fall under MITI guidelines," acknowledges Yoshizaki, "but we are regarded as a foreign-based company, and our product is equivalent to an import. TI, even if it could be assigned research projects on the part of MITI, would not because TI is very jealous of its own properties. Also, TI has no real research facility in Japan."

Nor has it been able to gain recognition as a truly Japanese company despite a work force of 4,700, almost all of whom are Japanese. "It is very hard to get the top graduates from the best universities," admits Yoshizaki. " 'Why don't you go to TI,' the professor recommends to a student. But his mother and father say, 'Texas what? That's not a Japanese company.' In the competition for talent, we are a not a first choice. Electronic engineers are still in short supply. And there are more companies going into chip manufacture, Toyota, for example, making it even harder."[16]

Still, TI fares very well against Japanese competition, and in 1985 one of its plants won the Deming Prize for outstanding quality of its products. Its success also convinced Sony it may have been somewhat precipitate in leaving the arranged marriage. In 1980, Sony decided the profits in the industrial IC market were so huge and the research benefits the manufacturing technology provided so great, it made even a marriage of convenience look like a love match. And so Sony is returning to the altar, this time a most eager bride. The blushing groom, doubtless thrilled by the dowry, is AMD, one of the many Silicon Valley companies spawned by Fairchild Semiconductor.

And like so much of the Japanese semiconductor story that bears a "Made in America" stamp, there is even a Fairchild parallel to be found in Japan. In the 1950s, a small company named Kobe Kogyo began to explore this new technology. A subsidiary of the Kawanishi Machinery Works, Kobe Kogyo was created during the war to manufacture vacuum tubes for the military. "Kobe Kogyo had several excellent engineers, and after the war they continued to manufacture vacuum tubes and then cathode-ray tubes for the then just developing television market," Professor Hajime Karatsu, the director of the Research & Development Institute at Tokai University, who speaks fluent English, told me as we slurped our way through a couple of bowls of noodles in Tokyo. "At first, they had a very high level of technology and quickly gained a large market share. But since they had been supported exclusively by the military, they really did not know how to deal with the commercial marketplace. Still, they went into the manufacture of transistors, and their product was quite excellent. But soon the large Japanese companies entered the market and competition began to erode Kobe Kogyo's market share. And while the level of engineers at Kobe Kogyo was quite high, their managers and salespeople were not top quality. Finally, the company failed, and its engineers went to Sony, to NEC, and to Fujitsu. Then the company itself was taken over by Fujitsu."[17]

Japanese companies at that time, in the late 1940s and early 1950s, were rebuilding rapidly, and American technology and American ideas provided much of the impetus. But Japanese industry was burdened with the reputation throughout the world of producing shoddy junk. The label "Made in Japan" was synonymous with flimsy wind-up toys and an array of cheap imitations of Western products that broke almost immediately.

American occupation forces were particularly distressed by the poor quality of Japanese telephone service. Calls were difficult to put through, often disconnected, and the lines so filled with static and noise that understanding what was said on the other end was akin to holding a conversation next to a pavement-ripping jackhammer.

But the Japanese in fact possessed a vast reservoir and tradition of skilled workmanship and quality engineering, all of which had been diverted to the support of its military machine. Faced with the prospect of imminent starvation and with no natural resources, the Japanese and the American military government realized that the manufacture, for export, of high-quality products was essential to Japan's survival. Beginning with the telephone, in May 1946, the U.S. military government ordered the Japanese telecommunications industry to apply American quality control methods to its manufacturing process.

At about the same time, a group of engineers and scholars known as the Union of Japanese Scientists and Engineers (JUSE) was formed. They asked General MacArthur to intercede with American industry to provide advice. Thus, in 1948, three WE engineers began to tour Japanese factories, holding seminars on quality control. They preached a doctrine known as statistical quality control that had first been introduced in the United States during World War II. The idea of quality control had first been articulated by Walter Shewart, an engineer at Bell Labs. His book, *The Economic Control of Quality of Manufactured Products,* had been published in 1931, but until the outbreak of war, only AT&T had ever applied its techniques to manufacturing. Then the U.S. government insisted that everything it purchased meet statistical control standards. So successful were the use of these standards that the statistical methods developed and utilized by the United States were classified as military secrets until the end of the war.

After the WE engineers' visit, the Japanese government passed an industrial standards law that established a set of criteria that all

Japanese products had to meet. JUSE then formed a quality control research group composed of people from industry, universities, and government. Its aim was to study and spread quality control throughout Japan. They began in 1949 with publication of a magazine devoted exclusively to quality control for all factory foremen. And then, in 1950, JUSE invited Dr. W. Edwards Deming, an American physicist, to teach an eight-day quality control seminar in Japan.[18] In height and manner, he is an imposing man (more than six feet three inches tall), courtly in bearing, and addressed by all but his most intimate friends as "Dr. Deming." A statistician with the U.S. Census Bureau before the war, Deming has been a professor of management at New York University's School of Business since 1946. Here he first began to teach his principles of quality control. His message then and now is simple and dramatic: "that continual improvement of quality decreases costs, captures the market and provides jobs. "Deming's way," explains the supplementary notes to the seminars he still gives at the age of eighty-five at NYU and Columbia University's Graduate School of Business and to corporations, "requires top management to foreswear emphasis on the quarterly dividend and on short-term profits and paper profits. Instead, management must adopt constancy of purpose to stay in business by delivering service and product that will help man to live better materially and which will have a market. Management will thus protect investment and will, by growth, create more and more jobs as time goes on."

The tall, imposing American apostle of quality control had a galvanic impact on the Japanese and was immediately elevated to an almost godlike status. A year after his lectures, the Japanese created the Deming Prize, a quality control award that companies compete for with the kind of fierce determination more often associated with life-and-death issues. And perhaps for the Japanese, the reputation for quality is a matter of survival. So important do the Japanese view the Deming Prize, it is presented on national television each year with as much fanfare as are Hollywood's Oscars.[19]

The commitment to quality is so all-pervasive it permeates the entire nation. November is officially set aside as "Quality Month" in Japan. November is also when the Annual QC Conference for foremen is held. It is the most important of the more than a hundred QC Circle conferences held every year throughout Japan. Everyone is invited to participate, especially teachers. At these conferences, they are kept abreast of the latest techniques and urged to introduce basic quality control concepts into high school courses.

The almost religious Japanese devotion to quality control is thought by some scholars probably to be rooted in a Japanese movement called *sekimon shingaku,* which emerged in the early eighteenth century as an outgrowth of centuries-old Chinese teachings. *Shingaku,* literally "heart learning," attempted to codify the teachings of Shinto, Confucianism, and Buddhism into a doctrine aimed at the moral edification of common people and especially the then rapidly growing merchant class. *The Kodansha Encyclopedia of Japan* declares:

Shingaku followers not only believed in the study of human nature, but also involved themselves deeply in the propagation of their beliefs. Two means of propagation were the *dowa,* an informal lecture that addressed concrete problems regarding family, employment, religion and personal relationships; and *sein,* a type of poster, occasionally illustrated which conveyed in verse a simple moral lesson. These were issued from some 150 centers that were established all over Japan. Although their numbers were never very large, since the teachings were largely limited to the then rather small merchant class, followers also contributed significantly to mass education by publishing *shingaku* tracts and holding special talk sessions for children. Involving themselves actively in *terakoya* (village schools) they were enthusiastic supporters of academic and moral education.[20]

Today, there are more than 2 million people in over 125,000 quality control circles, groups of workers composed of no more than five or ten people, registered with the JUSE. But like virtually all foreign products and ideas, the Japanese modified statistical quality control to suit their own needs and style. "Statistical quality control was difficult for the Japanese to understand," says Professor Karatsu, who, as an engineer at NTT, was sent to the first quality control seminar in 1948. Karatsu, who became an early convert to quality control, began preaching it at a number of Japanese factories that manufactured telecommunications equipment for NTT. In so doing, he caught the interest of Kounosuke Matsushita, the founder of Matsushita Electric. Two years later, he joined Matsushita and eventually became a managing director and member of the company's board of directors. He now serves on an advisory council to MITI as well as a consultant to the U.S. Defense

Department on their VHSIC project. "Like so many other things, we modified the idea of quality control, adapting it to Japanese ways."

"The Japanese learned much from American experts such as Deming, Juran and Feigenbaum," wrote Robert Cole, a professor of sociology and business administration at the University of Michigan and an expert on Japanese work organization and quality control practices. But they also evolved a system of quality improvement that is distinctively Japanese."[21]

Instead of involving only quality control experts, as in America, all production workers were incorporated into the process. "Production-level quality control must be backed up by the efforts of other departments," explains Professor Karatsu, "including market research, design, marketing, and after-sales service, in order to be effective. This concept of involving all members of the enterprise is referred to in Japan as total quality control."

Cole states:

No American in the early '50s was talking about quality improvement as a participative system of management that involved all employees and all departments. This was a Japanese innovation. They borrowed many basic principles from American experts—for example the faith that higher quality would lower costs. But even as they did so, they developed ingenious ways to make them operational that were not anticipated by their American mentors.

"Some people in the United States tend to think that they originally invented TQC," says Dr. Karatsu, and the Japanese simply imitated them. This is a misunderstanding. In the United States, QC is only used by QC specialists, while in Japan, QC activity involves all of the employees in the factory. Thus the QC method in the United States and Japan is actually quite different."

One of the best-selling books in Japan since it was first published in 1970 by JUSE is *QC Circle Koryo, General Principles of the QC Circle*. It describes the purpose of QC Circle activities as follows:

1. To contribute to the improvement and development of the enterprise.
2. To respect humanity and build a happy, bright workshop which is meaningful to work in.
3. To display human capabilities fully and eventually draw out infinite possibilities.

The emphasis on human dignity and achievement is deliberate and essential to understanding how the Japanese have managed to produce superior products. For one thing, participation in quality control circles in virtually every Japanese company is always voluntary. Sometimes the meetings are held during working hours, sometimes after work, but the participants are always paid.

There is perhaps no field of manufacturing where quality control plays a more important role than in the production of ICs, a complex process that involves more than 150 steps. Here yield is everything. The cost of production of each chip on a wafer is identical, but the number of usable chips that finally emerge at the end of the process determines the defect ratio. The lower the defect ratio, the higher the yield and the lower the final cost of production of each IC.

"There are two basic approaches to improving IC reliability," wrote Tamotsu Goto and Nobukatsu Manabe, both process engineers in NEC's IC Division in the March 13, 1980, issue of *Electronics* magazine.

One screens out failures by strict inspections, the other tries not to build failures in the first place. In the U.S., for example, the term quality control is often used as a synonym for inspections, and strict and frequent inspections then come to be regarded as good quality control, though of course they raise costs.

Japanese leaders of quality control take the opposite tack. They feel the highest reliability is achieved by building quality in; for if failures are held to a minimum, yields go up, costs come down and inspection becomes almost redundant. They sometimes even ask, "Is the quality control in your company so unsuccessful that you need such strict inspections?" Reliability is built in by always trying to fit the product design to the capability of the manufacturing process.[22]

To fully understand the role quality control circles play in Japanese semiconductor companies, I flew to Silicon Island, as the Japanese have dubbed Kyushu, the southernmost of Japan's islands. Here, in the shadow of a mountain chain dominated by Mount Aso, a series of volcanic peaks that enclose a huge twenty-four-kilometer-wide volcano, are several of the fifteen IC plants that produce 46 percent of Japan's semiconductor output. They were attracted to Kyushu by the low real estate and building costs, the presence of a highly skilled, educated population, and an abundance of pure water that bubbles up from a number of mountain streams through layers of lava that filter the waters before joining to become the Kuma River.[23] On the outskirts of the city of Kumamoto, surrounded by green fields of rice, is one of three NEC plants in Kyushu. This, the first and largest, was set up in 1970 by Masao Suzuki, who is now the retired chairman of the board of NEC Kyushu. It was Suzuki who first named Kyushu Silicon Island, in the 1970s, when many of the chip makers began to move there.

He greeted me at the plant entrance where a flagstone lobby leads to a carpeted floor. Here, at the very entrance to the factory, in accord with the traditional practice in all Japanese homes, I had to exchange my shoes for slippers. In this instance, they were green, had a Pierre Cardin designer label, and were about four sizes too big for me. As I followed Mr. Suzuki to the conference room, I had to take two steps inside those enormous slippers for every one I advanced. Finally, falling farther and farther behind, I simply shuffled along as fast as I could behind my host.

Now sixty-three years old, Masao Suzuki is a cherubic-looking man of great energy and enthusiasm. A rarity among Japan's technocrats, Suzuki is virtually a self-taught engineer, a high school graduate with no formal advanced training. Still, his education at that time, when Japan was seemingly winning the war, made him more important to industry than to the army. He joined NEC immediately after graduation in 1943. Recognizing his outstanding engineering talents, NEC moved him into their semiconductor division in 1957. Two years later he was sent to the GE plant in Syracuse, New York, to learn their IC manufacturing technology. The IC was barely two years old, and GE was producing discrete ICs composed of perhaps half a dozen transistors.

Syracuse in January, when Suzuki arrived, was, as always, bitterly cold and snowy. His allowance for hotel, food, and incidentals was fifteen dollars a day. He found a hotel for twelve dollars a day, breakfast was a dollar, and lunch and dinner were usually provided by GE engineers, who took a liking to the good-humored, eager, young Japanese engineer.

Ten years later, Suzuki was given the task of setting up NEC's first plant in Kyushu. Today it turns out 28 million semiconductor parts a month, microprocessors, and 1-megabit and 256K-megabit memories for industrial and consumer products. The Kumamoto plant is huge, covering 84,900 square meters. Seven buildings contain offices, manufacturing facilities, warehouses, and dormitories for about three thousand men and women workers.

Suzuki took me upstairs to the second floor of the factory building, to the diffusion line. In a locker room we pulled on blue coveralls, hoods, and masks and exchanged our slippers for white cotton shoes. In America these are known as "bunny suits." Then we stepped into an antechamber and took an air shower for two minutes. Jets of air hosed the coveralls, blowing away any dust or particulate matter that might have settled on them from the time they had been lifted from protective envelopes just prior to our pulling them on. It is just one more effort to reduce contamination inside the clean-room environment where the chips are actually made.

Finally, the shower ended, a door hissed open, and I stepped onto the diffusion floor. It is a vast room filled with widely spaced rows of computer work stations, wafer steppers, test stations, ion implantation machines, and all of the complex and expensive machinery required to produce ICs. The hum of mighty compressors, recirculating 12,000 tons of air every hour, was punctuated by the incongruous sound of Stephen Foster's "Camptown Races," played by chimes. The sound grew louder, and I had to leap out of the way of what looked like a filing cabinet disguised as a modern-day Dodgem car, a squat rolling box about four feet high, two feet wide, and five feet long. It was but one of twenty robots that roamed the plant floor carrying wafers from mask makers to photo resist coating stations to steppers to ion implanters, etc., on their way to becoming finished chips.

This and virtually all of the newer Japanese fabrication plants are highly automated and make greater use of robots than do even the newest American and European chip factories. Indeed, Mitsubishi boasts that its "fully automated Saijo factory is a major accomplishment toward the goal of the unmanned plants of the future." Other than maintenance engineers, the so-called front end of the plant, that area where the silicon wafers are imprinted with the chip designs and cut apart into individual chips, requires no people.

At the NEC plant, in a vast expanse of machines and robots, I counted only five people, most of whom simply observed the production process, looked infrequently into microscopes, or checked computer monitors and made an occasional note on a clipboard. We paused at one of the steppers, a rectangular box about eight feet high and four feet wide, with a window at about eye level. Inside, a series of lenses moved above a six-inch wafer. A spot of ultraviolet light touched the mirrored surface of the wafer and then moved on. Dot, dot, dot, repeated again and again, burning the pattern of 580,000 transistors, circuit lines, and other elements into an area on the chip about the size of my pinky nail.

There were thirty of these wafer steppers, all the most advanced design yet produced by Japan's Nikon Camera Company. Most of the equipment here is Japanese manufactured, and about 60 percent of it is built by other NEC or Sumitomo companies or subsidiaries. It is typical of the Japanese industry where each semiconductor manufacturer is but one part of a giant company that makes not only the end products into which the chips will be assembled but also most of the machinery used to manufacture the chips. And if the company does not make a particular machine, it will more than likely buy it from a rival Japanese company. Nikon, for example, is part of the Mitsubishi zaibatsu. Indeed, the only American chip-making machinery I saw at NEC Kyushu was a number of ion implanters, machines that add conductive properties to silicon wafers. These were built by Varian, an American company that also manufactures them in Japan in a joint venture with NEC.

From the diffusion floor we went upstairs to the assembly and testing area. Here there were more people, but again, most of the actual labor was done by machines. All of the assembly and most of testing are automated. It was most impressive, but what followed was even more so, and herein may lie the difference between the Japanese and American approach to production. We went back into the locker room, stripped off our bunny suits, and walked back out to the administrative section. We stopped before a glass-walled office. Above the doors were the letters TQC, followed by a string of Chinese characters. This was the Total Quality Control Circle office. Inside, seated around a small conference table, were five young women and a young man. All were in their early twenties and worked in inventory control. This was the South Wind Total Quality Control Circle. They looked to me more like the dance committee of the senior prom. In fact, as in most Japanese factories, the majority of the younger workers here lived in company dormitories right on the plant site. They had just come off a working shift and were being paid overtime for this meeting.

Spread before them was a large chart divided into four parts. The

Japanese call the chart "fish bones," and it looked not unlike the skeleton of a fish. Each member of the group had a number of small adhesive-backed labels with Chinese characters written on them. One of the girls used a magic marker to draw an arrow from one position on the chart to another. Two other girls spoke up immediately, then stopped as they realized both were talking at once; then both again started talking at the same time, stopped again, and laughed. They did not, however, consider the problem they were dealing with to be a joke. Their responsibility was to make up inventory sheets listing the types of products, their quantity, quality, and so forth. Occasionally, the computer printout would be incorrect. The group was meeting in an attempt to solve the problem.

Later, I met with two people who worked here, and their comments were revealing. Chiharu Okabe is twenty-four years old, a pretty, intelligent young woman who, unlike most of her coworkers, lives in Kumamoto with her parents. She has worked for NEC for six years, since her graduation from high school. Her job is to supervise the operation of the ion implantation machinery. She finds her work interesting and important. "It is important," she told me, "because if I make a mistake it will affect the job done by all the other people before me, and that will make their work to have been in vain."[24]

Okabe's working day is not unlike most factory workers. When she works the early shift, she arises at 4:30 A.M. and drives to work in her own car, a thirty-minute ride. The shift begins at 5:30 and ends at 1:30 in the afternoon. She works four days, then has two days off, changes to the afternoon shift, and works another four days. Woman in Japan are not permitted to work at night, and so she then switches back to the early shift. It adds up to twenty-one working days each month, for which she is paid $1,600. Like most of the workers here, she received three months of on-the-job training. She eats lunch in the company cafeteria, where meals cost an average of 100 yen, about $.75.

For the young men and women who live four to a small but comfortable dorm room, rent is 500 yen a month, about $4.50. And like college, there is a busy social life, parties, sports events, picnics, etc., for any who care to participate. Okabe occasionally attends these but more often goes to Kumamoto City to the movies or dancing with her boyfriend, who also works at NEC in the equipment-maintenance department.

Toshimitsu Kohtaka is thirty-seven years old, an engineer, a graduate of Kumamoto University. He has been with NEC for twelve years and has risen to the post of supervisor in the Production Engineering Department. "I analyze all of the defective chips and attempt to learn why they failed and so improve the yield." He still finds his job a great challenge. "The production process is a new technology, and so there is a lot I still must learn." His work day begins at 8:15 in the morning and runs to 5:00 P.M. He is up at 6:00 A.M., has breakfast, and bikes to work in five minutes. Married, he has three children. After work, he spends his time reading, swimming, and being a good husband and father. In addition, he is taking an evening course in quality control techniques given at the local high school.

Twice a year the company evaluates his performance and, based on his productivity, pays him a bonus in addition to his monthly salary. With the bonuses, he earns about 5 million yen a year, approximately $37,000.[25]

Salary and bonuses are not the only motivations that drive Japanese workers. Culture, tradition, and even religion play an important role in molding the Japanese work ethic. Professor Hajime Karatsu discussed this idea and others in a series of full-page advertisements labeled dialogues in the *New York Times* and the *Sankei Shimbun*. A highly visible, outspoken scientist, a sort of Carl Sagan of Japan, he is often quoted in the media in both Japan and the United States, especially on the differing technological standards in those two countries. His fellow discussants were Japanese novelist Ayako Sono and American authors David Halberstam and Robert B. Reich. "In Japanese we have this word *mottainai*," said Karatsu in the dialogue of July 14, 1987. "What a waste, what a shame, would be a rough translation. For example, if you make a million microphones and only one or two are defective, it means that almost all of the material that goes into production comes out as product. If there are many defective products, we have to discard them. What's the percentage? We measure in terms of PPM—parts per million. We once used percentages, but for the past twenty years components manufacturers have figured in terms of PPM.

"The very idea of producing a defective product is *mottainai:* It's a waste. It probably comes from Buddhism."

"This is different from just being stingy or avoiding making useless products so that neither you nor your company suffers an economic loss," added Sono. "It's against aesthetics not to use everything to its full potential. To some Japanese, the word *mottainai* has almost a religious connotation."[26]

This attitude, along with other cultural traditions, has been as important to Japanese success in the semiconductor industry as MITI-sponsored research, trading policies, low-cost financing, and vertical integration of its semiconductor companies. Indeed, it is the Japanese worker and his attitude toward his work and his company that are Japan's secret weapon in the Chip War.

4

Why Japanese Garbagemen Run

Early one hot September Sunday morning, I left my hotel in the Shibuya section of Tokyo and witnessed what to me was a most amazing sight. A shiny blue garbage truck, about half the size of the graffiti-covered behemoths I am used to seeing in New York, was parked at the curb. Two collectors were heaving bags into the open back of the truck. Then, to my astonishment, the two men literally sprinted down the street while the truck slowly followed. By the time the truck arrived at the next pickup point, the collectors had already lifted the trash and were waiting to throw it into the truck. As soon as they had, they immediately ran to the next pickup point down the street. It was apparent that this hustling, eager, even cheerful approach was the accepted method of garbage collection in Japan.

Later that day, I met with my friend Akio Akagi. Aki is the science correspondent for NHK. He has traveled about the world covering everything from the Apollo moon launches at Cape Canaveral to the nuclear plant disaster at Chernobyl. He has also written several books about science and technology and is one of the most respected journalists in Japan. About six feet tall, Aki in other ways is almost prototypically Japanese. He is a third-generation product of the Meiji Restoration. "My grandfather was a Tokugawa samurai, but one of the lowest class, called *ashiegaru,* which means literally 'light foot.' That was the infantry.

"After the Tokugawa were overthrown, he had great difficulty finding work and in fact was only occasionally hired as a scribe. In order to make ends meet, my grandmother worked as a seamstress, sewing clothes. By denying themselves a great deal, they managed to send my father to college in Wakayama, just south of Osaka. After graduation, he went to work for the Yasuda Trust Company in Osaka. At that time, Osaka was the financial center of Japan, and he entered the main office there as a trainee, a bank clerk. Eventually he was promoted to manager of the Kyoto branch, where I was born in 1932. That was just before the Japanese army invaded Manchuria. My father was promoted again to Nagoya and then to Tokyo. My father developed a taste for Western culture and education and spoke English and French, and he encouraged me to study foreign cultures. He never pushed me, but he would have a book on his desk and say, "You may find this interesting to read." So he encouraged me to look at the rest of the world."

Perhaps that is why Aki majored in English at Tokyo University and speaks it fluently. He has lived and traveled abroad and has, I believe, much more of an international view than do most Japanese. For one thing, he is convinced that the United States, as the leader of the free world, must remain a strong and technologically and scientifically superior nation. His convictions are strongly held, even passionate. Often he will seize upon an idea with such intensity that his eyes will squeeze shut and his command of English will be limited to a single word repeated four or five times until suddenly he slips back into gear with a rush of fluency that borders on the poetic. Aki and I were classmates in graduate school at Columbia almost twenty-five years ago.

We have visited one another several times over the years. Our assignments have taken us to each other's home countries, and we have tried to keep each other abreast of American and Japanese developments in science and technology. Aki has been invaluable in helping with the research for this book and has provided me with insights into the Japanese character and culture I might otherwise never have gained.

I told Aki what I had seen. "Why," I then asked, "do Japanese garbagemen run?"

At first he laughed. "Garbagemen always run in Japan; it is the way they do their job." Then he realized that there was an important point to make about the Japanese work ethic that might answer some larger questions about Japan that most Americans and Europeans never understand. His answer might also demonstrate what I believe is Japan's most important weapon in the Chip War—the effectiveness of Japanese workers.

"Let me think about my answer a bit," Aki said finally. The next night, we met for dinner, and he handed me a typewritten answer to the question:

"Why does a Japanese garbage collector run? Probably because there is a prospect for him to be promoted to a higher position, say, foreman of garbage collection, if he works diligently. You might say the foreman of garbage collection is still a garbage collector. However, in Japan, people don't think that way. For instance, Japanese people will respect a veteran garbage collector who has worked for thirty years more than a young engineer who has just graduated from a university (though this is a somewhat extreme example). The reason is the veteran garbage collector is a man of achievement and the backbone of society.

"The Ministry of Education acknowledges skilled craftsmen as *mukei-bunkazai,* intangible cultural treasures. Some are old ladies who are wonderful at weaving and dyeing, etc. All of them are respected as *ningen-kokuhou* (living national treasures). Their deaths are reported in newspapers to commemorate their achievements. This merit system does not work, however hard the ministry pushes, unless the general populace respects a person's skill, which is acquired only by diligent work. So a good garbage collector runs even if his manager does not watch his work.

"Let's take another example. In the Honda Company, every year all of the newly recruited university graduates are deployed in the assembly lines for more than six months after they complete the orientation course given at the company's training institute. This is not a unique example. In a railway company, new employees who are expected to become directors of the company within ten or fifteen years begin their careers as assistant train conductors or ticket collectors at stations.

"Of course, these young graduates are worse at these jobs than the foremen and quite inferior in performance compared to the laborers.

But they are supposed to learn from them. By such training, the young cadets of Japanese industry will recognize the importance of the foremen's work and will keep their respect for workers after they are promoted to a higher position. A young research engineer who has been recently assigned to his new job will greet the workers with whom he spent a six-month apprenticeship on the factory floor and from whom he learned what floor work involves. He will not order workers to make things with just a list of specifications. Rather, he will discuss the matter with them and listen to their advice. This human relationship, involving mutual respect among workers and engineers for the importance of each man's role, is one of the social bases of Japanese quality control.

"In other words, unless there is some merit-evaluation system, some incentives for individual initiative in every bracket of society, people would not work diligently. In the U.S. you have many success stories about millionaires but very few stories that praise ordinary people's diligence. Not everyone can become a millionaire, but many people can become socially respected by aspiring after some model character as the backbone of society.

"And that is why Japanese garbagemen run."

The garbagemen are not the only ones who run in Japan. Much of Japan's success in the Chip War can be attributed not only to the undeniable abilities and business acumen of its people but to their incredibly dogged work habits. In the offices, board rooms, and executive suites of Japan's giant corporations and government ministries, directors, vice-presidents, salesmen, accountants, engineers, and bureaucrats also run. The lights in Tokyo's office buildings burn far into the night as the modern Japanese executive works longer hours and harder than his American and European counterparts. Around the office buildings, taxi's wait, like schooling sharks, to snap up the departing office workers when finally they quit for the night. The streets in the business districts and the office buildings are honey-combed with coffee shops, snack shops, and sushi bars whose customers arrive in streams after eight and nine o'clock at night to have a drink, late supper, midnight snack, and convivial meeting with men they have just spent the last twelve or fifteen hours working with. Nor is it a time of relaxation. Business is definitely the preferred subject, until finally, almost reluctantly, the so-called salarymen tear themselves away and take what is usually an hour and a half or longer train ride home.

John Burgess, the Tokyo correspondent, for the Washington *Post,* writes:

He is hailed as an industrial warrior, the driving force behind Japan's economic success. He is also ridiculed in cartoons and commercials as a wimp who lives in terror of the boss's glower and chews antacids by the case.

He is as much a part of the Japanese cityscape as neon and sushi bars. He is found in dark suit, imported necktie, short hair parted on the left. No beards or moustaches. Accessories are standard too—pocket calculator, leather briefcase, commuting pass, business cards, pornographic comic book for long subway rides.[1]

Most of all, he is mass produced. The 'salaryman,' as the male white-collar worker is called here, is what most of the 280,000 young men who graduate from universities each year quickly become.

The good salaryman devotes himself body and soul to the company. If the company thrives, so will he. He loves his wife and children, but in a pinch he can be counted on to put the office first.[2]

Nor is such workaholic behavior limited to the home-office glare of Tokyo. Bill Taylor, a New Hampshire computer consultant, grew up in Japan, the son of missionaries. He left in 1963 at the age of eighteen to attend MIT. Some years later, he returned to Japan to sell software. "I was studying at a company hotel near the plant, in Fuchu," he recalled for me one day on a visit to New York. "There were about a thousand programers there, sitting at these little desks about two feet wide and eighteen inches deep with about a ten-inch aisle between them, writing code. They would show up in the morning at seven-thirty to get a head start on things, and everybody is still beavering away at eight, nine, ten o'clock at night.

"One evening I came back from dinner about seven o'clock, and the place is empty; there's only a couple of hundred people around. I ask, 'Hey, what's this, a plague?' And the guy says, 'No, it's a company rule; everyone must go home at six o'clock on Thursdays. Management doesn't think it's a good idea to work twelve hours a day, six days a week.'

"I asked, 'What about all these guys?' He said, 'If your project is very important, you can get special permission to stay late, but that can only be granted twice a month and can only be granted by a high-level signature.'

"And sure enough there's some turkey with an arm band walking through the place making sure all these workaholics have permission to stay."[3]

The Japanese work ethic is not cast off once the worker leaves home. And for many of the younger executives several lengthy tours abroad are part of the climb to the top. Koichi Shimbo, the forty-two-year-old chief representative for NEC's U.S. communications office in New York, has followed a fairly typical career trail. Born and educated in the suburbs of Tokyo, he graduated from the University of Tokyo with a degree in foreign studies, majoring in Spanish. He joined NEC in 1968 and after his initial training was sent overseas. "I spent more than fourteen years in our Latin American marketing division," he told me. "We lived in Nicaragua and Guatemala, from where I traveled to Mexico, Honduras, Panama, Colombia, Costa Rica, Venezuela, Brazil. As a marketing representative, I worked as an account executive to local governments, selling and servicing telecommunications, especially in those countries that were far behind the international standards in terms of telecommunications. When I started in Central America, for example, there was less than one telephone per hundred people. Now they have almost ten per hundred people because of our efforts. There was heavy competition there from Europeans and Americans. I worked to establish especially a network in those five Central American countries. We were introducing almost brand-new, the most advanced, technologies. Now Central American nations are one of the most advanced communications-accessible regions in the world."[4]

Now in New York, he lives in the affluent suburbs of Scarsdale with his wife and fourteen-year-old son and six-year-old daughter. His son spends more than an hour each way traveling by bus to The Japanese School of New York, a private school in Queens, one of the outer boroughs of New York City. The school is run by the Japanese Ministry of Education for those overseas Japanese who can afford its $2,280-a-year tuition. Shimbo spends less time traveling to and from his job, but like his counterparts in Japan, he puts in long, long hours. When his day, which started in the office at 8:30 in the morning, is finally finished and the last fax has been sent to Tokyo, he heads for Grand

Central Station. Here he catches the 10:30 P.M. train to North White Plains, the closest stop to Scarsdale. On board are hundreds of other Japanese New Yorkers, heading home on what they call the "Orient Express."

After a few more years, Shimbo will be transferred back to Japan to continue his hard work and doubtless his rise toward the upper-management levels of his company. But hard work in Japan is not fueled simply by a desire to get ahead. Loyalty to the company is a major Japanese trait. "The Japanese company is like a family," declares Professor Hajime Karatsu of Tokyo's Tokai University, "and treats its employees as such. Like the crew of a ship, when a ship is struck by a storm, they all work together. Since Japanese workers rarely leave their company, they are bound by ties of loyalty and family to always work together."

Company loyalty is all encompassing. Everyone from the chairman of the board to the meanest sweeper gives it tangible expression when all gather at the start of each shift for calisthenics and a rousing chorus of the company song. The Matsushita Electric Company's song is a virtual paean of loyalty to nation, company, and teamwork:

> For the building of a new Japan,
> Let's put our strength and mind together,
> Doing our best to promote production,
> Sending our goods to the people of the world,
> Endlessly and continuously,
> Like water gushing from fountain;
> Grow industry grow, grow, grow,
> Harmony and security,
> Matsushita Electric.[5]

Like so much else in Japan, company loyalty has deep cultural roots. NHK's Akio Akagi believes that much of the ability of both individuals and institutions to work together stems from Japan's agricultural traditions. "The cultivation of rice is a major factor in establishing the thinking of Orientals," he argues. "Rice is quite different from wheat and other Western crops. Rice requires much more work to grow. Rice cultivation requires very much more minute attention and care than does wheat cultivation. Rice needs great amounts of water, so every farmer must cooperate to get water to irrigate the paddy, digging canals, arranging the time to get the water. During the dry season, watering the individual rice field is a major problem, and so every farmer must participate in an organized schedule to draw the water. Even today every village has its own farmers' meetings to discuss the problem of water.

"Therefore, the dependence of Japanese farmers upon each other and their knowledge of the importance of the group are much more basic to our working ethic and working culture as a result. In America the fields are huge and the farmers are much more independent."[6]

In a small, crowded country, the rugged individualism that is so conspicuous a part of the American cultural heritage is almost anathema to the Japanese. The group rather than the individual is paramount. "If you look at our mythology, we talk about the Lone Ranger and Davy Crocket and Wild Bill Hickock, and all those guys were essentially one-man deals," says Bill Taylor. "In our mythology, any single white hat riding into town can blow away any number of black hats. In Japanese mythology, they talk about the seven samurai; that's about the smallest heroic group I know. There were forty-seven *Ronin*. . . . All of their mythology centers around groups. And as their fundamental thinking does not admit to an individual able to do anything significant, their natural ability to work together for the greater good is readily understandable."

Not only do the Japanese work well in groups, but they are often shifted around within a factory or plant, assigned to different jobs, so that each worker knows and appreciates what his counterpart does. This then adds to the ability to control quality from the ground up. "To organize workers in this fashion and maintain high standards of workmanship," believes Akagi, "there is a kind of mind-set that evolves from a mastery of *Go*."

Go is an ancient Chinese game of strategy, perhaps four thousand years old. Despite its reputation as among the most intellectual games ever devised, the most complex and exciting game in the world, it never gained popularity much beyond China, Japan, and Korea. Go is played on a wooden board that is divided by nineteen vertical and horizontal lines, creating 361 intersections. There are only a handful of rules, and they are quite simple. Two players, black and white, alternate their stones or pieces, on unoccupied intersections of the board. The aim of the game is to surround the opponent's stones and thus capture them.

That's all there is to it, but the number of possible play sequences is a staggering 10^{761}, that is, 10 followed by 761 zeros. Chess, by contrast, has a mere 10^{120} possible moves.[7]

"Unlike Go," points out Akagi, "pieces of chess each have their own role or assignment. Go pieces have no preset role, and the game is therefore inherently more complex. In a sense, chess play is limited by the role of each piece; chess pieces can only be moved in certain ways. There are no kings and no queens and no pawns in Go, so it's more closely allied to the Japanese system where there is no social distinction between bosses and workers on the job and no difference between the pieces on a Go board. If you go to a British factory, by contrast there are many grades of canteen. This one is for workers, this one is for executives, this one for foremen. In a Japanese factory there is just one huge dining room. The plant manager, the executives, and the workers all eat in the one dining hall."

Some experts also look to the stratagems used in Go as an arcane explanation for the Japanese success in selling their chips in world markets. I thought to pursue this Eastern mystic mumbo-jumbo approach to business strategy at a meeting with the sales manager of one of Japan's largest chip makers. And so, in an offhand manner designed to catch him off guard, at the conclusion of the interview I asked whether he played Go. Throughout the interview an interpreter had been translating my questions into Japanese and the answers into English. Occasionally, the sales manager would interrupt the translation to correct the English answer being given or begin to answer the question even before the translation had been completed. His English obviously was excellent. Thus, when I asked if he played Go, he did not even wait for the translation but instead responded almost immediately in English. "I love the game, but unfortunately my work takes up so much of my time, I can only get to play nine holes about every four or five weeks."[8]

So much for arcane Oriental subtleties based on the ancient game of Go. I laughed and explained I meant Go, not golf, and the idea that it provided a cultural background for Japanese business strategies, whereupon he laughed. But the incident pointed up one of the great tolls the Japanese have paid for their success. They have little leisure time, and the pressures and competition of corporate life seem to be affecting the health of many Japanese executives. A survey of 500

Tokyo salarymen in August 1987, conducted every year by the Dai Ichi Kangyo Bank, revealed that three out of every four men polled said that there was something wrong with them somewhere. Almost half complained of failing eyesight or tired eyes; twenty-nine suffered from fatigue or general lethargy; 23 percent had stiff shoulders. They diagnosed the cause of their ailments as age, stress, overwork, and lack of exercise.[9]

Despite the complaints, more than half of those responding said they had not taken sick leave during the past year. In the boardrooms of Japan's major corporations, the toll is even more devastating. At the same time the Dai Ichi Kangyo Bank survey was made, *Time* magazine reported the sudden deaths of the chief executives of at least twelve major Japanese companies.

The immediate causes of death ranged widely, from pneumonia to heart attacks. But many Japanese are convinced that the real killer was endaka, which means a strong yen. The 40 percent rise in the value of the Japanese currency since September 1985 has made the country's products more expensive abroad and stalled its vaunted export machine. As companies have increasingly suffered slipping sales and profits, corporate leadership has become more stressful—and possibly deadlier—than ever.

Time went on to quote a number of doctors on the effect of stress on Japanese executives:

Japanese managers, most of whom work for a single company during their entire careers, have an extremely close personal identification with the fate of their firm. Says Dr. Tomio Hirai, a psychiatry professor at the University of Tokyo: "A Japanese obsession with perfectionism puts pressure on executives. As a result they tend to overstrain themselves." Such strain peaks during times of economic hardship, according to a study of Dr. Okada. He found that the incidence of heart attacks among Japanese managers was nearly four times as high during the oil crises of 1974 and 1979 as in the high-growth period of 1966–68.

The unhealthy effects of job stress are made worse by the workaholic life-style of typical Japanese executives. They have relatively little time for their families and even their after-hours social encounters are usually work related. Says Dr. Yasuo Matsuki, director of Tokyo's Shin Akasuka clinic: "Top executives have to attend a party or two after work almost day in and day out. As a result, they end up eating high-calorie food and drinking a lot.[10]

If life at the top is tough, it is seemingly not much easier a little lower down the executive ladder. While most Japanese earn as much as, or even more than, Americans for the same jobs, their standard of living is much lower. Consumer expenditure data for both countries in 1984 (the most recent figures available) showed a median monthly household income of $1,600. The difference in prices for the same items, however, is startling. The cost of even the most mundane of goods and services can be staggering to the average Japanese. A driver's license costs $2,000; auto inspections, $600. In Tokyo, even if you can afford a car, which costs about 15 percent more for the same model in Japan than the United States, you could not get an auto registration for it unless you can prove you have off-street parking.[11]

Michael Kahan, an associate professor of political science at Brooklyn College and a consultant on Asia, wrote in the "Letters" section of the August 7, 1987, *New York Times:*

These and other examples, have a common root: the conscious, long-term and concerted policy of the Japanese Government (monopolized by the same political groups since the end of World War II) to discourage consumption, to punish personal indulgence and to reward only those activities that earn hard foreign currency. Thus, Japan has fewer automobiles per capita, fewer miles of paved highway and higher food costs than any other industrialized nation in the world. It also has the poorest housing.

Most Japanese live in tiny two- and three-room apartments or only slightly larger houses. More than 25 percent of Japanese homes have no indoor plumbing. Despite the often stated excuse of high population density, the fact is that not a great deal of the usable land in Japan has been used for housing. Only 47 percent of metropolitan Tokyo's available land, for example, has been developed. Much of the surrounding countryside consists of suburban farms that are so valuable that their land is worth $230 and more a square foot and is rising in price at 50 percent a year. By comparison, Park Avenue real estate sells for about sixty-five dollars a square foot. The farmland, however, is taxed only on its agricultural value, which is quite low. Local government subsidies and a traditional attachment to the land keep these small farmers from selling out to developers. Instead, many borrow against the value of the land and put the money in businesses and other investments. The end result is that most Japanese must live far from their workplaces. New employees of the major Tokyo-based firms are often housed in company-owned dormitories in cities as far away as Yokohama simply because no affordable housing is available any closer.[12]

It is also impossible to buy very much furniture or large appliances, since there is simply no room to house it. Thus, spending patterns in Japan are dramatically different from those in the United States. The average Japanese household spends only $1,077, while its American counterpart spends $1,581 per month. The Japanese spend $250 a month less on autos, $14 less on home furnishings and appliances, and because of the much smaller dwellings, $41 less on utilities such as gas and electricity. They spend more money than Americans on such things as personal care, food, and education.[13]

The seemingly well ordered structure of Japanese family life is beginning to feel the strain of the unrelenting work ethic. Young people are no longer as certain of the need to work unceasingly and forgo all luxuries, as were their parents. Sumiko Iwao, a professor of social psychology at Keio University in Tokyo, calls today's young Japanese the "no-hunger" generation. "Our generation was always hungry and had to work hard to live," she notes, "but now, the young have everything and can relax."[14]

Women, too, once the most submissive members of the Japanese family, are beginning to show the first signs of rebellion. "Japanese middle-class women once revered their hard-working, ambitious husbands, but the modern generation demands emotional satisfaction as well as social status," wrote the highly respected feminist author Seiko Tanabe in the June 1987 issue of *Chuo Koron,* a Japanese magazine. "The result is a quiet feminine revolt against uncaring, egocentric spouses."

Tanabe went on to describe a casual social meeting with the vice-president of a big corporation—a man she expected to be charming and urbane.

What a shock when his entire conversation consisted of bragging about his alma mater. His equally brilliant son, he pointedly added, had also attended the same school. The preening arrogance of this corporate peacock! I thought to myself, "If a dope like this can make it to the boardroom, you would have to be brain dead to fail in business."

Tanabe was born in 1928, and raised in an era when Japanese women were taught to respect and defer to men and honor their authority.

Thus I was all the more disappointed to discover the truth: The giants were actually pygmies.

Ineffectual elitists make inept fathers and husbands. What happens when women, unable to tolerate male vanity and emotional immaturity any longer, lose respect for their husbands, and abandon the role of mother and homemaker?

Many unloved women will end the charade and walk out. The shock reverberating from that is potentially more traumatic than Japan's defeat in World War II.

Social critic Kenichi Takemura, quoted in the *Wall Street Journal,* uses World War II terminology to describe the woman's role in the modern Japanese economy. "We call our homes and wives 'flattops.' Men are the fighter planes. We need the flattops to take a rest. But flattops never accompany fighter planes to the front."

In the fall of 1987, the *Journal* had assembled a panel of Japanese women to explore the problem. Plaintive quotes such as this were typical: "My husband rarely comes home before 11 P.M. and hardly ever eats dinner with our three sons. Isn't it a strange life, where night after night everyone works so late?"[15]

Still, a woman's role is not merely to sit home and wait for her husband. She is charged with overseeing the education of the children, and that is, in Japanese eyes, as important a job as there is in Japan. They are called *kyoiku* mamas, "education mothers," and they study with their children, wait in lines for hours to register their children for the competitive examinations they must take even to enter the right nursery schools, and then wait even longer hours while their children actually take the exams.

"Much of a mother's sense of personal accomplishment is tied to the educational achievements of her children," says *Japanese Education Today,* a study by the U.S. Department of Education. Those achievements are made in what may be the most demanding and successful educational system in the world today—a system that has produced what the study calls a "learning society of formidable dimensions."

That same educational system can also claim much of the credit for

Japan's remarkable economic success. "Nothing, in fact, is more central in Japanese society or more basic to Japan's success than is its educational system," says former ambassador to Japan and Harvard professor, Edwin O. Reischauer. A 1982 study comparing U.S. and Japanese manufacturing capabilities, by the New York Stock Exchange, attributed Japan's greater productivity to its educational system. "The single most important factor in Japan's extraordinarily high productivity," noted the study, "more important than quality circles, techniques of management or the partnership between business and government—is the high quality of Japanese primary and secondary education."[16] Japanese students consistently score higher on international tests of science and mathematics than the students of any other nation and much higher than American youngsters. Since the testing began in 1964, Japanese thirteen-year-olds have placed first or second in almost every category of mathematics skill tested, every single year.[17]

Data from the Second International Mathematics Study, which tested arithmetic, algebra, geometry, statistics, and measurement among the students of twenty countries, showed that Japanese seventh-graders scored highest in all five subjects. By comparison, American eighth-graders ranked average or below average in those same categories.

To determine why Japanese and other Asian youngsters did better than the rest of the world in these areas, a team of researchers from the University of Michigan and the University of Chicago studied first- and fifth-grade mathematics classrooms in Japan, Taiwan, and mainland China. The differences, they discovered, show up as early as kindergarten; hence, the researchers looked for cultural differences that might account for the marked differences in academic performance. And there were. American mothers and children, for example, placed greater emphasis on ability as the key factor in achievement. Chinese and Japanese mothers and their children thought that effort was the crucial factor. The research team concluded:

Motivation for academic achievement may be enhanced to the degree that students, as well as their parents and teachers, believe that increased effort pays off in improved performance. The willingness of Chinese and Japanese children, teachers and parents to spend so much time and effort on the children's academic work seems to be explained partly by this belief.[18]

Thomas Rohlen, a professor of education at Stanford University, gives the Japanese educational system very high marks. "The profoundly impressive fact," he says, "is that it is shaping a whole population, workers as well as managers, to a standard inconceivable in the United States." Rohlen, who spent a year in the high schools of Kobe, found that the greatest accomplishment of Japanese primary and secondary education was not in its creation of a brilliant elite but in producing a high average level of capability in its graduates.[19]

Dr. Makoto Kikuchi told me very much the thing. "Here in Japan, at the start of the Meiji era about 100 years ago, the leaders were very astute, much better than the present politicians in Japan. They planned the future of Japan, basing it on the same fundamental primary and secondary education throughout the country. That took a long time to change the fundamental intellectual structure of the Japanese people. Because of that we have a much more concentrated grouping of intellectual achievement, but unfortunately we do not have the sort of elite you do. But neither do we have a long tail of inadequately educated people."[20]

At this point, Kikuchi drew two bell curves on a paper to illustrate the spread in intelligence among the two populations (Fig. 1). One curve, representing the intelligence level of the Japanese population, fits inside the other, the intelligence curve of the U.S. population.

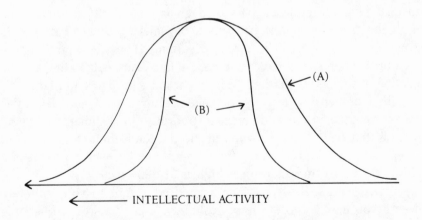

INTELLECTUAL ACTIVITY

"There are not very many outstanding elites pushing way ahead, nor are there very many people who are intellectually far behind. In the United States there are many far-ahead elites. I noticed in my visits to Bell Labs, MIT, and RCA that in most cases I could deal on the same level with the scientists I worked with. But occasionally one or two persons out of twenty or thirty whom I encountered were outstanding elites, so much so I felt perhaps the fundamental structure of their brains might be different from mine. You have a small group of elites who are actually guiding your society, but at the same time you have a long tail of less than normal intelligence."

This tight concentration of intellectual levels is, Kikuchi believes, one of the most important reasons for Japan's industrial success. "This clustering of intelligence," he insists, "has special meaning. People can understand how to do things without being given any special instruction. The lower end of your intellectual level must be given a special instruction book to accomplish an industrial task. We don't need that. That is very good and effective for a technologically based society. That is why we have made such rapid progress in industry to catch up."

Producing that high and homogeneous grouping of intelligence is achieved only with great effort on the part of Japan's children and their *kyoiku mamas*. The examinations, which begin as early as nursery school and are aimed ultimately at securing entry into the best colleges in the nation, which in turn guarantee job opportunities in the top companies in Japan, are savagely demanding. The "examination hell" that all Japanese children pass through is composed of batteries of multiple-choice questions covering a bewildering array of details in every subject covered in the curriculum. "The questions," notes Stanford's Thomas Rohlen, "focus on the fine print, demanding a close reading of textbooks that are densely packed with dates and facts."[21]

The competitive ferocity of the examinations has led to the establishment of thousands of *juku*, tutoring establishments or cram schools that every Japanese who can afford it sends their children to for at least several hours each day after regular school is over. Jukus are not cheap; the better ones can run as high as $7,000 a year. It is money the Japanese spend willingly. " 'Give a good education to your children even if you have to go to the pawnshop,' is a popular proverb among Japanese parents," Akio Akagi told me.

Many youngsters do not finish the school day and juku until

eight-thirty at night and have been away from home since breakfast. "Its difficult," admits Masato Hidoh, the assistant director of a Tokyo *juku*. "But whether they want to be here or not is beside the point. They must, in order to pass exams."[22]

Reporting on the examination system in an article on Japanese education, the July 18, 1986, *Science* magazine noted:

In Japanese fashion, the whole process is becoming more efficient each year. Before a real exam is taken, students pretest themselves, so as to have a true estimate of their ability in each of five areas. The results, known as *hensachi* or "deviation values" are calculated by testing companies and portend a bright or dim future. These *hensachi* become more and more accurate, because the companies reward their clients for sending back results on the actual tests and their college acceptance records.

So accurate is the system's predictive power that students waste no time applying to those schools whose *hensachi* "price" seems too high. The sort themselves into the leagues in which they are most likely to succeed based on the testing companies' past records. One cannot afford to waste chances, because one has few. Students must travel to the major universities to take their "second stage" exams on their own grounds, on one of two days in February. The national test, which comes earlier and is run by the government, is more important. It is so critical that in 1984, 204,000 students (called *Ronin,* a term for roving scholars of another era) had spent at least a year preparing for it. They were making their second, third, or fourth try, hoping for a better score.

Japanese children attend school 249 days a year. The American school year, by contrast, is only 180 days. But the two systems cannot be compared simply on a time basis. "Education is not to be measured merely in terms of years," notes Professor Reischauer:

The intensity of the educational experience counts too, and here Japanese education on average rates well above the American, except at the university level. The school day is longer, the school week is five and a half days and the school year is broken only by a short summer vacation of little over a month in late July and August, a New Year's holiday and a break before the start of the school year at the beginning of April. Discipline within schools is firm and the children devote prodigious efforts to their studies. In addition, they are assigned daily homework from the first grade on.[23]

Jared Taylor, computer consultant Bill Taylor's younger brother, is the West Coast editor of *PC Magazine* and attended Japanese public schools from kindergarten on through high school. In his book *Shadows of the Rising Sun* he writes:

For many Japanese, their most vivid summer memory is of plowing through piles of homework. On Japanese television there is a word association game show, in which the players try to help their teammates guess a specific word by giving them one-word clues. On one show, the designated word was "homework." The player thought for a minute and then gave his clue: "Summer vacation." His teammate immediately answered, "Homework."[24]

If homework and tests are the two grinding constants of the Japanese educational system, *kanji,* the ideograms that form the Chinese alphabet, is the bedrock upon which it rests. The Japanese language actually consists of three separate alphabets. Kanji, of course, which provides the bulk of the characters used in the written language, was imported into Japan in the fourth and fifth centuries along with Chinese classics, philosophy, and learning. Then the Japanese, as they still do, adapted it to their own needs—a process called *wakon-kansai* (Chinese talent with Japanese spirit).

This meant reducing an alphabet that contains upwards of 50,000 characters. The second alphabet, *katakana,* is used to express foreign words, such as the names of countries or people, words of foreign origin that have found their way virtually unchanged into the language, such as radio or dollar. It consists of a mere forty-eight characters, extremely simplified versions representing phonetically the sounds of specific syllables within the original Chinese ideographs. Katakana evolved over the course of about three hundred years to meet the needs of a vigorous and burgeoning indigenous literature.

At about the same time, still another set of characters called *hiragana* was developed. Hiragana is essentially a cursive script, more formal than katakana, that expresses the sounds of a select group of Chinese ideographs. It, too, contains forty-eight characters and was designed originally for use by ladies. Katakana was considered a male alphabet. In addition, the Japanese use the twenty-six letters of the Roman alphabet, which they call *Romanji,* and Arabic numerals. The result is a staggering galaxy of written symbols that must be learned in order to be literate in Japan.[25]

"Japanese children," Akagi told me, "are brought up surrounded by Chinese characters. They must be able to read them or they will have difficulty buying anything or even traveling. The more Chinese characters they learn, the wider their world becomes, so they are eager to learn them. At the end of six years of primary school, Japanese children will have learned 996 Chinese characters. Even under full-employment conditions, Japanese workers cannot get a job unless they can write at least one thousand Chinese characters. To read a Japanese newspaper, one must know about two thousand Chinese characters. Every day, garbagemen are required to report on their day's work in Chinese characters, as are factory workers. Therefore, every garbage collector and factory worker in Japan encourages his children to have at least basic literacy and numeracy."

As an example of the level of numeracy achieved in Japan, Akagi points to such simple daily tasks as shopping. "In a Japanese shop, if one pays with a large bank note, the cashier will make the subtraction mentally and give back exact change at once. In contrast, in Western countries, cashiers usually add money to the price until the total equals the number of bank notes paid. Japanese children, as soon as they are old enough to carry money, become aware of the necessity of calculation. If they cannot judge at once whether the change received is correct or if they carry calculators when they go shopping, they will be laughed at, even held in contempt by their friends.

"By simple arithmetic, any floor worker in a plant can deduce whether his product has deviated from specifications. As a simple example, by measuring the sizes of his product and dividing the total of sizes by the number of his products or drawing a simple bar graph, with the sizes on the horizontal axis and the number of products on the vertical axis, he can know the average size of his product or the statistical distribution of his product's size. In every Japanese plant, a visitor can see those bar graphs drawn by workers. Sometimes they are decorated with paper flowers or red ribbons celebrating their achievements in quality control.

"The mere exercise of discipline from the management above will not develop this kind of initiative among the workers. They are enjoying the result of their efforts just by simple arithmetic or by drawing a statistical graph. They feed back the results of their work as part of their working method. This is a basis of Japanese quality control activities. Let me emphasize again that Japanese quality control is not oriented toward raising a worker's moral. It has nothing to do with

Confucianism. Rather, it is closely related to the Japanese standard of numeracy. In fact, though it is simple, it is a mathematical technology exercised by ordinary workers."

Numeracy, numerical literacy, may actually be aided by the need to learn kanji. A study done by Irene T. Miura of San Jose State University, in California, demonstrated that a distinct advantage in understanding and manipulating numbers could be traced to Chinese ideograms. She notes:

Place value is inherent in the number language.

For example, the number 11 is read as ten-one, 12 as ten-two and 22 as two-tens-two. The numbers 13 and 30 which when spoken, sound similar in English, are entirely different in the Asian tongues: 13 is spoken as ten-three and 30 as three-tens.

A standardized test was given in several different experiments to determine the mathematical skills of youngsters in the first grade in mainland China, Korea, Japan, and the United States. The results showed that the Asian children, expecially the Koreans and Japanese, have a greater conceptual understanding even at this early age. Miura says:

Their ability to think of more than one way to show each number suggests greater flexibility for mental number manipulation. Socialization accounts for some of the differences in math achievement between Asian countries and the United States, but there also appear to be differences in the basic mental representation of numbers affected by language characteristics.[26]

In addition to its influence on mathematical ability, kanji may also have some effect on the problem-solving abilities of the Japanese and even on pattern recognition. Those three capabilities are essential in science and engineering, and I asked Dr. Kikuchi if he thought they might also be part of the explanation for Japan's great success in the semiconductor technology. "An experiment done about ten or fifteen years ago by a researcher at London University tried to measure the perception of children around the world," he responded by way of explanation. "He discovered a very high degree of pattern recognition for Japanese and Chinese children. A tremendous difference from Western children. Now this probably relates to the ideographic pictorial patterns of Chinese characters. Whether that relates to semiconductors is still in question for me. These kinds of things are more related strictly to the fundamental skills needed in scientific activities. I am not quite sure if it can be correlated to integrated circuits. The miniaturization is not so important for us, but how to treat the problem is more essentially related to that capability."

For all its presumed benefits, kanji has been a major barrier to computer usage in Japan and perhaps a goad to the development of chips with ever greater memories. One of the main problems has been the need to output Chinese characters to the monitor and printer. Whether the huge appetite kanji has for memory inspired the Japanese to design ever more memory into a chip or the inexorable expansion of chip capacity made possible the addition of Chinese characters to the computer's array of communications tools is a question akin to the primacy of chicken or egg. In fact, the solution simply called for the availability of great gobs of memory. With it, the Japanese were able to design a system that inputs phonetic Japanese by touching a Roman-alphabet-labeled key that also carries a hiragana character. Those words can then be expressed in kanji simply by pressing a special key. Software selects the appropriate Chinese characters, but to be outputted to the monitor, additional memory, stored in a ROM, is needed. These ROM chips are programmed to display the Chinese characters visually. There is also the additional need to parse the sentence into proper Japanese grammar, which is an extraordinarily complex problem, calling for still more memory.

In addition, English-language software for such purposes as data bases and spread sheets require an additional software interface and, of course, additional memory to elicit the appropriate kanji and hiragana characters. This also explains why the sale of American PCs in Japan are much lower than Japanese PCs which have solved the problem of outputting kanji with a keyboard containing only a few more keys than on a standard English-language computer.[27]

The Japanese educational system is an interesting amalgam of reforms instituted during the Meiji Restoration and Douglas Mac-Arthur's reforms aimed at promoting literacy and thereby, in his mind, improving the chances for democracy to succeed in postwar Japan. The result has been spectacularly successful. The Japanese are probably the most literate nation in the world, with a rate over 99 percent. Moreover, 94 percent go on to and finish high school, and 37 percent continue through college.[28]

Entrance to a top-flight university is the ultimate goal of every Japanese youngster and kyoiku mama, and that means selection to one of the nine universities established by the Meiji Restoration. Dotted across the length of Japan, from Hokkaido in the north to Kyushu in the south, nine imperial universities were created between 1877 and 1918 and constitute a sort of Japanese Ivy League. The jewel in the imperial university crown is Tokyo University. In Japan it is the equivalent of Harvard, Yale, and Princeton combined. Next come the rest of the former imperial universities, with Kyoto and Osaka the most prestigious among them. There are also dozens of national universities that were established in each prefecture after the war by merging and upgrading existing higher schools and technical schools. There are also a great number of private universities that were part of the prewar higher educational system, and two of them, Keio and Waseda, rank just behind Tokyo University in prestige.[29]

Less than twenty minutes from downtown, on the site of what was once the Daimyo of Tokyo's castle, is the campus of Tokyo University, a collection of sand-colored Gothic buildings laced with tree-lined walks. Founded in 1877 when the Meiji combined three shogunal academies, Todai, as it is affectionately known, was originally designed to provide bureaucrats for the rapidly growing government ministries. Today Todai serves virtually the same role. Japan's last prime minister, Yasuhiro Nakasone, and his adviser, Chief Cabinet Secretary Masaharu Gotoda, were Tokyo University graduates. The new prime minister, Noboru Takeshita, bends rather than breaks the tradition. His alma mater is the private but prestigious Waseda University. The old-boy network quickly reasserts itself, with Todai alumni holding the posts of finance minister, foreign minister, and the head of the Post and Telecommunications Ministry. The director general of the Defense Agency, the chairman of the Executive Council and of the Policy Research Council, and the leaders of Japan's Communist party are all Todai graduates. The bureaucracy that runs Japan is also almost uniformly headed by the men of Todai. Of the twenty vice-ministerial posts, eighteen are held by Tokyo University graduates.

Most of the government bureaucrats come from Todai's law school, which, unlike American universities, does not turn out very many practicing lawyers. Still, some law-school graduates do practice law, but even then, their ultimate goal is to serve in the government. Of the top twenty bureaucrats in MITI, for example, eighteen are graduates of Tokyo University. The Supreme Court numbers twelve of its fourteen associate justices as Tokyo University graduates, while the chief justice is a graduate of the law school at Kyoto University, which ranks just behind Todai in the eyes of the Japanese elite. At both schools, one of the most influential courses offered is in public administration, which stresses the government's role as a manager of the economy and as a protector of and adviser to business.[30]

It is no exaggeration to say that the Japanese establishment consists mainly of graduates of Tokyo University and these people are to all intents and purposes running the country," wrote education expert Masayasu Kudo in the October 1987 issue of *Business Tokyo.* "Their influence is felt throughout the entire political, bureaucratic and financial world, and they control three fundamental branches of government—justice, legislation and administration."

In the Japanese corporate world, the graduates of Todai control virtually all of the country's major concerns. A survey of company presidents showed that of corporations with capital of over $6.6 million, Tokyo University had the most, with 718 presidents, followed by Kyoto University, with 474.[31] Not only are these companies headed by the men of Todai, but virtually every key manager, scientist, technocrat, and bureaucrat I met while researching this book seemed to belong to that exclusive club.

The top companies are replenished by the graduates of Todai, Kyoto, Keio, and the other major universities each year by a rite as unvarying and as fierce as the player wars that once raged between the rival National and American Football leagues before they made peace some twenty years ago. Then teams would not merely offer staggering sums of money to sign outstanding college players; they would "baby-sit" them in isolated motels and resorts in order to hide them away from competing teams until they were safely signed. Eventually, with the merger of the two leagues, an orderly draft was restored, and players who were selected by a team had no other option but to sign with that team.

The Japanese corporate recruiting process falls somewhere between the frenzy of the football wars and the slave-market style of the draft. A gentleman's agreement between universities and corporate Japan bans recruiting of students save for a thirty-day period beginning about the middle of September. Then the companies open their doors to students entering their senior years and ply them with blandishments. Students from Todai and the other top schools receive an average of, according to this year's surveys, 2.79 job offers. Companies, however, are not permitted to sign the following year's graduates to employment contracts until the end of the thirty-day period, about October 15.

As many as 40 percent of the students, however, had already signed secret employment agreements and already knew with what company they would in all likelihood be spending the next thirty-five years of their lives until retirement. The prospects from Todai are contacted by phone and interviewed on a one-to-one basis by senior company personnel managers. During the summer vacation, many companies take their prospective employees on junkets overseas to keep them from the blandishments of rival corporations and the government.[32]

"Every trick in the book is used to encourage students to join private companies," reports Masayasu Kudo. "The main reason the companies want to hire Todai graduates is because they want to maintain close relations with government officials, most of whom prefer to deal with graduates of their old school—Tokyo University. To employ Todai graduates is to establish a conduit between business and central government."

But even the graduates of the most prestigious universities, who apply for jobs at the two dozen or so major companies, such as NEC, Mitsubishi, Toshiba, and so on, must pass yet another examination. For most, it is a mere formality; the years of testing have prepared them to pass almost any test devised by man. And herein lies one of the major weaknesses in the Japanese educational system. For despite its acknowledged contribution to Japan's industrial and economic success, there is great dissatisfaction with the system as it now exists. The educational pace is standardized; everyone from Honshu to Hokkaido takes the same courses at exactly the same time. There is a great emphasis on rote learning and passing the fiercely competitive examinations that many believe are inhibiting creativity and individuality in Japan.

"Before long we will be able to create robots who take entrance examinations" is the bitter comment of Naohiro Amaya, a former vice-minister of MITI and now a member of the prime minister's Ad Hoc Commission on Education Reform. "These robots will compete with real boys and girls and they will score higher. This means that we are educating people to be inferior to machines.[33]

Much of the criticism is also aimed at the Japanese universities. After the intense competitive pressures that have strained most Japanese students to the breaking point and in some cases beyond, the universities often serve, for the first two years at least, as little more than a breather. "As a result, the first two years have become a relaxing period during which students frequently cut classes, devote much of their time to clubs and other pleasurable activities that they had to forgo during the grueling period in upper secondary school when they were preparing for university admission,"[34] acknowledges the education committee's report.

In addition, Japan's universities offer very little graduate education. In fact, there is limited funding available either from the government or industry for graduate education. Advanced degrees do not improve one's job prospects, and most of the major corporations provide their own advanced training or fund higher education for certain employees at U.S. and European graduate schools. According to Monbusho, the Japanese Ministry of Education, Japan had only 20 percent as many graduate students as the United States. Nor do Japanese universities play a significant role in scientific and technological research. In fact, they are largely removed from the dynamic of the R & D process that is so characteristic of American universities. The reasons for this are many. Again, funds are very limited. Virtually 80 percent of all R & D is paid for by industry, and virtually none of that is spent on Japanese campuses. Ironically, corporate Japan spends most of its university-earmarked funds in the United States.

Then there is the matter of faculty. At the best of the Japanese universities, the state schools, which attract the majority of the top students, the professors are essentially civil servants, employees of the education ministry. As such, they are proscribed from accepting consulting fees from industry. This serves to underline the traditional separation of commerce and academe in Japan.[35]

"The contribution of Japanese professors to industry and science is quite different from that of the American and European system," acknowledges Dr. Kikuchi. "When MITI explores a possible project, they form a steering and evaluation committee. These contain professors from the universities, and so they have a large input. Unfor-

tunately," he adds with a laugh, "Japanese professors are like encyclo-
pedias; they are stuffed with information but not so creative. But today
we need more creative work from the university, and the professors in
and around Tokyo are interested in doing more."

Creativity is indeed the question Japan is now grappling with in its
efforts at educational reform. The kind of leaping, new ideas that spark
whole new technologies such as Americans are wont to burst out with
is not an area the Japanese have ever excelled in. Of fifteen major
developments in semiconductor technology, for example, fourteen
were developed by American researchers and only one, the tunnel
diode, was invented in Japan.[36]

That invention, which won the Nobel Prize for physics for Reona
Esaki, is only one of five Nobels won by Japanese. The most recent, the
Nobel Price in medicine, was awarded in 1987 to Susumu Tonegawa, a
graduate of Kyoto University. At first, national pride was puffed up by
the award, but it was soon acknowledged that Dr. Tonegawa had left
Japan years before and his discoveries had been made at the Basel
Institute for Immunology in Switzerland and at MIT, where he now
does his research and teaching.

"Almost certainly," editorialized the *Yomiuri Shimbun,* Japan's
largest newspaper, "Tonegawa could not have achieved his great work
if he had not been studying abroad. It remains true that no matter how
talented a person may be, if he or she is highly individualistic it is
difficult to achieve full potential here because of the nature of Japanese
society."

Even Tonegawa has faulted the rigidity of Japanese education,
pointing out that any free-spirited researcher must leave Japan to
thrive. "Universities here," he says, "are too authoritarian, too focused
on seniority, too inflexible to let young researchers pursue unproved
but potentially innovative ideas."[37]

For almost the entire postwar period, indeed for virtually all of its
history, Japan has looked either East to China for much of its
traditional learning and culture and West to the United States and
Europe for science and technology. As a result, Japanese creativity has
for the most part been channeled into improving upon the ideas of
others. "I classify creativity into two categories," declares Dr. Makoto
Kikuchi, who can lay claim to being one of the most creative scientists
not just in Japan but probably in the world today. "One I call

independent creativity and the other one associated creativity. Western people demonstrate independent creativity in that they usually try to find a new target. We start from the target-oriented point, that is, associated creativity. We are very skillful at this kind of activity because we are able to modify or improve the original idea or systematize the original idea. We don't mind starting from a hint given by another person."

Kikuchi's favorite example of this is the single-minded determination of Sony's Masaru Ibuka to build the world's smallest all-transistor radio. And so, using the Bell Lab invention and applying it to the target of producing a transistor radio, in 1953, Sony engineers, who probably knew virtually nothing about transistor physics, developed something called the ground-junction transistor, which markedly improved the frequency characteristics of the transistor and in fact made the all-transistor radio possible.

"With the target or goal clearly in mind, Japanese creativity is at its best. Western people don't like to follow someone else's hint because it means they are losing their identity. We don't care about that because we are a homogeneous people. The fundamental structure of our society is like small stones connected together by strong spring wires. People are not quite independent of each other. But in Western society people are heavy stones loosely connected with very weak spring wires. If one of the stones in the Western model vibrates, the other stones are relatively unaffected; they stay still. That is individualism. But in our structure, if you vibrate one of these stones, everything begins to oscillate in unison in a short time. That is a consensus. We don't care about losing our identity.

"An extreme example of this is the bee society. I have never talked to bees on this matter, but in that society each bee has no individual identity. At the other extreme is the U.S., where people have a very strong sense of identity and individualism. When I was a schoolboy and our house was burned out twice during the war, we were told, 'You are living for the emperor; you have no other identity or purpose.' A person is valuable only because of the contribution he can make to society. In this way we have much less degree of identity.

"That was the historical situation. But after World War II, Mac-Arthur came to Japan, and we were very much influenced by Western society. One important idea was that we should respect human rights.

So we are now moving closer to Western society. But still, we are closer to where we were. Our land is small, we have a much higher density of people, so we must be respectful of each other. That is a necessity, and it has influenced the fundamental nature of our society; we cannot escape from that."

For most of the postwar era, catch-up, not creativity, was the prime driver of Japanese industry and technology. "There is a reason Japan made such rapid industrial progress. When I started my research after university, the distance between the United States and Japan was vast," admits Dr. Kikuchi. "Both of us were climbing the mountain, and the United States was always ahead of us, so we could always see your back. This is extremely important for catching up. That saved the budget for doing R & D, saved the time, saved manpower, and increased the efficiency for catching up.

"Compare the pioneer to the second climber; the second climber always has the higher efficiency for R & D. For example, in the 1960s there was a great deal of excitement about electronic cooling based upon the Peletier effect, which is, if you flow electric current through a junction in one direction, the junction is heated up. If you reverse the direction of the current, the junction is cooled. There was the hope that in the next ten years all the air conditioners and heaters would be replaced with devices using the Peletier effect. There was tremendous work on R & D in the U.S., and in three and a half years they spent almost $40 million and nothing came out of it.

"We learned that approach was not effective and did not pursue it and so saved a great deal of time, effort, and money. So we could always see the direction not to take. Now, in certain fields we are doing quite well; in others, we are still lagging."

"What the Japanese are good at is developing a technology base," states Sheridan Tatsuno, a senior analyst at Dataquest, the largest market-research firm covering the semiconductor, computer, and telecommunications industries. Tatsuno, an American-born Japanese, spends half of his time in San Jose in Dataquest's home office and the other half in Japan watching developments and trends in the semiconductor industry there. "It's a corporate treasure for the vertically integrated Japanese companies that produce everything from air conditioners to fans, to heavy equipment and computers and semiconductors, in which you keep gathering information that will be of some use to some division within the company. Generally, they tend to pick up specific information along the way that will serve a specific corporate need. The Japanese are the ultimate pragmatists; they don't believe in reinventing the wheel. They believe that technology by definition is free flowing and despite all the tariffs and government restrictions, it's very hard to restrict the flow of technology. It basically goes over the wires electronically or into the heads of people, and short of imposing a Russian-style system, where you cut off the migration of people and ideas, there is no way to limit or stop the flow of technology.

"Given that scenario, the Japanese have positioned themselves right in the middle of the flow of technology. They subscribe to all the Western publications; they choose industries to protect, areas of technology to explore or target, specific and next-generation industries to compete in, and then they invest. Contrary to popular belief, the Japanese do not like to compete head-on in existing technologies because market shares have usually been divided out by then. They like to leapfrog into next-generation technologies where there is a probable need for it. So they will draw technology road maps, and that's what they did for the memory chips. It's done by a consortia they create of government people, academics, and industry. They all know each other. It would be as if you had all your Harvard MBAs and Stanford MBAs and government officials and Silicon Valley people living in one city and they all went to the same schools. By default you would have this cross-fertilization of ideas. In the United States, because it's so big and spread out, we don't get that cross-fertilization of ideas that occurs naturally during the evenings in Japan."[38]

It also flows in a very directed fashion from MITI, where the combination of government and industry working together is perhaps the ultimate Japanese genius. MITI is perhaps the best known of Japan's government agencies and often thought by foreigners to have almost supernatural powers of technological prophecy and omnipotent control over Japan's giant industrial corporations. But MITI's success is as much due to technology targeting as it is to the basic Japanese ability, if not compulsion, to work together for the greater good.

MITI bureaucrats and decision makers are neither technologists nor scientists. They are rather experts at pulling together advisory panels of specialists and gaining a consensus that results in a clear direction and course of action. "MITI people are all laymen when it comes to science and technology, so they never hesitate to ask questions about the nature of the technology," says Makoto Kikuchi patiently, acknowledging the fact that it is perhaps the one most asked by foreigners. "Many times I am telephoned by MITI bureaucrats. What's gallium arsenide? What's the future of VLSI? What's important? They get enough comments from informed people, opinion leaders, and many different aspects are gathered, because they don't hesitate to ask these simple questions. Once they get all information as they can get, they finally try to make a plan, and in this way VLSI was supported."

5

MITI, Machines, and Microns

The MITI building is white brick, inset with long, thin rectangular windows. A highly polished slab of orange-and-black-flecked granite outside the building is carved with the kanji characters for "Trade, Industry, Ministry." On the other side of the walk is a stainless-steel sign with the English letters "Ministry of International Trade and Industry." This is one of the two buildings MITI occupies on the Chiyoda-Ku in Tokyo. Flanking it and across the broad avenue are other government buildings, but none hold the almost mythic sense of power and omniscience of MITI. For from the bureaucrats who work inside this building has come the direction and policies that have set the Japanese economic miracle in motion. From steel to textiles to autos and now microelectronics, MITI has guided Japanese industry to economic megasuccess. In the late 1950s, MITI made one of the most crucial decisions in economic history by establishing a new target for Japanese industry—the knowledge industry. In a now famous position paper they stated:

The spirit of basing national development on technology should be our aim in the 1980s. . . . The basic course of knowledge intensification during the 1980s should be to increase the value added products through technology intensification. . . . International specialization between Japan and advanced countries will also become possible as a result of the growth of industries where Japan has unique creative technologies. . . . Possession of her own technology will help Japan to maintain and develop her industries' international superiority and to form a foundation for the long-term development of the economy and society.[1]

MITI's plans to date have been almost unerring in their ability to select areas of economic growth and demand and guide Japanese industry to it. In fact, there is probably no other government agency or bureau anywhere in the world quite like MITI. To learn more about it, I started by meeting with Ryozo Hayashi, the chief of MITI's seven-member delegation in the United States. "Be sure and ask for Ryozo Hayashi," he told me on the phone. "Of the seven of us here, three are named Hayashi. We are the Smiths of Japan."

Hayashi, in his early forties, is a career bureaucrat who spends most of his time in Washington. His official office, however, is in New York, in a glass-walled Sixth Avenue skyscraper where MITI keeps a profile so low as to be almost invisible. There is no listing for MITI in the telephone directory, and their offices are located within JETRO, the Japanese Export Trading Organization, a quasi-government agency used to promote exports.

Here I was escorted into a modest-sized, almost cozy room with eight leather armchairs grouped in a square about a large glass-topped coffee table and told that Mr. Hayashi would be with me shortly. I would later learn that virtually all Japanese companies had similar rooms where senior executives met with guests or visitors. Rarely did they meet with outsiders in their private offices.

Hayashi at first glance seems an unlikely bureaucrat. Long-haired, with long sideburns, he looks more like a holdover from a sixties commune, and even his background suggests a free-spirited approach to life that is not typically Japanese. His antecedents and education were, however, up to a point, almost typical of the shyu sai, the superior talents of Japan.

"I am from Kobe," Hayashi told me. "My father is a professor of law at Kyoto University. The function of law is important in Japan, but not the practice. I attended Kyoto University, and my graduation certificate was signed by my father. I joined MITI in 1970 and went to Harvard Law School in 1975 and '76. Then I fooled around for a year before joining a law firm in Washington. I returned to MITI in 1980 and was at the American desk for three years and then the electrical industries division for two years. All the technical knowledge I have, I picked up on the job. The concept of MITI is for us to apply practical wisdom and technological know-how to come up with an intelligent decision. MITI is run by generalists who are expected to know enough about the departments under them to evaluate projects in terms of the real world."

Hayashi's job in the United States is primarily to act as a liaison between the Japanese and American governments and industries. "For example," he explains, "I spend one-third of my time making speeches around the country and meeting the business people who sometimes have a complaint and who sometimes try to find some contacts in Japan, and we help with that. There are other organizations who do that, such as the consul general and other trade associations, but certainly when they want to see MITI people, we help them. Sometimes they want to know what is happening in Japan and how the Japanese view American industry."[2]

Hayashi explained that MITI is organized into eight bureaus and four agencies. The agencies correspond loosely to U.S. federal agencies and are largely regulatory bodies. In a nation with virtually no natural resources, it is only natural that the largest is the Natural Resources and Energy Agency, the counterpart of Department of Energy here. There is a much smaller entity called the Enterprise Agency, which corresponds to our Small Business Administration. The Patent Office, which is playing an increasingly important role as Japan's technological base grows at an awesome rate, is a separate agency within MITI and quite independent in the way in which it functions. The fourth grouping is the Industrial Technology Development Agency. It is responsible for the operation of all the national laboratories in Japan.

Staffing is on the lean side. Every bureau has around two hundred people working for it. The Energy Agency has more than five hundred, and the Patent Agency, by contrast, is huge, with more than two thousand people.

"One of the characteristics of MITI is that we have fairly strong control by the secretariat over policy coordination among the agencies and bureaus," explains Hayashi. "That means that no one agency or bureau is going to go its own way heedless of the effect of its actions on overall national policy." One effect of that control that directly effects the semiconductor industry is a national technology and trade policy that has enabled Japan to focus its efforts on such broad goals as "development of the knowledge industry" and narrower objectives such as the VLSI project.

"Closer to the policy-making machinery are the four functioning bureaus and three industry-specific bureaus. Of the four functioning bureaus, two are on the international side. They are the International Trade Policy Bureau and International Trade Administration Bureau. The policy bureau is probably analogous to the State Department, while the second is more like the Commerce Department, which implements things like export-import control. The other two functioning bureaus are the Industrial Policy Bureau, which, as the name says, coordinates industrial policy, and the Industrial Location Bureau, which implements the allocation of space and locations to industry in Japan."

But it is the industry-specific bureaus that have gained MITI its reputation for seeming omnipotence in the marketplace. They are the Basic Industry Bureau, which includes steel, petrochemicals, and aluminum; the second is the Consumer Goods Industry Bureau, which is responsible for textiles, housing, and related materials; and the third is the Machinery and Information Industry Bureau, which contains the computer, semiconductor, machine-tool, and consumer-electronics industries.

For the first two decades after the war, MITI made virtually all the industry-wide decisions in Japan. Recognizing the need for a coordinated national industrial policy as a matter of economic survival, the large Japanese companies followed MITI directives and even hints on what to produce, at what cost, in what quantities, and how much to charge. In addition, MITI controlled the national purse strings for the foreign currency that was essential to capitalize the then staggering Japanese economy.

"The mode of operation has changed dramatically in the last two or three decades," advises Hayashi. "The tight control over foreign currency the companies in Japan needed was exercised by MITI when it was a scarce resource. In those days following World War II, financing was very tight. At that time, MITI controlled the allocation of money. Also, the access to information was fairly limited, and MITI controlled that, too. One of MITI's main jobs was information gathering of foreign science and technology. Now the private companies have become larger, and we no longer have any problem of foreign currency; therefore, we do not have direct control over the companies' activities. While the money was once a precious resource for the private companies, it is no longer as important. The function of MITI has changed from very direct control to a kind of guide. As long as we demonstrate imagination and foresight, then the companies follow our recommendations, but when they think that MITI is acting against them, then they ignore us." The net effect is that MITI suggestions and guidelines carry the same sort of weight as does an executive department regulation in this country. It can be ignored, but only at great peril.

MITI's proposals have in fact been occasionally ignored in the past, but not its plan to thrust the Japanese semiconductor industry into the technological forefront with the VLSI project. VLSI was the generational leap forward, orchestrated by MITI with the vowed goal of gaining technological parity in computers and semiconductors with the United States. Although the computer goal was not realized, the success of the VLSI project in semiconductors has had a dramatic effect on everything from market share to manufacturing and has made Japan a dominant power in the Chip War.

The VLSI program was born ultimately of rumor. Although MITI had by 1971 announced to anyone who would listen that the information industry—computers, semiconductors, telecommunications, and all related technologies—was to be the focus of Japan's science and industry for the next two decades at least, no specific program had been created to design the next generation of chips.

Instead, MITI had established the goal of making Japan competitive in the computer marketplace. The target was to develop an IBM-compatible, mainframe computer. The Japanese computer industry, despite government subsidies and protection of its domestic market, simply did not have a competitive machine. To bring Japan's computer industry to a level of parity with the rest of the world, in 1966, MITI organized six of the country's major electronics firms into three teams—Fujitsu-Hitachi, NEC-Toshiba, and Mitsubishi-Oki. Assignments were specific: Fujitsu-Hitachi was to develop IBM-compatible mainframes; NEC and Toshiba, medium- and small-scale Honeywell-compatible computers. Mitsubishi and Oki were to design small computers for scientific and factory use.

Then, in the beginning of the 1970s, IBM brought out a revolutionary new design, the System 370. These new computers were so much more powerful than anything else in the world that several of IBM's major competitors decided it was easier to quit the field than fight. Giant American corporations such as RCA and GE simply got out of the computer business. The Japanese industry panicked. MITI thought the best approach was to reorganize the entire computer industry into two even more gargantuan companies. "I think that having six computer companies in Japan is definitely too many," one MITI vice-minister told the Commerce Committee of the Lower House of the Diet. "We should use administrative guidance to make two or three firms, not six, to unify and effectively use our engineers and thereby contribute to the development of Japanese computers."

The companies, needless to say, resisted a merger, but instead reorganized into three cooperative groups. The Japanese computer industry stayed in the game thanks to the MITI-organized computer program. "Because MITI started providing research grants and made different companies get together for cooperative development of new machines, for the first time, Japanese makers were ready for battle," said Fujitsu chairman Taiyu Kobayashi.[3]

Then, in 1975, the rumor swept the Japanese computer industry. "IBM was on the move again with yet another revolutionary design for a totally new type of computer." Such a computer, the Japanese reasoned, would have to have a high-density chip as its base. At this point in time, IBM computers were built with boards studded with dozens of discrete ICs, each containing only a few hundred transistors. Mirihiko Hiramatsu, then head of MITI's electrical division, used the rumor to mount the VLSI project. Central to the program was the decision to leapfrog existing semiconductor technology and develop a high-density chip, one that would contain a minimum of 100,000 transistors and circuits. The original company pairings were reorganized. Oki was dropped from the group, and NTT, the government-owned telephone and telegraph company, which had already begun its own VLSI development, was brought in and along with MITI formed a private organization, the VLSI Technology Research Association.

The program officially began in 1976 and ran through 1979. MITI put up more than $132 million in so called *hojokin* loans. Hojokin are "interest-free, refundable if and when" government subsidies. In fact, the hojokin for the VLSI project have never been repaid. The six firms, however, provided matching funds over the same four-year period of almost $200 million of their own money. The project set up three cooperating laboratory groups, a central research lab that was to develop the basic VLSI component and manufcturing technology and two others that were to develop computer applications for the new chip. The degree of cooperation among what were normally fiercely competing companies was remarkable. But each company knew that its selection to the VLSI program guaranteed not only favorable government treatment but the opportunity to cash in on the new products that would be coming out of the massive research effort.

The new Central Laboratory was established at NEC's plant in Kawasaki, a suburb west of Tokyo. It was headed by Yasuo Tarui, who had been the director of the semiconductor-device section in a MITI lab. Tarui was given the right to select literally anyone from the companies in the consortium to work at the lab, and he picked 100 researchers, engineers, and physicists—those he believed to be the cream of Japan's research talent.

The lab conducted basic research in every aspect of semiconductor technology—microfabrication, crystal technology, semiconductor design, process technology, test and evaluation techniques, and device design. The companies also benefited greatly from the exposure their top researchers had to the advanced technology and thinking being done in the Cooperative Lab.

At the research labs, tasks were divided among several working groups, each taking a different approach in the hope that one would come up with a workable solution. Each of the groups was led by one company but included researchers from other companies. One of the major problems was to develop electron-beam and X-ray-beam exposure devices to draw the ever narrower lines that would be needed for the very dense VLSI chips. Three different research teams, for example, tried seven different approaches to the problem of getting an electron beam to draw a fine enough line on a silicon wafer. Finally, the Toshiba-NEC group succeeded. "The timing of the project was critical," said Sakae Shimizu, senior managing director of Toshiba. "There was no electron beam, and we needed a breakthrough to get ahead. The firms did not have any of the equipment for producing VLSIs, such as the electron beam or testing equipment."[4]

But American semiconductor equipment manufacturers did; not

only electron-beam equipment but everything needed to make small-featured, densely packed chips. Indeed, fully 80 percent of the VLSI budget was spent on state-of-the-art U.S. equipment, most of which was not being incorporated into American chip maker's fabrication lines at that time. It was *wakon yosai* (grafting the talent of another culture, in this case Western, onto the Japanese spirit) yet again, the same strategy that has served the Japanese so well since the sixth century. The electron-beam technology came from Perkin-Elmer (P-E), a diversified high-tech company headquartered in Norwalk, Connecticut, that manufactures everything from analytic instruments to chip-making machines, superminicomputers to avionics instruments that guide missiles and passenger planes. In 1977, P-E produced its first electron-beam machine, MEBES, the manufacturing electron-beam exposure system and the Japanese promptly bought it.

"How did they get their start?" Chuck Minihan, the just-retired head of P-E's semiconductor equipment manufacturing division, asked me rhetorically. "They bought the third machine we made, and then they bought number four. They were shipped to Japan, put into their development laboratories, and everybody had access to them. And NTT then came up with technology so they could make their own."[5]

A burly, blunt-spoken Irishman, Minihan remains a consultant to P-E and is on the board of SEMATECH, which, not unlike the Japanese VLSI project, hopes to develop new process technologies for American chip makers to utilize in the Chip War. Under his guidance I saw the newest version of the MEBES electron-beam machines being assembled at P-E's Hayward, California, plant.

Selling them to the Japanese and competing against them all over the world will be extremely difficult. Today 80 percent of all new purchases of semiconductor manufacturing equipment in Japan, about $3 billion worth, is from Japanese companies, and those same companies have made major inroads in world markets as well. And that may be the most significant impact of the VLSI program. It gave the Japanese the process technology needed to manufacture the most advanced chips in the world.[6]

"To compete in the increasingly torrid race for market share," declares industry expert Dr. Gene Gregory of Sophia University in Tokyo, "chipmakers have had to constantly upgrade practically all the 100 odd processes involved in integrated circuit manufacture to assure reliable mass production using higher precision required by design rules steadily approaching submicron line widths."[7]

"The Japanese did not invent their way into their present dominant position. They basically got both their equipment start and their design start by buying our technology, installing it in special evaluation centers, and letting everybody in the industry come in and use it to learn how it worked. And then they replicate and improve on those processes," says Chuck Minihan with anger tinged with admiration. "They seem to be absolute masters at taking this original technology which we invent and we develop and improve the hell out of it in such a way that it's better and it's cheaper. And then they turn around and sell it back to us at a price that is so low, it just drives the Americans out of that business."

The Commerce Department concurs. Its analysis of the VLSI project, published in its 1985 report, *A Competitive Assessment of the U.S. Semiconductor Manufacturing Equipment Industry,* stated:

Some of this [U.S.] equipment was used for production of semiconductor prototypes, but much was dismantled and analyzed by technicians in an effort to fabricate SME [semiconductor manufacturing equipment] equal to or superior in performance to U.S. equipment.

Out of this phase of the VLSI research effort came the technology used by Nikon to produce its commercially successful wafer steppers, Canon's contact and proximity printers, advanced E-beam and X-ray lithographic equipment and dry etching technology. As important as these were, other benefits were felt to be of equal or greater significance. These programs served as on-site training for Japanese engineers and scientists whose efforts could be constrained by the commercial interests of individual firms.[8]

Going head to head, most American equipment makers have not fared too well against the suddenly highly competitive Japanese equipment makers. "We've lost a good part of the equipment business," admits Washington economist and semiconductor-industry consultant Bill Finan. Sitting in his wood, steel, and plastic modern office not far from the Washington Monument, I had the odd sensation of attending a wake as he ticked off the segments of the industry that were now controlled by the Japanese. "Silicon production is virtually gone with the exception of Monsanto, quartz masks are pretty much coming out of Japan exclusively, ceramic packages out of Japan, lead frames more and more. Chemicals have not gone heavily over to the Japanese yet, but the potential is there.

"Something a lot of people overlook is architectural engineering skills in structures. If you were to say where is the most preeminent capability in designing and executing clean-room technology, a lot of people would say Japan. When Texas Instruments constructed DMOS 4 down in Dallas, they utilized a Japanese architectural engineering firm. Other American companies that wanted to develop and construct a state-of-the-art facility also had to use Japanese capability.

"That's an area a lot of people overlook, but if you look at the nature of VLSI and ULSI technology, structure is becoming an integral part of the capability of the facility itself. You just can't go out in your garage and make these things. You've got a very, very specialized facility. If you look at the materials, there are a number of segments that are wholly produced by the Japanese; in the architecture and engineering of clean rooms they're very advanced. An area that's related to that is production-line integration, bringing in and interfacing different types of equipment. The protocols are complex, and the Japanese have had more experience there, so the tendencies for that will be to use a Japanese firm. The American situation is that you have to use a number of smaller entities and try to integrate them yourself. Whereas with the Japanese you can call on one engineering firm to take care of the integration of all the subparts."[9]

"Equipment makers are losing market share faster than the semiconductor producers are," acknowledges BRIE's (the Berkeley Roundtable of International Economics) Michael Borrus.[10] "In 1979, when the VLSI project officially ended, U.S. semiconductor equipment makers had over 75 percent of the $1.4 billion world market," according to the U.S. Commerce Department. By 1983, although the market grew to more than $3.3 billion, the U.S. share had dropped to two-thirds. The European makers' share, at 8 percent, small to begin with, fell to 7 percent. Japan's semiconductor manufacturing equipment companies by contrast increased their share of the world market from 24 to 32 percent. By 1987, the Japanese had raised their market share to 40 percent.

Nikon Camera, for example, which came quite late to the ball, began making optical steppers, in effect, photolithographic engravers that repeatedly print the circuit design of a chip onto the silicon wafer, in 1976. They are now the world leader, with 1987 sales in excess of $240 million. The next closest competitor is Canon, a Japanese company that is rising rapidly. Following closely behind Canon is GCA, an American company that was once the world leader in steppers.[11]

How did Nikon suddenly blaze into the forefront of a field in which virtually every technological advance has been made by American companies? I went to their headquarters in the Fuji Building in the heart of Marunouchi, Tokyo's bustling business district, to find out. Here I met with Shoichiro Yoshida, the managing director of Nippon Kogaku, as Nikon is officially known. Yoshida is in his middle fifties, an elegantly groomed, handsome man with a sleek head of gray hair more usually associated with male models in Cadillac ads. Yet another Tokyo University graduate, with a degree in precision engineering, he joined Nikon in 1956. Four years later, the company was building Japan's largest reflecting telescope, based on his design.

"As you know," he acknowledged, "MITI guided the chip makers to form the VLSI consortium. They ordered us to develop some equipment. We were one of many equipment manufacturers involved. We got the order from them, and we just developed an optical stepper for VLSI."

He insisted, however, that Nikon used its own technology and built on that in designing and developing steppers. In fact, Nikon has been a world leader in cameras and lenses for the past twenty-five years, with an extensive background in many of the fields needed to build the optical steppers, measuring devices, and masks used in semiconductor manufacture. Still, the ability to see it all pulled together in an already working American optical stepper cannot have hurt.

"All our technology has been developed by Nikon," Yoshida nonetheless responded to my question. "We entered into the semiconductor industry by designing and building the laser interferometric X-Y measuring machine. Before that we had developed by ourselves a ruling engine. This is a very ultraprecise machine. Its main purpose is to do the linear grids for spectrophotometrics. The space between lines is less than a micron, in an area five by ten inches. That's equal to the feature sizes on 1-megabit chips. In these machines we gained invaluable experience to develop machines for the semiconductor industry. These technologies are all incorporated in the new steppers."[12]

The Japanese do more than improve on existing equipment. Professor Gene Gregory offers still another reason for the Japanese equipment makers' success.

Not only have Japanese equipment builders dramatically up-graded their technology for VLSI production, reducing and often eliminating the advantage of American makers, but they have also shown a greater readiness to make modifications in their equipment to meet customer needs. And, since competition in the market for semiconductor devices is determined largely by superior manufacturing strategies and systems, Japanese chip makers tend to be exceedingly exigent in their requirements for equipment modifications to their particular production system specifications.[13]

"One of our research goals," admits Yoshida, "is to follow up on the requirements of our customers, which is for finer lines and spaces and much more accurate alignment systems and so on."

"Customizing equipment for Japanese needs is very important," admits John Suzuki, the Japanese president of P-E Japan. Suzuki is a dapper, urbane man with an international outlook nurtured by his father, a professor of English at Kyushu University, on what is now dubbed Silicon Island. Surprisingly, even after graduation from Tokyo's prestigious private Waseda University, Suzuki spoke no English. But after working for a Japanese trading company in the States for ten years, he is of course now fluent in English. Then, in 1977, he joined P-E Japan.

In the dining room of the Maronouchi Hotel in Tokyo, we had what amounts to a Japanese version of the power breakfast—rice, fish, some vegetables with a rather piquant flavor, and miso soup and tea—while he told me of the difficulties American manufacturers face doing business in Japan. "One of the reasons most American manufacturers were not so successful in Japan, they did not comply with customers' needs. For instance, a customer might ask that a power supply be located not on the right- but on the left-hand side of the machine to conform with the room layout. That is a simple change and does not require a major design change in the equipment. But even such minor requests in the past were denied by American manufacturers."[14]

Even more egregious is the seeming inattention paid to equipment after it has been purchased. Dr. Chang Soo Kim is the senior managing director of the semiconductor group of the Korean electronic giant

Goldstar. His experience with American equipment manufacturers has been less than encouraging. "I set up two fabs," he told me, "so we had to buy a lot of equipment. And since I was trained in the U.S., I bought over 95 percent of the equipment from the U.S.

"As much as I like to buy from the U.S., the distance involved and the attitude of support and service are not as good compared with the Japanese. When we bought equipment from the U.S., it was shipped in packages, and we would come in and we'd dismantle and assemble it. Many times we were missing a component, or a wrong component would be mixed in. So when those things happen, then you have to order the missing part or bring in new things and start all over. When those kinds of things happen, the delays in setting up that equipment cost a lot of money for us. A month delay, for example, in setting up something, when you translate that into lost opportunities, it's a huge amount of dollars. So naturally we get very upset.

"Finally, they'd send technicians to intall the machine. And the only way to relieve our frustration is to talk to these people. And these technicians will say to us that this is not my job and my job is to install it. Whoever made the mistake, it's not me, but it's our shipping department, so don't yell at me. I'm not going to do anything about it; you just tell our headquarters.

"There are, of course, some good, very competent Americans who will provide good support, and we still buy a lot from them. The Japanese, on the other hand, will not do things like that. First of all, the incidence of having things come in here with mixed components or wrong components is very small. On the other hand, when we complain to the people from the factory about that, about Japanese equipment, the attitude is that Japanese technicians or salesmen or engineers or whatever will first of all apologize to us. He believes he represents the whole company to us; even if he's the lowest-level technician, he takes the responsibility and calls the factory and gets things up here as soon as possible. Or they dispatch other people directly from the factory with the component to come in and set it up for us."[15]

As a result, a much larger percentage of equipment than the Korean's would actually prefer will be purchased from the Japanese. "Now we, of course, are trying to set up another fab, a big one," announces Kim. "We will spend something like $100 million. Out of that $100 million, probably one-third will go to a Japanese equipment manufacturer."

Americans, according to John Suzuki, are apparently also indifferent to doing the customizing the Japanese expect when they buy equipment. "Japanese customers often asked for different colors on the covers of the equipment or to be certain that emergency cutoff switches were installed in specific places to meet Japanese safety codes," explains Suzuki. "Even simple requests such as this were often ignored by American equpment makers. In some cases, it was more a failure of communications. American companies often did business with the Japanese through a distributor, who failed to convince the Americans of the importance of these requests."

The Japanese practice of customizing not only helps sell a machine; it makes for very close cooperation between maker and user, with concomitant benefits for both. The Department of Commerce says:

The Japanese semiconductor manufacturer can permit scientists and technicians of its tool making affiliate to have access to the production process for extended periods of time without fear of losing proprietary secrets. This situation is in contrast to the arm's-length relationship between U.S. equipment-makers and chip-makers. It is only recently that a closer relationship of trust and collaboration (though on a generally informal basis) are becoming common in the U.S. In the past the lack of opportunity to test and adjust machinery under operating conditions prior to marketing resulted in a situation where it may have required "six service people working weeks or months" to de-bug a piece of U.S. equipment and getting it to work properly. On the other hand, . . . a Japanese machine arrives with one or two service workers who, in a couple of days, set it up to spec. . . .[16]

Bill Finan agrees. "Once the equipment is on site," he advises, "it requires close cooperation to transfer the know-how from the vendor to the company in terms of operation of the equipment. More and more customers are demanding that the equipment maintain certain standards of reliability over a long period of time. So more and more you don't just put the equipment on site and walk away. Dealing with a number of vendors, this cooperation can be very complicated in trying to integrate a line. Very large companies like IBM, Motorola, TI, and such have the capabilities in house to work with a multitude of vendors and modify the equipment, if necessary, to integrate it into a line. Smaller firms don't."

The large Japanese firms, however, were positioned to take advantage of the new VLSI technology. Coupled with highly motivated workers, quality control programs, an already highly skilled and experienced global marketing force, and the fiercely competitive sales approach that had been honed over the decades by domestic competition, the ultimate effects of the VLSI program were stunning.

More than a thousand patents came out of the research, and the technology developed proved to be critical in enabling Japanese firms to make major leaps to 64K RAMs, 256K, and now the 1-megabit chip.[17] The ability to churn out these memory chips, the so-called jelly beans of the industry, in vast numbers and dump them in the United States and elsewhere in the world literally turned the pecking order of semiconductor producers upside down. In 1976, when the VLSI project began, the ten leading merchant producers, according to Dataquest[18] reports of worldwide sales, looked like this:

COMPANY	COUNTRY	SALES IN $ MILLIONS
1. Texas Instruments	USA	655
2. Motorola	USA	462
3. NEC	JAPAN	314
4. Fairchild	USA	307
5. Hitachi	JAPAN	267
6. Philips	NETHERLANDS	266
7. National Semiconductor	USA	249
8. Toshiba	JAPAN	236
9. Matsushita	JAPAN	177
10. RCA	USA	174

Eleven years later, in 1987, the world ranking according to Dataquest looked like this:

COMPANY	COUNTRY	SALES IN $ MILLIONS
1. NEC	JAPAN	2,613
2. Toshiba	JAPAN	2,257

3. Hitachi	JAPAN	2,157
4. Motorola	USA	2,025
5. Texas Instruments	USA	1,752
6. National/Fairchild	USA	1,427
8. Fujitsu	JAPAN	1,362
7. Philips/Signetics	NETHERLANDS	1,361
9. Matsushita	JAPAN	1,176
10. Mitsubishi	JAPAN	1,123

The almost bottom-to-top reversal of chip makers is largely due to the almost complete takeover by the Japanese of the memory market. "In the United States memory is given the affectionate term of jelly beans," explains Dr. Sheldon Weinig, an industry pioneer and CEO of Materials Research Corporation (MRC), a major producer of semiconductor materials and equipment. "The smart boys in the U.S. said, 'Who wants to make these jelly beans? It's boring; the prices are eroding. Let's make them offshore or get out of the business altogether.' So the Japanese got it almost by default, and they learned what Henry Ford learned. When you make enough of anything, you got to get smarter. So the next thing you know, the jelly-bean manufacturers knew more about manufacturing all kinds of semiconductors than the smartasses back in the States.

"Less than two years after the Japanese took over the 64K RAM, they developed the most efficient plants in the world, learned how to produce reliability and reproducibility, and suddenly they knew more about semiconductor processing than we ever knew. Now they had all of the jelly-bean capability—handling of wafers, operator training, clean rooms, automation, reliability—and guess what. You put the complicated chips into that setup and you find they make them faster, better, and cheaper than we did."[19]

The Japanese also learned that generational changes in chip-making equipment occur just as rapidly as they do in chip making. "In the semiconductor field the rate of improvement in technology is quite high," notes Tokai University's Hajime Karatsu. "Manufacturing equipment is often obsolete within three years. Therefore, the equipment has to be profitable as soon as possible. The Japanese makers see the DRAM, with its enormous demand, as the best and fastest means of returning their investment in equipment. Therefore, the companies targeted DRAMs for the purpose of replacing their equipment.

"This was about 1979, when the 16K DRAM was first introduced. This is the basic strategy of the Japanese manufacturer. This led to an oversupply of chips and a drop in price. With the 16K chip, the Japanese took a 40 percent share of the market. Then, with the 64K chip, they took a 70 percent share.

"The entire aim is to replace the equipment with the newest state of the art. The speed of advancing technology dictates that strategy. American chip makers made a big mistake in this case. When the price of the DRAM dropped, they left the market. Only TI stayed and built a new plant in Miho, Japan, to build DRAMs and then another in Dallas."[20] A second American company, Micron Technology, has also remained in the memory business, but their production is quite small compared to TI or the Japanese.

The Japanese not only did not leave the market; they sought to consolidate their gains by building more and more production capacity—especially during the inevitable downturns that plague the industry. "During each major industry recession over the last decade," says Michael Burros of BRIE, "Japanese firms have strategically gained market share through high capital investment that is beyond the resources of most U.S. merchant firms during a downturn. Their investments in 1974–76 permitted them to enter the U.S. market in the late 1970s. Their spending in 1980–83 permitted them to seize and extend leadership in commodity memory. Then, when worldwide demand picked up, the Japanese were able to supply the market with high-quality, advanced memory chips using the American techniques of forward pricing at low prices."

Japanese success in mastering the enormously complicated techniques of VLSI manufacturing pay off in higher yields. And profits in the semiconductor industry are sharply affected by yields. The production of an IC is one of the most complex and demanding of any manufacturing process devised by man. Moreover, it is enormously costly and technically challenging. And more is at stake here than mere profit. The ability to produce advanced memory devices provides critical access to the next technological step in chip design and production. To many experts, the production of memory chips, with their microscopic circuit features, is an essential technology driver for the semiconductor industry. "Only through large-scale production can the current generation of integrated circuits be improved," says Charles Dittrich, a microelectronics specialist at the Department of Commerce, "and the information needed to develop the next generation be generated. Bugs encountered and resolved in large-scale production are the key to developing the next-generation products."[21]

"DRAMs give you more advantages in a number of different places than you get from other parts," adds Bill Finan. "Principally you're going to get it in tuning a line. You've got to have a simple part to prove out a process and equipment, training people, getting operators to understand the consequences of their actions, integrating the line. We identified two or three other areas that we felt were clear couplings into processing know-how and advantages. There are some indirect things, too. It's good loading for plants, it helps you spread fixed costs, it generates cash flow. It's without doubt the biggest-volume part you have. If you look at it in terms of volume per family, memory is about a five- to sixfold factor greater than any other product area you can identify."

When TI looked to design their DRAM production facility in Dallas, they employed a Japanese architectural engineering firm. Then, to tune the line, they again turned to Japan. "We sent in what they called a SWAT team"—Hideo Yoshizaki, chairman of the board of TI Japan chuckled—"tens of people, managers, and engineers. I called them 442 after the Nisei Rigimental Combat Team that rescued a Texas regiment from the Germans in World War II at Bastogne. My file on the project was labeled 442."[22]

Less dramatically but with a similar approach, when IBM built its clean rooms at its advanced chip-making plant in East Fishkill, New York, they, too, had to rely on a Japanese firm for state-of-the-art construction technology. "We called in the Japanese principally because they had more experience in clean-room technology than any domestic contractor in bringing up a very clean facility," admitted Michael Attardo, the president of IBM's General Technology Division. "They had significant experience and more experience than any one domestically with clean materials, clean chemicals, and overall manufacturing technology. The requirements of cleanliness in the space program, for example, are an order of magnitude or two less than those required for the advanced semiconductor technology center. The company was Shimizu, and they basically designed the facility for us. We improved on that design, and basically we wanted to see what they had. We saw and had them design the facility for us; we enhanced that with our base knowledge. Then, after competitive bidding, we chose an American company named Walsh to execute the design."[23]

Despite the obvious benefits of producing commodity chips such as VLSI memories, most American companies have determined that the costs are simply too high and the profits too low. "The Japanese really don't have an inherent comparative advantage in memory," points out Bill Finan. "What's really driving the Americans out of that area is the fact that the profit margins are so lousy. There is no DRAM manufacturer as of 1987 who can say that they made money over the entire life of the product. Everybody has lost money on DRAMS."

Still, the Japanese continue to invest in capital equipment. "Japan-based semiconductor firms have been investing unprecedented amounts in capital plant and equipment in recent years," admits the SIA. "In 1983, Japan based firms outinvested the American firms for the first time, and will widen this gap through the balance of the decade."[24]

The hemorrhage of red ink that overwhelmed the industry in 1985 and 1986 reduced capital spending in the United States even further. According to Dataquest, American companies, for example, spent $1.55 billion, or an average of 13.1 percent of their sales, in 1986. That same year of deep industry recession, the top five Japanese semiconductor manufacturers spent more than $1.76 billion for new processing equipment, or almost 11.7 percent of sales. In 1987, however, as their sales mounted, U.S. companies actually outspent the Japanese, investing $1.9 billion to $1.68 billion.[25]

Massive Japanese spending for semiconductor manufacturing equipment actually began as a concomitant to the VLSI project. "From 1979 to 1985, capital investment in Japan jumped 15 percent, doubling in size as a percentage of semiconductor production," William E. Reed, executive director of SEMI, Semiconductor Equipment and Materials Institute, a trade group, told a Dataquest-sponsored industry conference in Phoenix in October 1987, "compared to a capital spending increase of only 5 percent in the United States. Capital spending fell in all regions in 1986, undoubtedly influenced by the prolonged industry recession."

That sort of investment was, to Reed, part of a strategy that paid handsome dividends. "But," he told the assembled industry leaders, "the overall trend toward increased capital investment in Japan not only reflects the Japanese business philosophy but also points out their strategy regarding the importance of market share. The Japanese purchase of capital equipment has enabled them to increase market share. They have found that for every dollar invested in capital equpiment, they will see a two-dollar return. They also have a strategic view and do not emphasize short-term returns on their capital-equipment investments."

Japanese companies can afford to make these massive investments for two reasons. Even though the equipment industry is structured much like that of the United States, approximately five hundred companies, most of them small, producing everything from rubber gloves to $4 million electron-beam mask makers, a scant dozen companies account for 75 percent of total sales. And of that dozen almost, all are either subsidiaries of Japan's electronic giants or belong to the same zaibatsu. Nikon, for example, is part of the Mitsubishi financial grouping, Anelva, a major maker of deposition and etching equpment, is a subsidiary of NEC.

The results of such alliances have produced a reordering of market leaders similar to the shake-up of the chip manufacturers. "The population of the top ten equipment companies," points out Reid, "has experienced a demographic shift that is very similar to the one seen among the top-ten device manufacturers. In 1982, the top-ten equipment companies included six U.S. companies, three Japanese companies, and one European firm. But by 1986 the top ten's makeup had nearly flip-flopped, with five Japanese companies, four U.S. firms, and one European company."

Indeed, only P-E retained its position as the world's leading maker of semiconductor manufacturing equipment. Figures compiled by *Electronic Business* magazine in their May 1987 issue[26] show the changes in sales leaders between 1982 . . .

COMPANY	COUNTRY	SALES IN $ MILLIONS
Perkin-Elmer	USA	162.1
Varian	USA	99.5
Schlumberger	FRANCE	96.4

Takeda Riken (Advantest)	JAPAN	83.8
Applied Materials	USA	83.4
Eaton	USA	79.9
Teradyne	USA	78.7
Canon	JAPAN	77.7
General Signal	USA	76.9
Nikon	JAPAN	57.7

and 1986:

COMPANY	COUNTRY	SALES IN $ MILLIONS
Perkin-Elmer	USA	264
Advantest	JAPAN	238
Tokyo Electron (TEL)	JAPAN	166
Canon	JAPAN	155
General Signal	USA	155
Applied Materials	USA	149
Nikon	JAPAN	148
Schlumberger	FRANCE	139
Varian	USA	137
Ando	JAPAN	136

Unquestionably, some of the Japanese success has come about as a result of their close relationships with chip makers. According to Sophia University's Gene Gregory:

Japanese semiconductor equipment builders tend to have closer ties with chipmakers, either as direct subsidiaries or in a client-supplier relationship that includes a two way flow of resources—men, money and technology. Although most equipment manufacturers sell across group lines, competing fiercely for customers throughout the industry, each has developed more or less intimate relationships with principal customers as a result of continuing cooperation in the design and development of successive generations of more advanced equipment.[27]

William Reid pointed out the benefits of just a few of those cooperative relationships between equipment and material suppliers and device manufacturers in Japan. "The Anelva/NEC venture has enabled NEC to move from the number-three position in 1982 to number one in 1986. The arrangement has also allowed Anelva to develop leading-edge sputtering systems for mass producing megabit VLSI devices. . . . . Hitachi has played an important role at Kokusai, funding much of their wafer fabrication research. Advantest has gained a technological edge over its competitors due to its alliance with NTT. And the list goes on."

An example of how this works can be seen as Japanese chip makers gear up to produce 4-megabit chips. Among them is Mitsubishi, one of the largest and wealthiest companies in the world. Mitsubishi Electric Corporation is but one of dozens of major companies within the Mitsubishi zaibatsu. It is headquartered in its own building in the Marunouchi district in Tokyo, just down the street from the Mitsubishi Bank Building, the Mitsubishi Heavy Industry Building, and two other Mitsubishi Company buildings. As I walked past these identical white concrete, glass, and steel boxes, for an instant I thought I was trapped on some giant Monopoly board, hurrying down Park Place or Boardwalk, the game's most expensive properties.

I was here for a meeting with Dr. Hisao Oka, the director of Mitsubishi's LSI Research Institute and the man directly responsible for setting up the line that would produce the 4-megabit chip. Oka is an electrical engineer and had joined MELCO, as the company is often referred to, upon his graduating from Tohoku University, one of the nine schools within the original imperial university system, in 1952. Among his string of impressive successes, Dr. Oka designed the semiconductors used in Japan's first bullet train.

"The development and production of 4 megabit will be an extension of our current line," he told me, "and of course the stepper requires an advanced version; however, this has already been introduced. So it will be an extension of the existing memory factory with a few replacements. That is not the case with the 16-megabit chip. That will require a total replacement. The equipment is purchased from outside, but in recent years the amount of equipment being produced under a cooperative development program is increasing. That is because we need to acquire equipment that is applicable to the advanced technology we need in our factories. By cooperative development we specify and place orders for specific improvement in our equipment. Since these companies are business partners, we cannot afford to give their names. They are for the most part Japanese companies. However, since the Mitsubishi group has a lot of corporations in its ranks, such as Nikon, etc., I think you can conclude that we will be jointly developing it with companies such as that."[28]

In the United States, virtually all equipment makers must sign nondisclosure agreements with their customers. This effectively prevents them from commercially exploiting what they have learned from working with a semiconductor company on a new-process technology. Despite these barriers, some American chip makers and equipment companies are forming alliances. P-E, for example, worked very closely with Intel in the development of its Micralign system in 1973, a projection printer that made possible the mass production of low-cost ICs. The Micraligner projected a high-resolution, one-to-one image of the circuit onto a wafer. Without it, chips would have to be contact printed directly on the wafer, which limits the life of the mask and greatly increases the possibility of contamination of the wafer. Since then, P-E has continuously refined and upgraded the Micraligner, working closely with IBM in subsequent machine generations.

Now, at a cost of $20 million, P-E has developed a new photolithography system called step and scan. Code-named Gamma Blue, in acknowledgment of the development contract for the system from IBM, the new system is a complex combination of scanning and stepping technologies that allow all the chip areas on a wafer to be exposed at the same time.[29]

"We've been quite successful until now in purchasing from domestic vendors," IBM's Michael Attardo told me. Appointed president of the General Technology Division as part of what IBM recently and euphemistically called a management restructuring, he is responsible for the development, production, and purchase of chips and chip-making equipment. Thin, dark, and balding, Attardo is at one and the same time a classic American success story and almost an antithetical corporate executive. Attardo grew up in a poor neighborhood in the

New York City borough of Queens. He commuted on the subway to Brooklyn Tech, one of the city's elite science high schools where admission is by competitive examination. From there he went to Queens College, part of the city's free university system, in a special program that was linked to Columbia University. The result was that he gained a BA from Queens and a BS from Columbia. "What made that attractive," recalled Attardo with affection, "is I couldn't afford to go to Columbia otherwise." He subsequently received a Ph.D. in materials science at Columbia and did his thesis work at the Brookhaven National Laboratory on Long Island before joining IBM.

Attardo is a tough-minded executive who thoroughly understands the importance to IBM of a strong American semiconductor equipment industry. "We have a vested interest in seeing that we have a viable domestic tool industry. And so we are trying to foster their development efforts. Alignment equipment, etching-equipment sputtering, measurement equipment, are all essential to the manufacture of state-of-the-art chips. The better alignment you have, the better your lithography, the denser the chip, the lower the cost, the higher the performance. And you can relate that to each piece of equipment. Measurement is extremely important in terms of yield, which flows back into cost. If you don't have good measurement equipment, you can't control your yields, you can't control your process, and so all those tools basically come back to the guy who can make it the fastest, the smallest, at the least cost."

But Attardo is not content simply to sit back and wait for equipment makers to knock on IBM's door. "We also have some joint development work with domestic vendors to enable them and us to get the best state-of-the-art tools. We kick in not only money but our skills and our people. Depending on the amount of intellectual input or resource that we invested, we would put in some restrictive clauses on the right of the vendor to sell that tool. For example, we would like to have the rights to the output of that vendor for a year. Then he can sell it to *SEMATECH* vendors if he wanted to. Those are the kinds of things we have in place, and they vary depending on our intellectual and financial value added to the tool."

In so critical an area, Attardo believes that Americans must learn to play by new rules. "I think the American equipment makers are technologically as advanced as the Japanese, but I don't think they pay as much attention to detail, I don't think they pay as much attention to tool reliability, the up time of the tool, the contamination levels in the tool, that the Japanese companies do. And I think they've got to learn to do that. We are presetting those parameters right now. We are having some difficulty in getting commitments to meet those specs from some suppliers, but before we sign any development contracts or any significant procurement contracts, we are insisting on their meeting those requirements.

"It's more a way of thinking. For example, we build our chips to last 100,000 hours in the field, and it doesn't cost us any more to do that. It's a design methodology. It's a design practice, a way of thinking; that's all it is. I think it's a cultural shock to some of these equipment suppliers. I look at a Japanese tool and its mean time to failure is 1,000 hours. I look at an American tool in our factories and its mean time to failure is 150 hours. And the difference is simply a question of design methodology, design practice. Looking at the material set that goes into the tool, looking at the tolerance the tool has to temperature variation, for example, the barometric variation. Those kinds of things are taken into account in Japanese tools, but not nearly as well in American tools. The Japanese use materials and tolerances that are more forgiving of environmental conditions."

What happens to production when a piece of equipment goes down? "When a machine fails," Attardo explains patiently, "it means you need to get a maintenance guy on the phone. It takes him an hour to get there, or a half hour. He looks at the machine for an hour; then he brings the tool back up. Then its got to be recalibrated, and I've lost three hours out of a fifteen-hour day.

"So tool reliability is really king. It's one of the most significant drivers of capital, of space, and of yield. And the guy who has got the edge there is going to be the winner. We are broadcasting this message to our equipment suppliers. We have specific teams that are going out to our suppliers and telling them what we want and why we want it. So there is an education process going on out there. And the message is definitely getting through. We have some companies that are very responsive but some that are fairly lethargic."

Most chip-making machinery has been improved over time to allow for the ever shrinking size of the components jammed onto a chip and to increase the throughput, the numbers of chips that can be pushed through a plant every month. Of the more than 150 processes involved in producing a chip, there are half a dozen major steps that begin with the making of a mask. Once the chip has been designed, each of its barn-floor-sized circuit diagrams must be photographically reduced to the size of a baby's thumbnail. As many as three hundred of these images are then etched onto a quartz plate, the mask. Each layer of the chip is reproduced on a separate mask, and each mask contains multiple images of the same layer. Thin wafers of highly polished ultrapure silicon, looking not unlike razor-thin mirrored pancakes, are coated in a furnace with a thin film of oxide. Layers of a photoresist, a light-sensitive chemical, are laid down on the wafer, which is then loaded with others onto a wafer boat that carries it to the stepper. The steppers are similar to photoenlargers, although in this case their negatives are the vastly reduced circuit diagrams etched on the mask.

The mask is then precisely aligned over the wafer, and a beam of ultraviolet light passes through the mask and strikes the surface of the wafer. The exposure "eats down" through the photoresist to the oxide film, forming on it the outline of the circuit. The process is repeated again and again, up to three hundred times on a six-inch wafer, covering it with tiny squares. The exposed wavers are then etched, in essence, developed just like film, washing or burning away the exposed areas of the photoresist and the oxide layers beneath it. The etching cuts a set of windows through to the pure silicon below.

Now the wafer passes to another machine called an ion implanter, which zaps electrically charged atoms of arsenic, or phosphorus or boron, into the silicon windows. These dopants convert the silicon to appropriate n-type, for negative, or p-type, for positive, junctions. This doping process is alternatively done in a gas diffusion furnace. The photolithographic stepping and doping processes are repeated for each layer the chip is designed to carry. The last mask provides the layout for the interconnects, the microscopic grid that joins the 100,000 or so transistors of the chip together. A thin layer of metal, usually aluminum, is then laid down on the chips by a machine called a sputterer. The sputterer fires molecules of an inert gas such as argon at an expendable target, such as aluminum, gold, or platinum, like a pitching machine hurling baseballs at a batter. In this case, the argon molecules nip off particles from the target and at the same time give them an electrical charge or ionize them. These ionized particles of metal are then electrically slammed into the surface of the chip in a very precise pattern, providing the hundreds of miles of microscopic wires that connect transistors and circuits together.

At virtually every step of the way, the chips being built on the wafers are tested. When finally finished, the wafers are then diced, or sliced, by diamond saws into individual chips, at this stage called die. The die are then mounted on lead frames, bonded with gold-wire electrodes and then molded into plastic or ceramic packages. The end result looks more like a plastic-armored cockroach with perhaps a dozen metal feet on each side than the chips that hold the economic and geopolitical future of nations in their convoluted, microscopic geometries.[30]

The entire complex process may take as many as 120 days from start to finish and is subject to a horrendous catalog of potential defects every step of the way.

Intel's Bob Noyce points out:

Defects can arise from many sources. The photo masks used may have pinholes in dark areas or opaque specks in areas that should be clear. Severe defects in the basic silicon crystal can make the circuit inoperative. Dust in the photo printing operation or in some other processing step that affects a critical spot in the circuit will cause failures. Errors in aligning successive photoengraving steps or lack of control over critical dimensions and impurity concentration will make the circuit inoperative. The correction and elimination of these defects is a difficult task and represents a major portion of the effort and expense of semiconductor device development and production.[31]

Much of that expense is generated by the effort to keep the processing area clean. A few motes of dust falling on a wafer can destroy most of its chips. And so the bottom line of all semiconductor production is yield, the number of working chips produced on each wafer. And the ultimate requirement for high yields is cleanliness. Chips are created in fabrication facilities, called fabs, or front ends. The heart of the fab facility is a clean room. NASA popularized the industrial concept of a clean room by assembling statellites in them. But compared to the levels of air purity demanded by VLSI chip production, NASA clean rooms were little better than brightly lighted dustbins.

The yields of Japanese IC plants are generally conceded to be among the highest in the world. "The way they did it was by reducing contamination," says P-E's Chuck Minihan, who understands clean-room technology as well as anyone in the semiconductor industry. An ex-fighter pilot and career air force officer, he has spent virtually his entire life in military and industrial electronics. While fighting in three wars, he still managed to secure advanced degrees in electrical engineering and held such important posts as commander of the team that launched the military space satellites at Vandenberg Air Force Base during the sixties and was head of the air force's space tracking network at Colorado Springs. In 1969 he retired after thirty years' service. He then spent three years with GE in their Missile and Space Division until P-E hired him away.

"Contamination is particles," he said to me, shoving his big hand out as if feeling for raindrops. "If I hold my hand out, the particles falling into that hand can be tiny, no bigger than a micron. And those little micron-sized particles fall on that micron-sized line, and if they get into the interconnects, that line is no longer a functioning line; that chip is lost. That's yield. How many useful die do you get off a wafer that's been through an entire processing line. The Japanese are in the 80 to 85 percent range. If you find a company in the United States that's over 55 per cent, I'll be very much surprised. We just never automated enough. We take smoke breaks, allow women workers to wear makeup."

The problems of keeping a clean room clean are enormous. Consider that a typical office may harbor 100,000–500,000 airborne particles per cubic foot. Just breathing, the average person spews out about a million particles every minute, and that's while at rest.

"If you go into a clean-room facility that is class 100," points out Minihan, "that means there are 100 micron-sized particles per square foot per hour. Now, if you go into a class 100 clean room, that is, we'll say of 500 cubic feet, after a smoke break, that room will go to class 1,000, simply due to the carbon that you exhale after smoking a cigarette outside the clean room. Class 1,000 of course is much dirtier; it has ten times more particles.

"Today, class 0 is the ultimate goal. If you go into American mask-making facilities, you are beginning to see workers in bunny suits with bootees, masks, the entire rig." It's an outfit some American workers, at least, seemingly resent. Koichi Shimbo, NEC's director of communications in New York, recalls the difficulty of getting American workers to dress for clean-room operations.

"NEC had bought a small chip-making company in Mountain View, California, called Electronic Arrays. Today it produces more than three million memory chips a month for the American market. But it took more than two years to make the company profitable. Electronic Arrays had always run at a deficit because their productivity was so low. The yield was very low due to high contamination. We took a number of steps to reduce contamination, including asking the employees to take off their shoes and replace them with clean-room slippers. It took more than one year before they learned to do it. That experience made us believe that quality is from the production. Americans believe quality is from testing and checking after production. We believe quality must be incorporated in production itself. We believe it, and we do it."[32]

American companies generally have more problems because their workers do not have the same fierce dedication to quality the Japanese do. "I asked a guy at TI why the yield at their Miho plant in Japan is greater than the yield in Dallas," Dr. Weinig told me by way of explanation. "He said because at Miho I've got 500 Japanese housewives and in Dallas I've got 24 percent blacks, 23 percent Hispanics, 26 percent Vietnamese, etc. In Japan I say I don't want anyone to smoke for one hour before coming to work because particulate matter comes off your lungs. I don't want anyone going in without washing your hands; you're only allowed one toilet break a day, because every time you egress the clean room you bring more dirt in.

"Can you envision my saying to the people in Houston, 'You can't smoke for one hour before going to work'? Or tell the women, 'No makeup,' because it is a contaminant? Try that one in Houston. You're dealing with an American mentality. When I deal with the heterogeneity of the U.S. versus the homogeneity of Japan, I say wait a minute, let's only look at problems that are solvable. I can't change the culture of the American worker. At least we haven't been able to do it yet. That means our yields will always be lower."

Thus, the Japanese permit no smoking breaks during a shift, ban the wearing of cosmetics, beards, and mustaches, and of course insist on the wearing of protective clothing. The result is a clean-room technology that is second to none. To me there is an almost laughable irony here. One of my most vivid impressions of Japan on my first visit in 1953 was of the astonishing contrast in the public health attitudes of the two countries. Many Japanese, I noticed, wore surgical masks to avoid spreading their colds to others. At the same time, with absolute unconcern, men would urinate in the streets.

When I mentioned this to Sony's Dr. Makoto Kikuchi, he laughed and attributed it to a mix of feudal behavioral norms and Meiji education that produced the most cultured people in Japan. "I remember walking on the outskirts of Tokyo with some of my father's friends," he told me. "They were men of culture, bearded, with flowing white hair, the standard-bearers of an education and cultivation inseparable from the Meiji years. But I was flabbergasted to see them expectorating in the streets and stopping to urinate on the hedges."

The Japanese still wear surgical masks to prevent the spread of cold-causing viruses, but those other flabbergasting practices I remembered are no longer in evidence. Nor are there many people to be found in most Japanese fab facilities. Mitsubishi's new Saijo factory on Shikoku Island, just to the northeast of Silicon island, is entirely automated. Its 650 employees are almost entirely maintenance people. Total automation does seem the way in which the industry is headed, but until the advent of VLSI technology, chip making was a labor-intensive process. American companies had responded to that situation by moving production and assembly offshore, where labor costs were much lower than in the United States, as early as the 1960s.

Many of the nations that had proved so attractive to the American garment industry offered the same compelling advantages to the semiconductor companies—a large pool of young, unmarried, illiterate women who could be paid twenty-five cents to a dollar an hour, often at piece rates. TI set up plants in El Salvador and several Asian nations. National Semiconductor, Intel, Motorola, and many other companies established plants all over Asia, in Indonesia, and in the Philippines as well as Mexico and Central America.[33] Dependence on low-wage offshore labor increased until, by the late 1970s, 80 percent of the chips produced by American merchant companies were assembled outside the United States.

"Work is tedious and stressful," declared the Pacific Studies Center's Global Electronics Information Project in 1980. "It requires intense concentration and pinpoint accuracy, constant coordination of hands, feet, and eyes—sitting in an arched position all day—and rapid movement. Microscopes, especially when used all day six days a week, can cause dizziness, eyestrain and tension, and deteriorating vision.

Few scientific studies have been done on the impact of scope work on Asian assemblers, but women who must have 20/20 vision to be hired usually end up wearing glasses after a few years. In Hong Kong, most assemblers over age twenty-five are called 'Grandma' because they wear glasses."

The move offshore did more than just provide poorly paid jobs to the women of the third world. It transferred American technology to the rest of the world and sowed the seeds of a competitive whirlwind that has shaken the U.S. semiconductor industry to its very foundations. Offshore manufacturing was done under a series of special tariff provisions that allowed American manufacturers to produce products overseas, bring the components back to this country, and pay duty only on the work that was done abroad.

"We began watching rising competition in the early 1970s," says Jack Clifford, director of the Commerce Department's Office of Microelectronics and Instrumentation. Clifford, an economist who has grown white haired in the service of his government, has been with Commerce as the industry specialist since 1963. An intelligent, hardworking bureaucrat in the best sense of the word, he has watched the American electronics industry literally give away its technological leadership while chasing after cheap labor. "In 1973, I had started to do a project on offshore manufacturing by U.S. companies, with its attendant introduction of technology to foreign facilities. We began to collect trade data on the effects of the 806 and 807 special tariff provisions that allowed U.S. manufacturers to produce overseas, bring the components back to the United States, and pay duty only on the work that was done outside the U.S. Examining those tariff provisions and looking at the countries where the trade flows in those particular areas were rising gave us an indication of when the offshore manufacturing started and the velocity of increase that took place. Then we began to watch the foreign content increase. That meant the foreigners were now beginning to put more and more into the product that was being produced. You could see symptoms or indications of the transfer of technology.

"We were looking at countries like Malaysia, Hong Kong, Singapore, Taiwan, Mexico, and Japan. At the time we started looking, Mexico was one of the major areas because it was border operations for people like Motorola, Texas Instruments, and firms in California who

had production-assembly facilities in Mexico. The Mexican government had made some special provisions to encourage a border
industries program in Mexico, and that caused that area to build up a
little faster. Companies could have U.S. management because the
facilities could be managed from just across the border; transfer of
personnel, travel, was all kept to a minimum. When you started dealing
in the Far East, many of those nations gave special breaks to U.S.
companies who would put up facilities over there, but the distances
involved were much greater, management control was more difficult,
transportation more costly.

"We watched the shift from the level of U.S. content, which would
have been maybe 60 or 70 percent at the beginning, drop to 45 and 40
percent, which meant the foreign manufacturers were providing more
native-produced content, for example, bases and headers, which we
had earlier shipped over there. They were beginning to pick up the
technology and build a production base that would eventually produce
semiconductors."[34]

But labor was less a factor in VLSI production than automated
state-of-the-art manufacturing technology. And so the shift to VLSI
technologies accelerated the loss of American dominance to the
Japanese.

TI's new DRAM IV plant in Dallas and their VLSI plant in Miho,
Japan, for example, each cost more than $100 million to build.
Although the Japanese now also have dozens of assembly plants in
countries around the world, they have always been reluctant to send
production offshore. And when they have, labor costs have not been
the determining factor.

Mitsubishi's Dr. Oka explains why. "As far as our company is
concerned, we feel we should not resort easily to the option of offshore
production. Because once you start offshore production, let's say you
assemble in one place and produce in another place, then your whole
corporate integrity becomes undermined. I think you lose the quality of
production that is required when you take that kind of approach. So
the decision to go offshore should be based on many different
considerations that far exceed that one factor of labor costs. Of course,
labor cost is one factor. As far as cost is concerned, the two basic factors
that come into place when we talk about cost are yield and design. So
labor cost is not the only consideration. So in view of this I believe we

will be able to meet the competition by automation to increase production at home."

That reluctance is now being overcome by the inexorably appreciating yen, which is forcing Japanese manufacturers not only to import lower-priced parts from the rest of Asia but to establish semiconductor plants in the United States, Mexico, South Korea, Taiwan, and elsewhere. The reason is the same one that initially sent U.S. semiconductor makers overseas—cheaper labor costs. At the same time, the Japanese continue to run their domestic plants full blast and continue to look to better manufacturing methods and automation to keep costs down.

In the 1970s, however, when American semiconductor firms were still dominant, the Japanese made automation a key factor in their attempts to catch up and at the same time maintain the entire production process, from design to assembly and test, at home. That need to automate led to the development of significant new technology by the Japanese equipment makers that gave them a sizable jump on their American competition.

"The reluctance of Japanese chip makers to send assembly processes to low-wage off-shore plants," declares the U.S. Commerce Department, "translated into a demand in Japan for more highly automated assembly equipment than was being used elsewhere in the world."

It is a well-practiced technique that helped Japanese semiconductor equipment manufacturers challenge the U.S. lead in the early 1970s. The Commerce Department notes:

Shinkawa, a Japanese firm, successfully adapted to the demand with such innovations as the automatic wire bonder (1972), thereby equalling the sophistication of machinery produced by U.S. firms. Another Japanese firm, Disco, was successful in mastering dicing saw technology. Once the United States' lead in assembly equipment evaporated, so did its domination of that segment in the Japanese market. In addition, Disco and Shinkawa were able to expand their overseas sales, penetrating the U.S. market for the first time in 1975 and 1976 respectively. Since that time, competition between U.S. and Japanese assembly equipment firms has been intense, with technological leadership and market share changing from year to year.[35]

Japanese equipment makers also made significant advances in testing devices and a number of photolithographic dry etching and ion implantation equipment in support of the VLSI project. The end of the VLSI program, however, did not end Japanese chip makers' investment in new capital equipment. The reason is simple. The Commerce Department reports:

In its efforts to surpass the U.S. device [semiconductor] shipments both in quantity and complexity of device, Japanese semiconductor manufacturers are not hesitating in replacing "obsolete" equipment (even though only two or three years old) with equipment reflecting the newest, although yet unproven, technologies. U.S. device manufacturers seem to be more reticent about incorporating unproven technologies into their processing lines. Japanese SME makers, therefore, are being given an extra impetus to improve existing technologies and develop new ones.

"The Japanese will always buy your equipment first," emphasizes Dr. Weinig. "They do not stand around asking how many people are using your equipment and what is their experience; they don't care. I developed a brand-new sputtering machine, a million dollars a copy, and the first sales were made in Japan. They put the money on the barrelhead; they always did."

As just one example of that Japanese willingness to spend money on the newest equipment, in October 1987, the Kayex Corporation, a Rochester, New York, company that makes silicon-crystal furnaces, reported the sale of its newest model, a monster capable of holding silicon in lengths up to ten feet. The furnace, for which an unnamed Japanese company paid almost $900,000, is capable of producing six-, eight-, and ten-inch-diameter single silicon crystals.[36] Although most new fab lines are capable of handling six-inch wafers, only IBM, as of this writing, has installed, in its Burlington, Vermont, plant, an eight-inch wafer line. To date, no one in the world has yet attempted to produce chips on a ten-inch wafer. But when the technology does arrive, the Japanese, and probably IBM, will be ready and waiting to take immediate advantage of it.

For the most part, the American philosophy is quite different, more of a "mix and match" approach to capital spending. The Commerce Department, in its "Competitive Assessment of the U.S. Semiconductor Manufacturing Equipment Industry," reports:

This consists of reducing production costs by using older generation equipment for less-exacting processes and restricting the use of expensive, state-of-the-art processes for tasks where they are absolutely necessary. In effect, this selection practice reduces the demand for the advanced equipment in which U.S. firms have had an edge (albeit shrinking), and increases the size of the market for less sophisticated technology where the Japanese equipment makers have been better able to compete on the basis of technical equality.

That situation may not last very much longer. The next generation of chips will require advances that may push chip-making technology beyond the borders of visible light. VLSI is already a geriatric case, and the industry is now looking ahead to ULSI and beyond that to GSI, or giga scale integration, when a billion transistors will be jammed on a chip. Viewed under a microscope, the circuitry of such chips is as labyrinthine as an Arab Casbah the size of North America. Here, compared to a transistor, a virus is the size of a battleship; a human hair looms as large as the World Trade Center and the Sears Tower placed atop one another.

Among the new technologies that may be used to build the next generation of chips is X-ray lithography. A number of companies are already well advanced, and some have moved into commercial production. "We are now coming out with an X-ray stepper," Nikon's Shoichiro Yoshida announced. "Sales are only to domestic companies so far. NTT and other Japanese companies are using it but do not want their names mentioned. We are not sure just what kind of semiconductor device will be produced by this kind of technology. Maybe 64-megabit area or much higher density area. We are not sure yet. It is one of the candidates, but nobody can say right now that it will be the one. But we want to be in the forefront, and so if X-ray technology is the one that will be used, we want to be there."

P-E is one of the very few American companies also selling X-ray steppers, but all of the commercial machines are based on conventional X-ray sources, which are not considered powerful enough to photograph the extremely small geometries of the ULSI generation of chips. What will be needed are X rays produced by synchrotrons, machines better known as atom smashers. And already a race is on to develop compact synchrotrons to provide X rays for steppers. Leading the field are the Germans. In West Berlin the Fraunhofer Institute for Microstructures Technology and a consortium of European semiconductor makers known as COSY Microtec GmbH have just built a synchrotron specifically for X-ray lithography.

The Japanese have also mounted a ten-company cooperative effort called Sortec, which has begun construction on a synchrotron center for X-ray lithography research. In the United States, the Brookhaven National Laboratory has proposed the establishment of a center to develop and produce commercial chips with feature sizes of 0.25 micron by X-ray lithography. The laboratory is the site of the National Synchrotron Light Source, the world's most powerful source of X-ray light for research.[37] IBM, P-E, and AT&T have already conducted experiments there. "We are running product right now at the Brookhaven National Laboratory with their X-ray source," notes IBM's Michael Attardo. "And we are seeing some very excellent results. It's got the advantage of allowing you to print devices with feature sizes as small as a tenth of a micron. It also has the potential advantage of being immune to defects, because the X ray penetrates low-atomic-number defects. In other words, a dust particle on a mask cannot block the X ray the way it can a visible light wave. In effect, the X ray doesn't see the particle.

"So one could say this technology has the potential to significantly improve yield and to make you more tolerant of contamination. So we think the technology is worth investing in. Given what's going on in Japan, we think we cannot afford not to look at it. And so we have work going on at Brookhaven right now, and we have also asked Oxford Instruments in England to build us a superconducting synchrotron ring. And that's going to go into our new advanced semiconductor technology center. We have no expertise in this area, and so we thought it would be counterproductive to try and do it ourselves. We looked around the world, and we made visits and requests for quotes, and Oxford turned out to be the best. Perhaps it was because of their expertise in superconducting devices and CAT scanning, which employs a high degree of X-ray technology."

At the same time, the Brookhaven National Laboratory is looking for $395 million to build a facility that would house a pilot manufacturing plant and an R & D center for ULSI chips. As of this writing, neither industry nor government has committed funds to the project. But here is a clear opportunity for American technology to burst out once again ahead of the world. The question is, will the opportunity be seized?

6

Bye-bye TV, VCR, and Just About Everything Else

MITI's success in targeting industries has become the focal point of trade frictions between the United States and Japan and inevitably has also become a factor in U.S. presidential politics. During his campaign for the 1988 Democratic presidential nomination, the Reverend Jesse Jackson used a standard opening gambit in addressing audiences across the nation. "How many of you own a videocassette recorder," he challenged. Everyone in the audience shot up a hand. "All right," he then demanded. "How many of you know anyone who owns an MX missile? Above the laughter he shouted, "We don't make VCRs in America, but we're making what there ain't nobody buying." Thus does the former radical preacher turned presidential candidate point up what he and many others believe is the misplaced emphasis of both government and industry on military research and not consumer products.[1]

In fact, the United States has global commitments that include, among other things, defending the nations that do make VCRs: Japan, Korea, France, West Germany, and Holland. Nonetheless, the Reverend Jackson's point is also a singular fact: The United States no longer has very much of a consumer-electronics business. "If by consumer electronics you mean products such as VCRs, CD players, stereo components, such as receivers amplifiers and turntables, cassette tape players and recorders, and AM FM radios—those are not made in the USA," acknowledges John Streeter, an industry economist at the U.S. Department of Commerce.[2] The explanation is simple: The U.S. consumer-electronics industry was won by Japan in another trade war a decade ago.

And since GE sold off its consumer-electronics business, including RCA's color television division, to a French company in July 1987, there remains only one major American manufacturer—Zenith—still producing color television sets in this country. And no one makes VCRs, CDs, stereos, Walkmans, Watchmans, or any of the other semiconductor-based consumer-electronics products.[3]

After the first early successes with the transistor, MITI, in the late 1950s, targeted consumer electronics as a major growth industry. Product lines were developed and refined; meticulous studies were made of the American market and tastes in cameras, tape recorders, television sets,and other consumer electronics. The result: What had once been a thriving American electronics industry, producing millions of television sets, high-fidelity and stereo audio systems, and other consumer goods had by the late 1970s lost most of its market to the invading Japanese.

Among the first targets were television sets. Bill Taylor, the New Hampshire computer consultant who grew up in Japan when the Japanese first began to build TVs, has reduced their marketing approach to a simple strategy that might have been drawn by a Go-playing Shogun. "The way they worked in television sets, they went to RCA and said, 'Look, we can sell you your low-end model, which you are selling to Sears, cheaper than you can make it.' RCA said, 'Great,' bought it from them, and put their label on it and sold it to Sears. After a couple of years, once the factory was humming along, the Japanese went to Sears and said, 'We can sell you this same set cheaper than RCA can and put a Kenmore label on it.' And Sears bought it from them. Then they went to RCA and said, 'Evil round eye, you are not buying the quantities you said you would.' The agreement with RCA was abrogated, and the Japanese had their own distribution going to Sears. Once they did that, they had the distribution channels, the shipping, and now understood the American market because RCA and Sears had explained it to them, and so they started selling direct. That's how Sony got into the business."[4]

Still, you cannot sell junk, and when the Japanese first began making television sets, they were decidedly inferior to American-made TVs. But the Japanese were the beneficiaries of a couple of genuine strokes of luck that they were quick to understand and turn to their advantage.

"The Japanese television sets in the fifties were not competitive with anything being produced in the U.S." explains Jack Clifford, the head of the Commerce Department's Office of Microelectronics. "The Japanese did an awful lot of offshore assembly for companies like Montgomery Ward and Sears, and a great deal of consumer-electronics technology was gained by that offshore assembly of U.S. components. By the early 1970s they had built a good production base that was pretty near the size of the U.S. consumer-electronics industry. They knew they would have to do the same thing with the components that went into TV sets, and so they had a choice, going to tubes, which they were working on at that point in time, or going solid state, to semiconductors.

"The U.S. at that time had a very large vacuum-tube manufacturing base still in place. People like RCA, General Electric, and Zenith and Westinghouse and others were somewhat more reluctant to switch from tubes to semiconductors. The Japanese did not have this huge tube-manufacturing establishment in place, and they were quick to recognize the importance of semiconductors to their consumer-electronics industry. So you began to get such things as portable radios, TVs, and tape recorders, all solid state, from Japan pretty early in the game.

"The Japanese also recognized the importance of transistorized radios, which we did not seem to pick up on for one reason or another. When somebody demonstrated a transistor radio to the board of directors of Texas Instruments they said, 'It's an interesting product, but what value does it have?' TI was not a consumer-oriented company. In Japan, most of the companies were vertically integrated; they produced their own components, they produced the TV sets, and they were quick to recognize the importance of the strong component base they would need to build their consumer-electronics industry.

"So they started producing their own TV sets with Japanese names and a very high semiconductor content. U.S. manufacturers had all this tube foundation, and they were reluctant to write it off. But solid-state sets were more reliable than tube sets, and that gave the Japanese a lead in selling sets in the United States. Even though we had developed tube technology to a high state at that time, semiconductors produced less heat, needed less power, and most important, reduced the size of the set. The picture tube in the set determined the size of the carton to be used. In the early stages of television, most picture tubes had a ninety-degree deflection angle. That meant the gun had to be farther back from the screen, and that made the carton larger. If you added an inch to the neck of the tube, it meant you had to expand the box an inch times the overall size of the box. Transportaton costs were based on cubage, not weight. The Japanese saw that if they increased the deflection angle they could reduce the length of the picture tube. That would bring the size of the box down by a couple of inches and thus decrease their shipping costs.

"They introduced the 110-inch deflection tube, which not only reduced the size of the carton; it made tabletop and portable television sets possible. U.S. set manufacturers responded not by switching to semiconductors but by abandoning the field to the Japanese. They said, 'Okay, we'll build large console sets; the Japanese can't compete in this area. We won't go toe-to-toe with them in the smaller sets; we'll just make the larger console models.' That worked for a while, but it turned out that the most popular size was nineteen inches. With the Japanese breakthrough on the deflection angle, the nineteen-inch tube became a table-model set, and you could buy that for one heck of a lot less than you could buy a nineteen-inch console, even if the guts were the same price. The furniture costs were much higher."

The inroads the Japanese made in the American market had a ripple effect that swelled to a roaring tide throughout their entire electronics industry. "In addition to semiconductors," Clifford points out, "by increasing the production base in Japan for consumer-electronic products, radios, TVs, tape recorders, VCRs, they increased the whole production base of their entire components industry. That meant that not only were their semiconductors more competitive in the United States but so were other components they produced, such as capacitors, resistors, etc., because of the sheer volume. We began to notice in our passive components industry increased competition from foreign components coming over here to be installed in U.S. sets and then the subsequent loss of the U.S. television market to Japanese manufacturers.

"There is a lot of concern on our part of the direction things are taking in semiconductors. It is frighteningly similar to TV sets. When you get into integrated circuits, it's a much larger problem than discrete components. The reason is the IC not only represnts the components that go into a finished product, but it represents a larger and larger proportion of the technology of that device, be it a television set, an instrument, or a computer. The IC is actually the technology of the device. In the old days, when you were wiring components together, a manufacturer could redesign a circuit using standard components and could differentiate his set from somebody else's. But when you are using ICs, the same ones everybody else is using, it becomes much more difficult to differentiate your product. So your competition through produce differentiation becomes much more difficult. In computers, everyone must be compatible. So that reduces the competitive advantage U.S. companies had when they could differentiate their products by circuit design."[5]

With fewer means of differentiating a product, quality, reliability, and price become the critical factors. And here the Japanese excel. Having fine-tuned the manufacturing process to the point where their chips were the equal of anything the Americans were making, and probably better in some classes, the next step was to penetrate the market and then capture market share. When the share of the market was large enough, the Japanese would dominate it no matter what the cost or the time it might take.

And that is precisely what happened in the DRAM market. These chips, minuscule collections of boxes that store information, may be likened to the brain cells of computers and the dozens of other electronics devices that need some form of memory storage. And like brain cells, they store within their own crystalline cells millions of evanescent electrical charges. Each tiny charge represents what computer jargon calls an information bit, a binary number of 1 or 0. All the ones and zeros, offs or ons, are a numerical code that can represent any type of information, the product of one plus two, the organization of the alphabet into words, or any other piece of information.

Memories had grown prodigiously from the 1K chips that could store 1,024 such bits of information, in 1970, to the 16K RAM, which had the capability of holding 16,384 pieces of binary data. The 16K memory chip first appeared in 1978 and was to become the focal point of a battle for dominance of the memory market between the United States and Japan.

Significant advances had been made in chip technology, and a new American innovation, a method of coating the layers of a chip with metal oxide silicon, or MOS, became the new standard for memory ICs. MOS technology, as are most U.S.-inspired breakthroughs, was swiftly transferred to the rest of the world, either through licensing or theft. Moreover, as had been typical of each previous generation, the cost per bit of memory dropped as the amount of memory on the chip increased. In 1970, a single bit of memory on a 1K chip cost about a penny. In 1978 the price per bit had dropped to five cents. That made a 16K chip the equal in cost to the previous-generation 4K chip. It was a bargain that created its own heightened demand and promised great rewards to the companies that could meet it.

"The most significant international market battle took place in the MOS memory market and centered on the domestic U.S. market for 16K MOS RAMs," Michael Borrus, James Millstein, and John Zysman of BRIE told the Joint Economic Committee of the U.S. Congress in 1982.

The price per bit equivalency between 4K and 16K dynamic RAMs [the point at which the cost of a 16K DRAM equaled that of a 4K chip] occurred in 1978, and with it came a significant and accelerating demand for 16K RAMs. On top of the demand generated by this crossover, IBM entered the merchant market with a huge demand for 16K RAMs to meet the memory needs created by the rapid market acceptance of its new series 4300 computer. The increasing demand for 16Ks was paired in the U.S. market with a significant production capacity shortfall. This stemmed largely from the failure or inability of merchant IC firms to invest in capacity expansion during the 1975 recession, and from their cautious investment policies following the recession. Here, of course, reliance on internal funds and equity markets constrained the business-strategy choices of U.S. firms. By contrast, the stable availability of capital for Japanese firms permitted them to engage in a rapid capacity build-up that could support their export strategy.

The results were almost inevitable. "Indeed, in 1978 and 1979 the major Japanese firms strode in force into the market gap created by significant undersupply in the domestic U.S. memory market," point out Burros, Millstein, and Zysman. "By the end of 1979, they had taken 43 percent of the domestic U.S. 16K RAM market."[6]

With distributor networks firmly established and their footholds in the American market secured, the Japanese moved swiftly to consolidate their success. In 1979, at the conclusion of the VLSI project, Japanese companies began limited production of the 64K DRAM. It was to be their knockout punch, but first they had to increase their production capacity.

"Beginning in the late 1970s, Japanese firms began a major buildup of semiconductor production capacity—a buildup which not only continued, but accelerated despite the onset of a world wide recession," accused the SIA in a 1985 position paper entitled "The Effect of Government Targeting on World Semiconductor Competition."

Most major Japanese semiconductor firms at least tripled their annual rate of investment in semiconductor production facilities in the five years 1978–82. Throughout 1980 and 1981, Japanese producers continually revised upward their initial projected investment levels—notwithstanding the fact that prices were falling rapidly and demand was below expectations. Much of this investment was directed toward production capacity for 64K RAM devices.

What the Japanese had done was to leapfrog an entire generation. By 1981, while most of the world's chip makers were producing 16K RAMS, six Japanese companies were into full-scale production of more than 20,000 64K chips a month. At the time, there were only three American merchant companies beginning to produce 64K DRAMS, and one of those soon dropped out. Everyone else had been caught with their technological pants down and were still making 16K DRAMS. The effect of this onslaught of new and far more powerful chips staggered the Americans.[7]

"With their early start," explains Jerry Sanders, the chairman of AMD, a major Silicon Valley chip maker, "Japanese producers began cutting prices sharply. 64K RAM prices dropped 96 percent, from an average 25 to 30 dollars per unit to about 8 dollars during 1981."[8]

Production of 64K RAMs was increased dramatically faster, according to the SIA, than the world market could absorb. "Japanese 64K RAM production capacity increased sevenfold from an annual capacity of 9.0 million devices per year in July 1981 to an estimated 66.0 million per year in August 1982," according to SIA figures.[9]

Eventually prices dropped below five dollars a chip as the Japanese firms launched an all-out assault on the American market. How did it work? How could the Japanese afford to sell chips at prices seemingly considerably below the cost of making them? The key was a strategy geared to obtaining market share, no matter the cost. Part of the effort, however, was subsidized by higher prices at home. Burros, Millstein, and Zysman explain:

Until late 1978, Japanese producers apparently used a two-tier pricing strategy. They kept RAM prices high in their controlled domestic market, thereby subsidizing their ability to offer lower prices in the U.S. market. . . . As domestic Japanese demand for 16K RAMs rose in 1978 and 1979, Japanese IC firms chose to let imports (mostly from U.S. firms) meet domestic Japanese demand. . . . This enabled Japanese producers to divert their own production to the United States in order to increase their share of the U.S. market.

The Japanese were also producing a higher-quality chip than were the Americans. Burros noted:

The Japanese firms also used the issue of higher quality as an extremely effective technique to help penetrate the U.S. RAM market. A number of U.S. consumers of Japanese 16K RAMs, notably Hewlett-Packard and NCR, have suggested that the failure rates of the Japanese product were significantly lower than those of U.S. devices. While U.S. devices met the quality standards of U.S. purchasers, there was unexploited market demand for higher quality devices. Japanese producers correctly appraised the U.S. market and used a higher quality penetration strategy to capture additional market.[10]

All strategies paled beside the ultimate willingness to beat the competition no matter the cost. And so armies of American salesmen spread across the land selling Japanese chips and armed with the knowledge that they could meet any price. Bob Flowers, a young engineer who had begun his career with Intel, went to work for Hitachi in 1981. "Hitachi had a sales philosophy called the top ten," he told me. "There are ten customers in the U.S., and then there's everybody else. IBM, General Motors, Kodak, Xerox, etc. To Hitachi, that was who you sold to. We focus on those guys, bend over backwards to give them what they want, and bring in another product designed for use in Japan and see if we can't also sell it here. When I was at Hitachi, there was so much money to be made, you could be a whore and be encouraged. You had an American salesman who was the eyes and ears of the Japanese in the U.S. marketplace. The Japanese didn't understand what it took to sell or establish a product, and the quality of the people they had took the path of least resistance. That was, I'm not going to sell the technical advantages of this product. I'm going to sell it on price, make my life easy and make money as well.

"You built budgets based on the transfer price of a product, which had already been established in Japan. We were then asked how much can you sell? We would come back with a forecast. In 1981, in my area of complete microchip computers, memory and CPU right on one chip, I would assess what their average demand would be for the budget period. I would provide competitive prices I could sell it at.

"I would tell them what I needed to sell it at; they would then adjust the transfer price, which we then marked up to the price we needed to sell it at, including our profit. If the market couldn't sustain the sales at the transfer price originally established, it would be changed to beat the competition. If my four-dollar, 4-bit microcomputer was losing to the Fujitsu version, I would lower the price. In those days it was more the Japanese companies competing among themselves, fighting for the same markets. They all had the same technologies, with the same sales approach and the same available marketshare."[11]

Indeed, it often seemed to the Japanese companies trying to crack the U.S. market that their chief competition was other Japanese companies. Long accustomed to the savage competition that marked their business practices at home, price cutting in the States seemed a logical extension of a traditional sales philosophy. "The overall sense I have," Professor William Egelhoff, a research associate at the Center for Science and Technology Policy at Rensselaer Polytechnic Institute, told me on a warm spring day in 1987 at his office in the financial district of lower Manhattan, "is that the competition between the large Japanese firms is intense."

Egelhoff had just published a comparative study of the business strategies employed by Japanese, U.S., and European semiconductor companies entitled *Business Strategies and Competition in the Semiconductor Industry: A Comparative Study Across U.S., Japanese, and European Firms.*

"It's an intensity that we are not familiar with. Therefore, it's not at all surprising to me that when they come into our markets we look upon them differently than they look upon themselves. It's like a high school football team getting on the field with a pro team. It's a different level of competition. And neither is bad. The pro people, the Japanese, will simply say, 'You guys don't have your act together; you're a bunch of amateurs fooling around.' The Americans say, 'You guys cheat. You play too rough; you're too big, too strong.'

"Part of the problem is we don't have the same definition of competition. So when we get in the same arena, everybody is yelling foul. Because we are not playing with the same rules. The Japanese way of competing is analogous to the way wars used to be fought. Both armies had the same strategy. We find a nice big field, your guys line up on one side of the field, my guys on the opposite side. It's as if we both had gone to the same military academy and are going to fight the exact same kind of battle. And we have very similar agreements. If it's raining we don't fight; we don't fight at night. The guy who wins is the guy who executes the strategy better. If my people are trained better than yours, if their marksmanship is better and my discipline is better, I'm probably going to win. The army that gets even a 10 percent advantage on each of those factors will probably be invincible, given equality of size. That's what the Japanese try to do; they try to get that 10 percent or 5 percent advantage. But they all try to do it the same way. They're all running down that same learning curve, trying to get higher yields on the same 256K chip and trying to get to the next-generation chip a week or a month ahead of the other guy.

"We, on the other hand, are used to competing through strategic maneuvering. Pepsi tries to position itself a little differently than Coke. Usually they don't go head-to-head. At times, they do, but that is not a respected tactic in American business. Going head-to-head means we both lose and the consumer walks away with the day. You get into a price war and the whole industry takes a bath. That is not admired in American business. What is admired is uniqueness. Digital Equipment doesn't take on IBM head-to-head.

"Basically, we have a game plan that says when you're outflanked, or outmaneuvered, don't insist on fighting to the death. That just isn't the way American business does it. I'm taught in business school to beat a hasty retreat, conserve my resources, and win in another segment of the market. So that philosophy goes a long way to explaining why Americans pulled out of the 64K and then the 256K DRAMs so quickly. You're taught to do that; when you are at a disadvantage and the foreseeable future looks no better, you're supposed to get out of that business.

"So when we see we have lost, we cut out and look for something else to do. I leave the field and leave the spoils to you. The Japanese, on the other hand, don't leave the field. The Japanese way is to regroup and try harder. So what you get is a lot of hard pushing back and forth. It doesn't drive anybody out of business, but it drives down prices and costs and is a boon for the consumer, who benefits from the intense competition."[12]

To the Japanese, fierce competition is the norm, but for many years, three of the giant companies had a major advantage in the Japanese market. Mitsubishi, Toshiba, and Hitachi were the best known and well entrenched. "Until the advent of the computer, Fujitsu and NEC were largely unknown to the Japanese people; they were primarily suppliers of telephone equipment to NTT," explained a Japanese electronics-magazine editor who asked me not to mention his name. "Then, as they invaded other markets, they became known as *nobushi,* "wild samurai." Mitsubishi, Toshiba, and Hitachi were much better known to the Japanese public and did not have to claw their way into the marketplace. They were known as *tonosama shobai,* "Lords of Commerce."[13]

The ultimate expression of Japanese competition can be seen in a section of Tokyo called Akihabara. Akio Akagi took me there on a sunny September Sunday, in much the same way I would take him to see such New York tourist attractions as the Statute of Liberty or the Empire State Building. Like the Diet, the imperial palace, and sushi bars, Akihabara is a tourist attraction, a monument commemorating and expressing the essence of Japanese competition. Akihabara, a crossroads for Tokyo railroads, was originally a major black market at the end of World War II. Today it is a ten-square-block area of modern stores laced with a winding labyrinth of stalls and shops that would do credit to an Arab bazaar. In them is the high-tech outpouring of Japan's electronic companies, everything from flat-screen plasma TV to 256K DRAM memory chips selling for as little as 200 yen ($1.25). Here young boys and men paw through bins of components, drop their treasures into blue plastic dishes, bring them to the cashiers, pay for them, and then take them home and assemble them into electronic wonders undreamed of by their parents. At the same time, the latest in personal computers, telecommunications equipment, and audio and video components are presented by Japan's mighty consumer-electronics and computer giants. Here market research is as simple as what sells. Here too, scientists and engineers from nearby research laboratories who cannot wait for technicians to assemble a suddenly urgently needed test device or specific gizmo collect what they need from a stall and rush back to the lab to assemble it.

Within the four- and five-story buildings that house the major electronic stores, displays are dazzling, prices firm, and, it seemed to me, high. In the booths and stalls outside the big stores, prices were also high, but negotiable. With Aki interpreting, I asked the price of a Sony telephone answering machine, which I had just bought at a New York discount store for $169. Here, at 138 yen to the dollar, the then going exchange rate, it was $220. I then did what no Japanese would do. I asked for a better price.

They came down by five dollars. Similarly, in one of the large stores I priced an Olympus microcassette recorder for which I had paid $59 in New York. Here it was $106, and there was no bargaining. It was virtually a mirror image of the Japanese electronic industry as a whole. At home, prices are high. Abroad, prices are lower and will in fact be cut to whatever it takes to sell.

A now infamous memo from Hitachi to its American semiconductor salesmen illustrates to just what lengths they were prepared to go.

> Quote 10% below competition.
> If they requote . . . bid 10% under again.
> The bidding stops when Hitachi wins . . .
> Win with the 10% rule . . .
> Find AMD and Intel sockets . . .
> Quote 10% below their price . . .
> If they requote, go 10% again.
> Don't quit until you win . . .
> 25% DIST[ributor] profit margin guaranteed.[14]

"For the most part it was the jelly-bean chips, memories that they could sell below U.S. producer's costs, and they definitely had better quality." Flowers told me. And it was illegal. American law and the international trade agreement known as GATT, for General Agreement on Trade and Tariffs, to which both the U.S. and Japan subscribe, specifically forbid the sales of a product below cost to eliminate competition. The process is known as dumping.

"Here was one of the classic problems in trade between the United States and Japan," writes Clyde V. Prestowitz, Jr., in *Trading Places, How We Allowed Japan to Take the Lead.* Now, with the Carnegie

Endowment for International Peace in Washington, Prestowitz had been counselor for Japan affairs, the principal policy adviser to the secretary of commerce on all aspects of commercial and high-technology relationships between the United States and Japan, from 1981 to 1986. In that position, he was a key negotiator for the United States on all of the semiconductor agreements reached during that period.

U.S. law frowns on price discrimination between customer or markets; but in Japan, where there is no tradition of antitrust, it is the rule. Thus, while the U.S. producers accused the Japanese of acting illegally, the Japanese thought they were just being good businessmen. They needed customers, and the easiest way to get them was to cut price. That did not seem to them unfair. However, when U.S. producers threatened law suits for dumping in 1980, the Japanese eventually reduced prices at home, while raising them a bit in the United States to achieve parity in the two markets. By that time, however, they had achieved the volume required to be competitive.[15]

"That's how they got into the major American user markets," continued Flowers. "Not with better architectures but with cheaper and better memories. And quality and price were the keys. The pricing structure was whatever it took to sell. For the American salesman selling for the Japanese, it was the path of least resistance. Do whatever you had to do to make as much money as possible with the least amount of effort. And the customer took advantage of it, too; he had to be profitable in an extremely competitive marketplace."

Virtually all of the other Japanese companies were following essentially the same rules, dumping chips below cost to seize the major share of the American market. At first, the U.S. chip makers did nothing about it. Business was just too good, for everyone. Then came the recession in 1984 and 1985. Computer makers all over the world began canceling orders. Other users also cut back on chip purchases. The Japanese firms, with their enormous cash flows, government support, and other product areas in which profits could be made, had only to hunker down and wait for an upturn. For U.S. makers, the reduced orders coming on top of their loss of market threatened disasters.[16]

Most American companies simply dropped out of the RAM market. Intel, which had invented the RAM chip in 1968, stopped producing memories in 1985. "We announced our withdrawal from DRAMs on October 9," George H. Schneer, vice-president for memory components at Intel, told me sadly. "It sticks in my mind like Pearl Harbor."[17] Motorola and National Semiconductor also dropped out of the memory business. Mostek, another American company that had, in 1976 and 1977, been the largest producer of 16K RAM chips in the world, didn't simply go out of the RAM business; they went out of business entirely in October 1985.[18]

"The resulting high losses experienced by U.S. firms," acknowledged AMD's Jerry Sanders, "can only serve to discourage future investment and participation in the RAM market. It's become a new ball game called 'You Bet Your Company.' And the consequences of disinvestment are serious. RAMs traditionally have generated technological capability that translates into leadership in other semiconductor product areas. The RAM experience represents the 'entering wedge' of an all-out assault on other semiconductor markets—repeating a pattern familiar to many other industries targeted by the Japanese."[19]

Most American memory makers simply dropped out, but some chose to fight. In June 1985, Micron Technology, a small Idaho firm, filed suit against seven Japanese companies, alleging dumping of 64K DRAMs in the U.S. market. At the same time, the SIA filed a petition asking the U.S. government to investigate unfair trade practices by the Japanese under Section 301 of the 1974 Trade Act. Just two months later, the U.S. International Trade Commission issued a preliminary finding that imports of 64K DRAMs from Japan were injuring U.S. producers.[20]

The Japanese assault was not limited to DRAMs. Another type of memory chip, called EPROM, for erasable programmable read only memory, began to flood the American market. EPROMs are used primarily to store programs in minicomputers. One type of EPROM, which had sold for $17.50 in January 1985, could be bought from the Japanese in November for as little as $3.00. Another less advanced EPROM was sold for less than it cost to test and package it. In September, Intel, National Semiconductor, and AMD filed suit against eight Japanese companies, charging them with dumping EPROMs in the U.S. market. Their losses on EPROMs were estimated at $200 million in 1985.

In December, yet another antidumping case was filed, this time by the U.S. government itself. This suit alleged that Japanese companies were dumping 256K DRAMs on the American market at prices far below fair market value.[21]

At the same time, the European Electronic Component Manufacturers Association also filed an antidumping complaint against Japanese memory makers. The Europeans accused the Japanese of selling DRAMs and EPROMs for less than half their cost.[22]

After all the filings and accusations had been made, a series of heated meetings between U.S. and Japanese trade representatives took place in Tokyo and Washington. The United States threatened the Japanese with stiff tariffs and heavy fines for violating U.S. fair trade laws. For six months the two sides argued. From the U.S. standpoint there was an almost religious fervor to the stand they took on the side of free trade—the economic equivalent of motherhood, apple pie, and the other American verities. Rather than looking for specific benefits from an agreement, the American position has been to argue in support of the general principle of free trade. The result has been only the haziest of commitments on the part of the Japanese and confusion as to just what the agreement really means.

"I think we could negotiate for a significant increase in sales if we had different policies and a different willingness toward our negotiations with Japan," Clyde V. Prestowitz, Jr., told me. "We must negotiate for something concrete. If we are prepared to negotiate for a particular level of sales or a particular market share in Japan, we could do that. There is nothing in the agreement that guarantees any 20 percent. It's a very fuzzy agreement. It's one of the things I dislike about the nature of our negotiations with the Japanese. The agreement says that the Japanese government understands the expectations of the American industry, that their share of market will increase. The wording of the agreement is such that the American government believes that it has a commitment to a doubling of the U.S. share in Japan. But the Japanese can and are saying, based on the wording of the agreement, that there is no commitment."

"It's not that we don't try to force the Japanese to make the commitments; it's more that we ourselves are loathe to spell out just what the commitment should be. To some extent, the Japanese outnegotiate us, but they do so because we handicap ourselves. We never allow our negotiators to negotiate for an explicit share. In this negotiation of semiconductors, our negotiators could never actually ask for 20 percent. And the reason was not because the Japanese didn't let them; the reason was because our own trade policy, our own thinking about economics, prevents us from asking for a market share. We think it's contradictory to the free-trade principles that we uphold. So we never allow our negotiators to ask for anything concrete. And so they are always in a position where they ask for market access. The Japanese say, 'You have market access.' And our negotiators say, 'Yeah, but we never sell anything.' The Japanese say, 'Well, how much do you want?' And the Americans can never respond to that. They often have to say we want significantly more. The Japanese say, 'That's fine. How much more?' And Americans say, 'Buh-buh-buh-buh-buhhh.' The Japanese say, 'How about . . .' and they name a number. Then you get into winking and nodding across the table. So we do get outnegotiated, but it's because we tie our hands behind our backs before we begin."

Finally, at midnight on July 31, 1986, just one minute before a U.S.-set deadline, agreement was reached between the two sides. In addition to the agreed-to terms that were to be published, there was a secret side letter that committed the Japanese to a doubling of the U.S. share of the Japanese market. But when news of the side letter leaked out, MITI denied it. Subsequently, they admitted that indeed there was a side letter to the agreement, but MITI insisted it contained no commitment to a specific U.S. market share.

"The interesting thing about the MITI denial," Prestowitz told me, "is that in the fall of '85, just prior to the time the U.S. government began to initiate the 256K dumping case, I was contacted by one of the top Japanese negotiators for a secret meeting in the Mayflower Hotel. After going through long hallways and double doors and so forth, we sat down and talked about what to do in semiconductors. This was a high-level MITI official, and he did not want to self-initiate. He said to me that MITI could issue administrative guidance that would stop the dumping and assure American industry a 20 percent market share. There are two interesting things about that. One is that MITI is now denying that they would ever consider such a thing. The second thing is that I went back to the American side and reported this, and the American side rejected the idea that we should negotiate for a 20 percent increase. And so, in the fall of '85, we specifically rejected a negotiating position that MITI said was acceptable to them.

"Then, in the course of '86, we spent a lot of time negotiating market access in which we finally, desperately, tried to get the Japanese, in the form of this side letter, to commit to a 20 percent market share. I believe they did commit, and this is the problem with these damn side letters. In the course of the negotiations, we spoke often of increased market access and what would be a benchmark. How do you know we have access? The U.S. side said 20 to 30 percent of the market would be commensurate with the American share elsewhere in the world and reflective of American competitiveness. And the 20 percent figure was thrown around a lot in the oral negotiations. The Japanese kept saying it depends on the efforts of American industry, and we can't guarantee a market share that depends on free trade. And the State Department, the Office of Management and Budget, the Council of Economic Advisers, and the Treasury Department also would say that's not free trade. But Commerce and U.S. trade representatives were prepared to ask for a 20 percent share.

"The end result is you get into elaborate use of code language. The side letter says that the Japanese government understands and welcomes the expectation of the American industry that their share will double. And the Japanese government will undertake efforts to help realize it.

"Now, in the code language of the negotiations, recognizing the expectations is understood by the Americans to mean the Japanese government is committed to seeing to it that the U.S. side gets a 20 percent market share. But the language doesn't explicitly say that, so you are in a situation in which the U.S. government, based on the context of the negotiations and its understanding of the secret code language of the negotiations, says we have a commitment. The Japanese government first denies there is a side letter, and then it doesn't tell its public of its commitment. So as far as the Japanese public is concerned, the American demand for an increase in sales in Japan is irrational and not understandable. Then, when it's leaked that there is a side letter, the Japanese government says, 'Ah, but there's no commitment to a market share.' On the basis of the explicit language of the agreement, there's not, so it's very fuzzy."[23]

In addition to that somewhat fuzzy agreement to increase American access to the Japanese market, the Japanese also agreed to stop dumping their semiconductor products in the United States and in third-country markets. The Americans agreed to drop their suits, which, had agreement not been reached by the deadline, would have set fines of almost $259 million against the Japanese companies.

The agreement was finally signed in September 1986, and MITI agreed to impose strict antidumping curbs on all Japanese semiconductor manufacturers. What the agreement did not forbid, however, was the sale of semiconductors below cost in Japan. This created a so-called gray market for chips. Access to the gray market enabled most Japanese companies to dump chips on the one hand and deny it on the other. There are three basic channels of semiconductor distribution used by the Japanese, I learned from my anonymous electronics-magazine editor. "One is the company's official sales office. Then there is a so-called *keiretsu*, which is a subordinate sales and distribution company. There are fifteen or sixteen keiretsu used by each company. Finally, there are *chusho*, small agencies at the very edge of the distribution system. The exporting traders worldwide get their products from those three organizations.

"The destination of products, especially from the major companies, can be identified, but not those distributed by the smaller independent keiretsu and chusho. Those chips often find their way into the gray market. The sales departments of the large companies do not directly dump. The small-scale distributors, the chusho, are most often the channel through which chips are dumped in a third country. "Often, when American purchasers approach Japanese makers directly, they refer them to American distributors rather than sell to them directly. They do not want to be implicated in dumping. It still goes on through small companies in small quantities, but I have no idea as to the number," explains the editor.

"There are also so-called dark distributors, American companies that sell semiconductors lower than the market price. According to rumor, they bring the chips to a warehouse in Vancouver, in Canada, and then take them into the U.S. from there. As a result, the major Japanese makers refuse to sell chips to an unidentified company."

But Japanese producers did sell millions of 64K and 256K chips at fire-sale prices in Japan through their chusho distributors, who packed them into suitcases and carried them to Hong Kong, Taiwan, and throughout Southeast Asia. American experts, for example, calculated the Japanese production cost of 256K DRAMs at approximately $2.50.

Transported by the suitcase brigade, those same 256K DRAMs were selling throughout Southeast Asia for $1.60–$1.80.[24]

Faced with such competition, some small American distributors have quite simply joined the act in order to stay in business. In its June 30, 1986, issue, *Forbes* magazine quoted one smuggler who had made two trips to Tokyo during the year, picking up 20,000 256K DRAMS the first trip and 40,000 on the second. Suitcases bulging, he flew to Vancouver, where chips enter duty-free. There he rented a car, changed from business suit to jeans and a sweatshirt, and drove across the border to Washington. The chips were then sent by air freight to his office in Southern California. The smuggler then backtracked to cover his trail by driving back to Vancouver, where he took a flight to California.

The gray market also spread to Europe, affecting not only European and American producers but even the European distributors of Japanese semiconductors. "It's a ridiculous situation," complained one European executive of a Japanese semiconductor maker to a reporter. "I can go to the gray market and buy parts there cheaper than I can from my parent company."[25]

U.S. complaints were louder still and directed to the Japanese government. And finally, seven months after the initial trade agreement had been signed, President Reagan proposed the imposition of sanctions against Japan for noncompliance. But the proposed sanctions created almost as much turmoil among American electronics manufacturers as it did among Japanese chip makers. For the low-cost chips were in fact a competitive boon to American electronics makers, while they were at the same time a disaster for U.S. chip makers. The Computer and Business Equipment Manufacturers noted that its members "source [buy] many component subsystems in Japan . . . and sanctions would inflict extensive damage on the U.S. information technology industry. The information industry depends on high tech components and other products included in the proposed sanctions list."[26]

So tangled have the relationships and dependencies become that in one case at least National Semiconductor Corporation, one of those suing for antidumping relief, found itself in the anomalous position of testifying against the new duties. National Semiconductor, it turned out, imported Hitachi computers through a subsidiary. The penalty duties, they claimed, "would damage National Semiconductor more than the Japanese semiconductor violations."

"You've got to work both sides of the fence," Charlie Sporck, National Semi's CEO admitted to me. "We very much supported the sanctions, but we were very much interested in applying those sanctions so that U.S. customers were not hurt. Including ourselves. We were not in favor of sanctions. We were trying to pick products that no other manufacturer would be affected by. And by and large that's where the sanctions ended up."[27]

In the end, the Commerce Department placed 100 percent duties only on Japanese-made hand-held power tools, eighteen- to twenty-inch color television sets, and portable computers. Ironically semiconductors per se were exempted, although all the sanctioned products used chips.

Electronic News columnist Jack Robertson wrote:

The outcome of the 2-day jam-packed sanctions hearing was a huge relief for most of American industry, which escaped the penalty tariffs on imported purchases. That however may be only a short-sighted reprieve. As long as American high-tech industry is so dependent on Japanese supply, as proven conclusively in the sanctions testimony, then it is only a matter of time before trouble can strike again.

Such vulnerability to your chief competitor in the survival fight for technological dominance isn't very reassuring."[28]

Nor are the dichotomies that split the American electronics industry. For the systems makers, the builders of computers, test equipment, medical diagnostic tools, and dozens of other electronic devices need low-cost chips to be competitive and show a profit. And they don't care where their chips come from. "The U.S. market is a high-price island," says Gene Norett, director of the Semiconductor Industry Group for Dataquest, which compiles world chip prices every week. "Anybody stupid enough to buy parts here pays a lot more than in Hong Kong."

Tariff barriers will just drive the cost of American chips even higher and send computer makers and other systems assemblers heading not for the hills but across the Pacific. "Our survey of major users found more and more people saying they have to buy or manufacture in Asia," said Norett. The American Electronics Association, which represents most American electronics firms, echoed the same sentiments.

"Because of the pricing differential, about 85 percent of 200 respondents in our survey," they reported, "are opting for offshore purchase or assembly, or they are just using chips from their backlog."[29]

Without resolving the real issues of why the Japanese were able to seemingly produce higher-quality chips at lower cost than U.S. manufacturers, the sanctions had the desired effect. In October 1987, the Commerce Department acknowledged that Japan had ceased dumping computer chips. "Our monthly monitoring of semiconductors now shows clear, firm and continuing evidence that third-country dumping has stopped," declared S. Bruce Smart, Jr., the undersecretary of commerce for international trade.[30]

In response, the United States dropped most of its sanctions, and a new surge in computer sales worldwide boosted sales of U.S. and foreign-made chips.

Of even more concern, if possible, than dumping has been the virtually closed market Japan has presented to the rest of the world. In a study by the national Academy of Sciences entitled "A High Technology Gap?," Academy president Dr. Frank Press asserted that compared with the rest of the industrial world, Japan's economic nationalism results in the lowest imports of manufactured goods as a fraction of total imports. "Virtually all American and European governments and firms that do business in Japan," he stated, "complain of barriers that put foreign firms at a disadvantage in marketing products and services."[31]

To the Japanese, protection of their industries is a matter of national survival, and that means protecting the home market. Philip Abelson, the deputy editor of *Science* magazine, in an editorial in the May 8, 1987, issue, wrote:

In our relations with the Japanese, we set a pattern shortly after World War II that persists to this day. To facilitate economic recovery in Japan we did not object strenuously to protectionist measures on their part. They have been skillful in erecting intricate import control restrictions on all kinds of goods and services, and especially those of high technology. We have not erected barriers of anywhere near comparable magnitude. One is reminded of a coin-flipping game in which "heads I win, tails you lose." The Japanese act as if the existence of these historic asymmetrical trade policies are guaranteed.

They are skilled at promising changes and in avoiding substantive implementation.[32]

For many years Japan had resisted U.S. pressure to lower the tariffs that protected their producers of semiconductors and other high-technology products or to allow foreign companies to build plants in those industries in Japan. "During this catch-up period," declared the SIA in a conference on global productivity in 1985, "U.S. semiconductor investment in Japan was restricted and the Japanese domestic market was protected by a system of import licenses and quota. U.S. companies were able to sell semiconductor products in Japan which Japanese firms did not yet make, but as Japanese companies developed an indigenous capability in a particular product area, U.S. firms' sales tended to disappear. Thus, while U.S. semiconductor firms enjoyed a strong position in all other world markets, they never held more than a foothold in Japan, rarely exceeding 10 percent of the market."

Finally, in 1974, a so-called liberalization took place. The Japanese market supposedly was open to foreign investment, and tariffs were substantially reduced. It had little effect, however, on increasing the sales of foreign-produced semiconductors, according to the SIA. "Thereafter, however, the foreign share of Japan's semiconductor market showed virtually no increase, reflecting the fact that Japanese semiconductor consuming firms—which were often also the leading Japanese semiconductor producers—pursued an unofficial 'buy Japanese' policy."[33]

The new trade agreement was designed not only to end Japanese dumping in the United States but to open Japan to foreign-produced chips. Just how open the Japanese market has become depended more on whose dictionary one was using than upon numbers. "When you say open, you get into a question of definition," declares Clyde Prestowitz, "In terms of the way Western economists define access, we have it now. There is no tariff, there's no quota, there are no funny standards or anything like that. The thing that prevents us from selling semiconductors in Japan and in quantity is partly the structure of Japanese industry and the relationships between suppliers and customers and producers in Japan. It is also partly the fact that this remains in Japan a high-priority national-development industry which in some ways could be compared with the Apollo project in the U.S. Therefore, there is a high degree of nationalism attached to it in Japan. So you need a different definition of access. My view is that to call a market open and you can't get any sales in it is not really open."

Indeed, one year after the agreement was signed, there was little evidence of increased sales by U.S. or other foreign producers in Japan. In its report to President Reagan on the effectiveness of the agreement in 1987 one year after its implementation, the SIA stated:

U.S. companies' share of the Japanese market is only one-half percentage point greater than the market share level at the time of the initialing of the Agreement in July of 1986, and it is almost one-half percentage point below the 9.4 percent market share held in the quarter prior to SIA's June 1985 filing of its Section 301 case.[34]

In other words, despite all Japanese protestations to the contrary, the United States had actually lost its share of what has become the world's largest semiconductor market. In September 1987, as the SIA report was being delivered to the White House, I went to MITI headquarters in Tokyo to get their response to the American accusations that the Japanese market remained closed to U.S. semiconductor

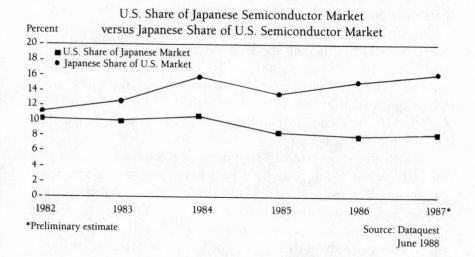

U.S. Share of Japanese Semiconductor Market
versus Japanese Share of U.S. Semiconductor Market

Percent

■ U.S. Share of Japanese Market
● Japanese Share of U.S. Market

*Preliminary estimate

Source: Dataquest
June 1988

products. On the fifth floor of the MITI building that houses, among other bureaus, the Industrial Electronics Division, I entered a large room bisected by shoulder-high file cabinets. With me was my friend Akio Akagi, who would serve as my interpreter. He gave the girl my name, and we were escorted to a tiny waiting area, four chairs tucked between another wall of file cabinets and a magazine rack. In the rack were full-color glossy reports by the giants of Japanese industry, NEC, Mitsubishi, Fujitsu, Toshiba, and the bible of the Japanese electronics industry *Nikkei Electronics*. The wait was very short; Tetsuo Matsui, the deputy director of the Industrial Electronics Division would see us. The deputy director of what is perhaps the most important division in one of the most powerful bureaucratic institutions in the modern world is a smooth-cheeked thirty-two-year-old who looks like a teenager. He ushered us to a tiny, round table beside one of the twenty or so gray steel desks that jammed the large room. We sat at the table in folding metal chairs that proved to be as uncomfortable as they looked. This, I subsequently learned, was virtually VIP treatment, for Matsui's office was one of the desks, with no room around it to fit another chair.

"In the past the market was closed," he admitted, "but now there is no import tax on ICs. The structure of the market here is very different from the U.S. The Japanese IC makers produce semiconductors for consumer electronics, but the U.S. semiconductor makers are not in consumer electronics but rather in the industrial and computer areas.

"U.S. makers, generally speaking, do not produce the type of ICs Japanese electronics makers need. There is also the question of price, quality, and the major problem of delivery. U.S. makers take much longer time to deliver their products here than do the Japanese makers. Despite U.S. insistence that quality and price are now comparable, MITI thinks that is not so."

After listing these standard arguments to account for the failure of the U.S. makers to do better in Japan, I was treated to a somewhat patronizing lecture by Mitsui. "Generally speaking, international exchange is welcome, but there is a tendency for the U.S. to provide design and the Japanese company to make the product. If you take that approach, we are afraid that the U.S. will lose the potential for future design by giving up production. You will lose the technical facility by losing the feedback between production and research. We believe that as long as we maintain our production capability we can continue to design and meet future needs. We are actually more concerned for the U.S. in that kind of arrangement, for we are, after all, allies."

Allies notwithstanding, the Japanese market remains a very difficult one for Americans to crack, and virtually all of the Japanese I interviewed had essentially the same response.[35]

"If we feel the product is good and the service is available, we are not averse to buying a product from the United States," Mitsubishi's Dr. Hisao Oka told me. "In this area of business there is day-by-day advancement, and so what used to be a good product in the past will not be good forever. There is very keen competition in this field. So a manufacturer to survive and win in this competition must procure the very best of equipment and semiconductors. So I believe that Japanese companies are happy to buy the very best products regardless of whether it's American or Japanese made. Especially so under the current trade frictions. We'd like to do everything we can to correct this trade imbalance.

"Let me also add, when Japanese manufacturers sell to the U.S., they learn and practice the U.S. law and follow U.S. business practices. They make their presentation in English; they will also rewrite their documents into English according to the American format. However, when American copanies try to sell their products in Japan, they bring along their English-language catalogs and will not change their business practices to suit Japanese style. And so they have a handicap because the Japanese businessman finds it difficult to understand the product as well as their way of doing business. If U.S. companies can correct this, we would more than welcome the introduction of U.S. products into Japan."

As an example, Dr. Oka points to SEMICON JAPAN, an annual trade show organized by SEMI. "The people the U.S. makers send to this can only speak English, and they only bring along English catalogs. If this company works through a Japan agency, the agency may provide Japanese translation of catalogs. If they are not, that doesn't happen."[36]

Dr. Oka's point may be true for some American companies but not all. Many have been in Japan for many years and understand the market thoroughly. Intel's Bob Noyce has little patience with the standard Japanese response that their market is not closed, merely misunderstood. "The cries that we don't understand how to sell in Japan, that's all bullshit," he snaps bluntly. "Essentially everything that Intel produces in Japan is sold in Japan. We are not using Japan as an export base. It's not as if we don't understand the market; it's just that we are a non-Japanese company doing business in Japan. We could do a lot better if the market were truly open. Our sales in Japan are about 1.5 percent of the Japanese market, they are 10 percent of the U.S. market, 5 percent of the European market. So it's not trade barriers, because the stuff is produced in Japan; it's basic cultural bias against foreign-owned facilities."[37]

The bias is often expressed as a practice called "import displacement." "U.S. firms have repeatedly encountered the phenomenon of 'import displacement' in Japan," reported the SIA in a 1985 report, *The Impact of Japanese Market Barriers in Microelectronics*.

That is, they are able to achieve some penetration of the Japanese market with a particular product so long as sufficient quantities of a competing Japanese product are not available, but as soon as Japanese firms can supply the product (at times simply a copy of the U.S. device), U.S. firms' sales drop sharply, frequently to zero.

One of Intel's major products was its 8080 microprocessor. First developed in 1972, it was used in a wide range of minicomputers and became the brain of the first PCs. Two other companies were also producing the 8080 under license from Intel, and all three were selling it in Japan. Sales in Japan reached a peak in 1979, with more than $500,000 worth of the microprocessors sold. But in 1975, NEC began producing small numbers of the 8080s. Intel lawyers will say only that the company never signed a formal licensing agreement with NEC for the 8080. In fact, in 1975, the protection of intellectual property rights, such as the design of microprocessors, was nonexistent in Japan, and so NEC probably had a perfect legal right to copy and make its own version of the 8080. By 1978, it had expanded production dramatically, and by the middle of 1979, U.S. sales of the 8080 in Japan virtually halted. By 1980 they had become negative numbers; in other words, advance orders for U.S.-made 8080 microprocessors by Japanese customers were canceled. Still, NEC continued to produce large numbers of the 8080.[38]

The SIA in its 301 representation to the International Trade Commission pointed out:

The U.S. firms' 8080 experience in Japan should be contrasted with the same three firms' experience in the world market, where sales tapered off gradually as the life cycle of the product came to an end and a new generation of microprocessors reached significant volume in the world marketplace. It is not clear what actually caused the sudden disappearance of U.S. firms' 8080 market. If Japanese 8080 quality were superior, seemingly some significant displacement of U.S. firms by the Japanese should have occurred sooner, since the Japanese 8080 was on the market several years prior to the 1979 displacement. Japanese firms should not have been able to underprice U.S. firms because they had been in the market for a shorter time and had produced lower volumes, so that U.S. firms' costs were presumably lower than those of Japanese firms. The evidence strongly suggests that 'buy Japan' propensities on the part of Japanese users were at least partly responsible for the disappearance of the U.S. 8080 market. It appears when NEC had built up sufficient capacity to supply the entire market, Japanese users quickly switched to the Japanese product.

The SIA also pointed out other examples of U.S.-manufactured semiconductors that had broad acceptance in Japan only until such time as Japanese producers began large-scale production of the same product. But having said all that, it must be recognized that the situation is simply not simple. The Japanese market is controlled not only by market forces but by cultural and business traditions that are not easily understood by outsiders. "We think about the Japanese market in Western terms, about it being open or closed," declares Washington economist Bill Finan. "Its not a black-and-white issue. A crude way of thinking about it is that the differential, whether it be price or performance, between an American firm and a Japanese competitor selling to a Japanese company has to be very significant. To the Japanese, our companies are really second-rate operations in the Japanese sense. They divide their companies up into three tiers; the first tier are the giants. They get the most favorable financing favorable treatment from the government, and they only consider a very few American companies to be in that league, say, IBM, Boeing, GE.

"Then there are the second-tier companies, which are outside the core group, Sony, Sanyo, TI, Motorola, would fit into that group. Those kinds of companies, the second-tier guys, won't be allowed to fail. Banks will prop them up, engineering graduates from the best schools will go elsewhere, but they will still get good people, but not the best. They are a little bit less favored in the system over there.

"Then there's everybody else. So if you are a company like Intel and AMD, or the companies we think are well managed, to the Japanese these companies are nowhere. So what it means is you've got to have hot technology, like microprocessor know-how, like AMD does, or Intel does. Or analog devices, where you're producing very specialized parts in an area the Japanese, who are very volume conscious, won't touch.

"By their standards, we don't try hard enough. By our standards, the market is closed. We've got to come up with a different way of thinking about it. I went in to talk to Mitsubishi to ask them why they didn't buy American devices, and they said the quality is no good. I asked for quantitative data to prove it. He said here and whipped out a piece of paper. The names of everything were in English, and it was the ranking in terms of quality of all their suppliers. Mitsubishi was ranked number one, because by definition they were the best suppliers to themselves. The number-one external vendor and the number-three vendor were both American. And I said, 'Wait a minute. Americans are rated one and three. How can you say our quality is no good?' He just looked at me. It's such an inculcated perception that even when being confronted with their own evidence they don't accept it."[39]

What the Japanese companies did accept, even sought, after the so-called liberalization was MITI guidance. In December 1974, on the eve of the lifting of all tariffs on imports of ICs with over two hundred elements, the Japan Electronics Industry Association submitted a list of requests to MITI, including the following:

"In view of the fact that the demand for ICs is growing rapidly not only in the electronics industry but in the automobile, watch, camera, and other new fields, it is necessary to deepen the interrelationships with these industries and enlarge the use of Japan-made products. Appropriate guidance that will assist both sides of production and demand is requested, because it is important that the demand for national products be enlarged."[40]

Then, a year later, when the last formal barriers to foreign semiconductors were removed, among the suggestions issued by MITI was the one that stated:

"If a Japanese model is on an equal level with a foreign model, the Japanese model should be selected."

Such formal and informal rules worked not only to keep out semiconductors but computers, consumer electronics, semiconductor manufacturing equipment, and virtually any other high-tech item produced in Japan that the Japanese chose to protect. Prior to liberalization, the Japanese not only effectively restricted the importation and sale of semiconductors but also the establishment of wholly foreign owned factories in Japan. TI was the exception, and virtually every other American company was effectively excluded from owning production facilities in Japan. Makers of semiconductor manufacturing equipment had long since realized they needed to manufacture in Japan in order to compete with Japanese equipment makers. One of the first equipment makers to establish a factory in Japan was MRC. MRC is one of the world's largest producers of sputterers, machines that lay down thin metal films on silicon to provide connections between transistors on the chip. Their entry into Japan was influenced probably more by MRC's president, Dr. Sheldon Weinig, than it was by the Japanese. But once there, the game was played by Japanese rules.

MRC was founded by Dr. Weinig, an industry pioneer who has been awarded the SEMI Award by the Semiconductor Equipment and Materials Institute, has been inducted into the National Academy of Engineering, and has been awarded the Ordre National de la Légion d'Honneur by the French government for his pioneering research in semiconductor materials. A metallurgist with the salty vocabulary of a longshoreman, "Shelly," as he is known throughout the semiconductor world, is a blunt-spoken pragmatist whose scatologically sprinkled conversation does not hide an almost encyclopedic understanding of the industry and the Japanese. His career is an "Only in America" story and helps explain his unique insights into the high-technology war that now embroils the semiconductor industry. Toward the end of World War II, Shelly graduated from Stuyvesant High School, one of New York City's elite schools for gifted youngsters. From there he went into one of the army's wartime specialized-training programs, which included a year at NYU and Rutgers University into which was crammed a four-year engineering curriculum.

"Although I survived, I was hardly an engineer," he says in self-deprecating fashion. "So I spent my two years thereafter in the army, at the Pentagon, and I came back out and went up to University Heights at NYU, where I took two bachelor's degrees, one in mechanical and one in metallurgical engineering in a relatively shortened period. Then I went to Columbia, where I did a master's and a doctorate and then became a professor."

University life proved to be too tame for Shelly, and after several years he left and in 1958 started a company he called Materials Research Corporation. "The thesis of this company," he explains, "is that one day the key to most advances in technology will be materials. Things will be prevented from moving forward because we don't have the materials to do it with. I like to say that I was just twenty to thirty years too early, but a lot of it did happen. Over the course of the next ten years, I worked on anything that was fun. I worked on nuclear submarines, on ANP, aircraft nuclear power, which never came to be, I worked on hip joints and poop joints—you name it and I was involved. I prepared the alloys that were used in the first manned-space-flight experiments. I screwed around in materials of every nature. I was a half-assed consulting company, half-assed government research company, trying to survive. I didn't quite have it all put together. I'm not that bright or that orderly that I sat down and made a ten-year plan. I made a ten-week plan and said if I could survive that long it will be fantastic.

"So a lot of it happens, and out of it comes a feeling that we spend more money pedigreeing dogs than we do on the analysis of the metals or materials that we study for years. The reality is that when I was a graduate student and a young professor I'd get a contract from Uncle Sam for Wright Field or the navy, and they'd say study the so-and-so properties of titanium, for example. Lo and behold, you would get a piece of titanium from somebody—your cousin had a piece, you had a brother-in law working at Titanium Corporation of America, you had a friend at Dupont—and somehow you'd get some metal and begin some research."

About a year after he formed MRC, Shelly was invited to a conference of materials research firms working under government contract. "It was unbelievable," he recalls. "I said how did you get those results? What did you use? It turns out I have 1 percent impurity of this, they have 1 percent of that. It was ludicrous. This led me to the belief that what we really needed was centers of excellence, places where we would learn about materials, pedigree them if you will, and thereby advance the science of materials qualification. So I go to the government and annoy enough people, and an outfit in the Defense Department, then called ARPA [Advanced Research Projects Agency], finally comes out with a program. I wrote the original work statement, and the first funding was $5 million, parceled out to half a dozen labs, Penn State on ceramics, RCA on 3-5 compounds, indium antimonides, gallium arsenides, and my own lab working on elemental metals.

"The idea was that each of these centers of excellence was to produce these very highly predigreed materials and supply them to researchers all over the country at a fee. The fee would then be used to propagate additional work, and ultimately the government would withdraw because these would become self-sustaining centers. We would be allowed to charge modest fees to maintain our progress. It worked real fine in my case. I started to produce some super pure metals; I start to characterize them up the nose and back again, and we begin to sell. The other laboratories refused to do the same. RCA developed a new 3-5 compound, and you go to them to buy some, and they say no way, this is ours, it's proprietary, we're going to make a fortune out of this.

"In 1963, we are now working on this program; President Kennedy has meetings with the Japanese, and he decides we will give them science and technology. This is the beginning of the great exodus of science and technology from this country. It was not promulgated by the Japanese coming here with their little cameras; we went there and delivered it to them. It was a good idea then. JFK appointed ten committees called U.S.-Japan scientific exchange committees. They covered everything from medicine to atmospheric research to the science of materials. I am appointed to this latter committee. I go off to Japan in '63. Kennedy unfortunately has been killed, but President Johnson goes ahead. I lecture over there for several weeks along with a number of other very prominent scientists, and Bill Pfann, the inventor of zone refining, an absolute necessity for the growth of germanium and silicon crystals pure enough to turn into semiconductors.

"My lecture in Japan is what I had told our government. You must have centers of excellence. You must produce people who will become the greatest authorities on X, Y, and Z. And they must feed this information to the world. The Japanese listen, and they write it all down, and they start the centers of excellence. They set up a number of laboratories, mostly under the aegis of MITI. They start to study materials, and they do become expert. At that time, we taught them many things, such as zone refining, characterization of materials, electronic measurements.

"I come back to the States and continue to run my little business making little bits and pieces of metal and selling it to Dr. So-and-so and Dr. So-and-so. I claim at this point I am the most published scientist in the world because literally thousands of papers come out and say the materials used in these experiments came from the Materials Research Corp. All refractory group metals such as tungsten, molybdenum, titanium, etc., and middle-range metals as well. I have taken the periodic table and made a list and said I'll provide them all. Some were 99.999 percent pure, some 99.5, some in single-crystal form, where all the atoms are aligned along a single direction. We did a lot of this with government support and from our commercial sales. It's a tiny company, a $2-million-a-year business. It's got maybe twenty to thirty kids working there, maybe half a dozen with Ph.D.s, some with master's degrees."

Then, in 1965, the IC became big business at TI and Fairchild. But the IC requires thin films of metal to be laid on the silicon to provide the connections for the increasing number of transistors that are being built onto the chip. The metal of choice is aluminum, a very, very pure aluminum. "Along comes Texas Instruments and asks to see some of my superpure aluminum," continues Dr. Weinig. "Aluminum in 1965 is worth roughly sixty-five cents a pound. I charge them $2,000 a pound, but this is aluminum that has been lovingly and carefully made. TI comes back in six weeks and says we would like to buy 500 pounds. I say at my present rate of production that's about a forty-year order. We're producing about a pound a month at this point, slowly zone refining it. Obviously I increased my production, and I am now in the thin-film-materials business.

"It was literally the old saw of the problem finding the solution; I was the solution. I was known as a research-materials house; I'd invented a business. I'm a big mouth, and I've lectured all over the world on the meaning of purity. I joke about pencil purity—that's where you change the numbers—or omission purity, where you only analyze to get the things you want, or traditional purity, where it is traditional to look only for certain impurities and not others. And in this period, 1965 to 1968, our popularity grows, but we are still a nickel-and-dime company, now doing several million dollars a year.

"At that point we become of interest to some of the sharks going around devouring the little guys—including a company called Teledyne, which is run by a metallurgist named George Roberts, who used to be a professor of metallurgy. He comes to visit us with an entourage, and our whole company is 20,000 square feet in a little building. They wander around and then come into my office and say, 'Call in your secretary and we'll give you a letter of intent to buy your company for X million dollars.' I don't have any idea of what the company is worth. I run it like an extension of the university, running students like a professor. The only difference is that instead of driving a Chevy I'm driving a Buick and instead of making $12,000 a year maybe I'm making $30,000.

"I'm too stupid to sell out, and I suggest that I have to think about this. They think I'm bargaining, and they increase the price. So I hold my first planning meeting with my people, and we sit down over a case of beer and ask, 'What have we got? Why do they want to buy this company?' We realize that we have a peculiar marriage of materials know-how and equipment know-how. We've built a lot of equipment to refine metals, to analyze metals, and we decide not to sell. We are now selling X-ray attachments for the analysis of metals to North American Philips, to Picker, to GE. We start to sell zone refiners to refine metals, but most importantly, we look around for a new technique that will maybe marry the materials and the equipment. We come up with a thing called sputtering, which was invented in 1852. It's a method of laying down a thin film, something that has now become a necessity for anyone who wants to build integrated circuits.

"In 1968–69 we came out with a sputtering machine. We sell our first one to GE for their semiconductor operations and thereafter became the world leader in sputtering machines. We also in the same period get involved with Gillette, with their platinum-plus razor blades, which are produced by sputtering. And we bcome a player in this game. We go public, and today we are a $100 million company with factories in Europe and in Japan and of course here in the U.S.

"Having gone to Japan in 1963, I become fascinated with it, as everyone must. Over the years I go back several times a year. We work through a small trading company, but we are just fooling around. But in the late '70s you begin to see the movement to Japan. I begin to realize the electronics industry is moving out of the U.S., and I get a little panicky, and I go to Japan in 1980 with the express purpose of opening a factory. But I make a decision not to fool around with the normal progression, which is to take the agency away from the trading company and open your own sales office. When that works, you open a service department, then a demo lab, and finally a little assembly plant.

"By this time I have operated a factory in Europe, so I'm ready. I should have done it ten years before, but I wasn't smart enough, and it wasn't until 1980 that the law changed in Japan to allow an independent to do it. I found nobody cared, neither the U.S. government nor the Japanese government, whether I came or not. I get a firm to do a public relations thing at the American Club in Tokyo. I knew these public relations people from the Semiconductor Equipment Materials Institute where I was a board member. I make a speech, and I say, 'What I have you need, and what you have I need, and therefore I don't want any roadblocks. I want signs of welcome.

"Three governors sent me letters, one from a prefecture called Oita, which is on the southernmost island of Kyushu, which the Japanese call Silicon Island. I actually visited five prefectures, and I settled on Oita. I go to Oita, and I negotiate a piece of land which is originally offered to me at several million dollars. I say I don't have that kind of money, but I have a plant in France, and the French government put it up for me. I don't think the Japanese would do any less. The result is the prefecture buys the land and gives me a lease to buy it within two years. I buy it for approximately $200,000, and it's now worth several million.

"The next problem is, how do I build a plant? So I ask them, 'How does a Japanese company build a plant?' The answer is 'They go to Japan Development Bank' [JDB]. So I say, 'Fine that's where I would like to go,' but they say, 'You can't; only Japanese companies can go there.' I say, 'But I am a Japanese company. I just opened a corporation last week called Nihon MRC.' I go to JDB, and I ultimately become the first American company to get a Japan Development Bank loan. The reality is the application schleps along for eight months and nothing happens. It's not that they're difficult, we don't understand them, and they don't understand us. It's two totally disparate accounting systems. They want a projection on the next hundred years; we're lucky if we have one for six months. Our balance sheets show items like deferred taxes, because we are allowed to have different schedules for depreciation for business purposes than for tax purposes. This is not understood in Japan, and they perceive us as crooks. Finally, I send a telegram to JDB, and it says, 'Dr. Weinig will arrive on Monday for finalization meeting.'

"They ask what is a finalization meeting? And I keep telling my secretary repeat the same telex. Don't change a word.

"I arrive in Japan on Sunday and am met by JDB and my own people. I say very simply, 'I can't devote my life to applying for money. If you don't want to give it to me, I'll do something else.' By Wednesday they ask for my application; they've got twenty-seven cases of data already from me, but this is the official document which I present to them. The guy thanks me and leaves the room. He comes back in five minutes and says, 'Come with me.' We go down the hall into a bigger office, where I meet the governor of the bank. He says, 'We have considered your application, and we have decided to grant it.'

"So, with the loan, I build the plant. It's not that the loan is so fabulous, but it gives you the stamp of approval. It's like having a kosher stamp on your ass. They just said you're a good guy, the government approves of you, and things just open up after that. Now, it sounds like I have a great success story, and I do. Within two years I'm making money. And so by '83, '84, I'm in the black in Japan. I've opened offices in Osaka, Tokyo, and of course the plant in Oita. I'm building machines there and selling mostly to the Japanese, but I am starting to explore selling to Korea, Taiwan, and mainland China. But not with a great deal of success. Those folks still want to buy directly from the U.S. That begins to change as the Japanese become more dominant. So in the '85, '86 period they change and begin to buy from Japan.

"Then two things happen. We start to sell in those areas, but we are selling in dollars, and the drop of the dollar against the yen doesn't differentiate between Japanese-owned plants or American-owned plants. All of a sudden my costs go through the roof. I have started to sell into Taiwan and Korea in dollar-denominated sales, which is the way they want them, but my costs in yen are going sky-high. Coupled with this, the Japanese are feeling the pinch and decide to turn off the faucet; they stop buying. They do it in a very interesting way. They say you must buy from your own zaibatsu. So Nippon Electric, NEC, who is my biggest customer, owns within its zaibatsu Anelva—my number-one competitor. That's difficult to understand in the U.S. but very much part of the game there.

"The next thing that happens is our business turns to zilch. The only buying is done within the zaibatsu, and if they can't get it within their own, they go to their cousin's zaibatsu and so on. Finally, if all else fails, maybe they'll buy from us. Prior to this we had been selling a million dollars' worth of machines a month in Japan. At this time, I'm selling $2 million a month in Europe. I'm coming up nicely, and boom, they turn off the faucet. At this point, I am in very difficult straits. In the U.S., when business is lousy, you can always hustle it a bit. By that I mean there is always some guy who wants a machine badly enough he'll get you twelve purchase orders for spare parts that miraculously add up to the price of one machine. There is always a guy who will figure out a lease arrangement; there's always a start-up company that has some venture capital for a machine. None of this exists in Japan.

"Nobody does twelve purchase orders to equal a machine, nobody does a lease buy, nobody does a start-up with fresh money. All the money is controlled by the major zaibatsus out of major manufacturers, and when they say that is it, the budget is dead, believe me, it is dead, and rigor mortis has set in. The only business I've got there is dollar denominated out of Japan, which I ship to keep my factory working.

"Then I pull off a Japanese trick. I go to the governor of Oita, and he arranges for me to transfer fifteen or twenty people to Toshiba, which is nearby. That's unheard of in the U.S. People come to work in the morning, and we say you'll be working in Toshiba tomorrow; we transfer them like chattels. I sent work from the U.S. to keep the 120 people I have left there working. I read the crystal ball, and I say to myself, Shelly, you're in very bad shape. The thing that you're missing despite your JDB loan, your friendship with Hiromatsu [the governor of Oita prefecture], and all these wonderful things is you don't have a marriage in the family; you are still a roundeye.

"So I began to work on a man in the U.S. by the name of Kondo, who is president of Mitsubishi International, their American company right here in New York. I get lucky. Kondo becomes president of Mitsubishi Corporation, which is the mother company, the big trading company. He buys 25 percent of my Japanese company. And that is an extraordinary phenomenon. First of all, when Mitsubishi buys you, they overcome one of the great problems; they eliminate the problem of getting good people. They now second the people to you; they lend me several important people whom I pay at my rate, while they continue to be paid by Mitsubishi. So if I want you but I can only afford to pay you $50,000 and Mitsubishi pays you $100,000, it's no problem. I send the $50,000 to Mitsubishi, and you continue to be paid by them at your salary, including all the benefits and loving care they usually provide their workers, and you work for me now for two or three years. It raises the whole level of the operation, because I was scrambling for lesser-level people. I was not getting the number-one-caliber engineer or physicist; nobody from Tokyo University was joining my company.

"The second thing it does is almost indescribable. All of a sudden, I'd get an order from NTT; I couldn't get an order from them to save my ass before. Then I get a second order and then a tremendous order from a large Japanese computer manufacturer for one of our latest machines for a million-plus dollars. Suddenly things are happening. After five straight quarters of losing our shirts, my Japanese company turns around and makes $30,000–40,000. We're now into the third quarter, and on opening day of the quarter we have enough orders already to be even. And yet you can't say that Mitsubishi sold this for me or that. They didn't do anything. But the association with Mitsubishi has certainly helped. And I'm beginning to teach the Japanese a few tricks. I got my first order in three parts that added up to a machine. I sold a used machine, which we've never before done; we traded one in. So we are starting to play in the American style and get some results."

Shelly also learned the Japanese style and practiced it as well as any Go player. "Before I got in bed with Mitsubishi"—he chortles—"they were selling one of our products, and my people there got hysterical. This product was a target, an expendable material used by a sputterer to lay down the metallic interconnects on integrated circuits. Mitsubishi Light Metals was selling a target that we had patented throughout the world but not in Japan. Nobody can patent in Japan. And they were beating us to death pricewise. I said go and talk to Mitsubishi, this is ludicrous, a company that big is not going to screw around in this nickel-and-dime business when they know we have it patented in the U.S. Even though it's not patented in Japan, it doesn't look good for them. Well, they wouldn't talk to my people.

"I go to Japan and invite the top dogs of Mitsubishi Light Metals to lunch. They arrive, they walk in, bow, and each one hands me his *meishi,* his business card. I notice one guy whom I knew but could not recall from where. So when he handed me his card, I flipped it over to the English side and read that he was from the Mitsubishi Bank. This son of a bitch has been trying to get our account over there for several months.

Now stop and think. I just arrived in Japan on Sunday; I sent out invitations on Monday for a Tuesday lunch and was fortunate to set it up. And this enormous company, the largest in the world, was able to communicate with the bank that they were having lunch with me. In my company, which is only a thousand people in Rockland County, if you were to call up, it would take three weeks to figure out who you are. These guys were able to do it in twenty-four hours.

"So that's number one. The bank controls, the bank is the core, the center, and the bank makes a difference in Japanese business to a degree we don't yet understand in this country.

"The lunch proceeds. I say, 'Gentlemen, it seems silly to me that a company the size of Mitsubishi would fool around with this little device that we have patented around the world, and although not in Japan, because it's not possible, still, that is no way to have a long-term relationship.' The guy from the bank says, 'We would like to have a long-term relationship, especially at the bank.' So we all shook hands and left. I opened an account at the Mitsubishi Bank, and they no longer sell my target, and everybody's happy. And that's Japanese-style business."[41]

7

The Little Dragons

They are called the Little Dragons, or NICs, the newly industrialized countries of Southeast Asia—Taiwan, Singapore, Hong Kong, and
South Korea. It is here that a new challenge has arisen for both the
American and Japanese semiconductor industries. With a highly
educated, low-paid, hardworking labor force, the Little Dragons of the
Pacific Rim attracted the semiconductor makers of the United States,
Japan, and Europe as a big, cheap, and efficient assembly plant for
packaging chips. As more and more companies built assembly facilities,
they transformed the Little Dragons into fire-breathing monsters—the
world's leaders in chip assembly. And by 1990, Southeast Asia will have
reached a capacity to assemble 30 billion units, enough to package
more than 80 percent of the entire world's semiconductor production.[1]
    But the Little Dragons were not content to stop there, for they
learned not only how to assemble chips into packages; they also learned
how to make the semiconductors that go into those packages. In the
process, they have moved into other high-tech industries and developed some of the highest-soaring economies in the world. From 1963
to 1985, while Japan's domestic economy was growing at a rate of 5.9
percent a year, Hong Kong's was growing by 8.5 percent; South Korea's,
by 8.7 percent; Taiwan's by 9.2 percent; and Singapore's was virtually
exploding at a rate of 9.7 percent a year.[2]
    So successful have the Little Dragons become that in January 1988,

191

President Reagan notified Taiwan, South Korea, Hong Kong, and Singapore that they had been dropped from the Generalized System of Preferences, a program that provides a special duty-free status to developing countries. In 1987, $16 billion worth of goods came into the United States under this program—$10 billion of it from the Little Dragons. It will not, however, be much of a brake on their surging economies. Nor will it affect the sales of semiconductors, since chips were not given preferential treatment to begin with. The tariffs they will begin paying on January 1, 1989, on goods now duty-free, will average about 5 percent, or $450 million a year. But their total sales to the American market last year were a record $60 billion. It is, however, a sign of how far they have come in a remarkable short time.[3]

The key to this surging growth has been the expansion of the manufacturing sectors of the four economies and the percentage of those manufactures that are exported, especially to the United States. The U.S. sops up almost 50 percent of Taiwan's exports, 40 percent of South Korea's, one-quarter of Hong Kong's, and over one-fifth of Singapore's. Taiwan's trade surplus with the United States in 1987, for example, was more than $18 billion. The U.S. trade deficit with the Little Dragons ballooned from $6.1 billion in 1981 to $37 billion in 1987, a remarkable 22 percent of the total U.S. trade deficit.[4]

Europe is also beginning to feel the heat of the Little Dragon's export blitz. The European Common Market deficit with the four Asian NICs had risen by 20 percent in 1987. At the end of the year, Korea lost its duty-free privileges in Europe not so much for its trade success as its refusal to protect foreign copyrights and patents.[5] The overall effect in Europe, the U.S. Congress, and the media has been to lump the Little Dragons together as a sort of matched set of mini-Japans. Yet they differ markedly in many ways and in the Chip War may in fact pose a greater competitive threat to Japan than to the United States and Europe.

"Southeast Asia is the place to watch, because they are doing to Japan what Japan did to us," enthuses Sam Korin, the director of IBM's Manufacturing Technology Institute in Thornwood, New York, a small, rustic community on the Hudson River about forty miles from New York City. Here IBM has established a center to create, develop, and assess new semiconductor manufacturing technologies. And from here Sam Korin is in a position not only to watch the world for new developments in technology but to seek trends in the forces that shape the semiconductor industry. Among the most important is the impact of the Little Dragons. "They're doing it on the basis of new technology platforms," he observes, "in which they didn't have to invest any effort or money to develop because it was already done for them by the United States, with Japan lately adding to it. Now they have a still higher knowledge base on which to build.

"The strategy here is so brilliant and so obvious, and I'm appalled that nobody understood it." In fact, lots of people do understand it; the KGB, the Japanese, and just about every competitor in the world except the United States understands the concept. "The strategy," explains Korin, "is that you borrow from the biggest free-research resource in the world, which is the United States. Through technical reports, through visits, through business interactions, acquisitions, and licensing, you get free research and technology."[6]

Among the Chinese especially ideas flow as freely and swiftly as family gossip. "There's a very close relationship between the overseas Chinese," a young Taiwanese engineer told me. "My uncle, for example, is at Lawrence Livermore (a U.S. national laboratory at the University of California, Berkeley), "and we talk whenever possible; we have a great many technical exchanges."[7]

"You analyze this technology," continues Korin, "and determine where the weaknesses lie in the products that are coming out of that research and where the leading-edge capabilities are technologically, logistically, and from a business standpoint. Then you take the products and value engineer them. In other words you correct for the deficiencies, and you find new market niches, what was missed by the initiating generator or developer of the technology."

In effect, the Japanese have done this for years, until they finally reached a level where they quite probably have less to learn from the United States than vice versa. "In the memory area," Korin points out, "the Japanese have dominated the chip business. Starting with our 64K RAM, they looked for the deficiencies, which were mainly in the yield. By virtue of the fact that they do things cleaner and more precisely, they were able to increase their yields and take over that market. We were preoccupied with developing the tools; that's another key point. They also took our tools; they bought all of our semiconductor photolithographic equipment, produced by American vendors, and enhanced them when they got to Japan.

"Now, having developed that knowledge base for free, they are developing their own equipment, which is superior to ours, and now we are buying their equipment. The interesting thing is the Southeast Asians are doing exactly the same thing. They are looking at Japan's capability and looking for flaws and deficiencies in that knowledge base and finding the niches and opportunities to create a market and build on it. Eventually, when they get up the learning curve, they will then start in doing their own development. They are initially doing it in the consumer-goods area, which is precisely what the Japanese did to us. They get into the easy things, such as consumer electronics, in order to build a base. You're beginning to see new names—Samsung, Multitech—which are Korean and Taiwanese. And when they get the foothold, they build on that base to get into the industrial product world."

Having achieved their foothold, two of the Little Dragons—Taiwan and Korea—have suddenly become a factor in the Chip War. Indeed, their challenge may be even greater for the Japanese than the Americans. That at least is what Sam Korin believes. "In my judgment, I think the war is going to be between Japan and some of its Southeast Asian neighbors. Taiwan, Korea, and pretty soon, Singapore and perhaps Thailand are already pushing hard. Look at what their universities are doing; that's the first clue. Look at the courses they're beginning to introduce. The Asian Institute of Technology (AIT) in Bangkok, Thailand, is a consortium of twenty or thirty countries in Southeast Asia that send their students to AIT. In Singapore there is Nan Yang Technological Institute. Taiwan was one of the first to set up a technology institute. At the moment, their emphasis is on programming and software, but they are migrating toward semiconductor manufacturing and to manufacturing systems and CAD CAM [computer-aided design and manufacturing] and a variety of technologies, so that we like to think we have the edge."

Taiwan

What we definitely do not have an edge on is hustle. Taiwan is a booming island nation. Taipei, the capital, is a city in a state of urgent transition, with modern high-rise buildings poking up alongside older Chinese-style four- and five-story apartment and office blocks. Traffic snarls every street. Hundreds of motor scooters and small motorcycles dart in and out, causing autos to stop suddenly, drivers to honk their horns and curse in rapid singsong Chinese. Although Asian, it has none of the neat, controlled air of Japan. Rather, it is an American industrial frontier with bold Chinese characters and a Confucian ethic.

I survived the ride from my hotel in the center of Taipei to the National Taiwan Institute of Technology (NTIT) without incident. The institute is on a sprawling twenty-five-acre campus dotted with poured-concrete-walled buildings in an architectural sytle that owes more to prison modern than university Gothic. Established in 1974 to train engineers and managers for Taiwan's rapidly growing industries, the Institute has 2,630 students and 269 faculty members. They come from an excellent school system after nine years of compulsory education and usually two years' training at any one of the thirty technical junior colleges dotted about the island. NTIT is the most advanced technical college in the country, with a two-year program leading to a B.S. in engineering. For those junior-college graduates who are already at work in industry, there is a six-year program leading to a B.S. degree.[8]

The importance of the institute was underscored for me by Chang Husan Lin, the young, energetic chairman of the Department of Electrical Engineering. "If they graduate from this institute, they have greater job opportunities. When they graduate from the junior techni-cal college, they have no bachelor's degree, but in this society they care about degrees, and that motivates them. The Confucian ethic also is very strong. Parents always want their children to gain degrees for the prestige it gives to them, the parents.

"Most of the people who graduate from here stay here. About 5 percent go to the U.S. for further study. In the vocational educational system they do not emphasize the teaching of English, so basically our graduates do not have enough English like the regular university graduates. The second reason is they get more practical training here, and so they are more welcome in industry. They do not need on-the-job training."[9]

What they do need is qualified teachers. Taiwan's sudden emer-gence as a mini-industrial power has sucked up virtually all its technicians and engineers and put them to work. "There is a shortage of teachers at the institute," admits Chang. "Most come back from the U.S., and the demand in industry in Taiwan now is so great we just cannot turn out enough teachers fast enough. An engineer here makes about 50,000 to 70,000 New Taiwan dollars; teaching he makes a maximum of about NT $50,000 (approximately U.S. $16,650)."

Industry in Taiwan actually got its start about twenty years ago, but there was not much demand for engineers for quite some time thereafter. Thus, many Taiwanese went to the United States for their education and their jobs. The situation is quite different now, with more money and more opportunity beckoning the technically skilled and trained Taiwanese. Just two years ago, for example, there were only two VLSI-related companies. Now there are about twenty, most of them small design houses. The reason is that the entrepreneurial nature of the Chinese is more akin to that of the Americans than the Japanese. And while a state-of-the-art chip-making plant can cost $200 million to build, a design house can be started with little more than an idea and an office over a bicycle shop in downtown Taipei.

But the government sees the need for production facilities to turn the creative dreams of young Taiwanese engineers and the technology acquired from the United States, Japan, and Europe into jobs, products, and trade surpluses. Hence, its emphasis on manufacturing. "In this institute and in the Electrical Engineering Department we emphasize manufacture of semiconductors rather than circuit design," explains Professor Chang. "One of the missions of the institute is to provide the practical training. Other universities in Taiwan are closer to the American model, with a heavy emphasis on theory. So at universities like Taiwan National University, the best one here, our Harvard, 90 percent of their electrical engineering graduates go to the U.S. And they tend to stay in the U.S., because their training is not industry oriented and they are more comfortable with the academic life in the U.S.

"In the past, the U.S. was like a paradise compared to Taiwan, but that is now changing. Most ordinary university graduates go to the U.S., but more and more are looking to come back now that there is more need for highly skilled engineers, more opportunities than ever before.

Chang is himself an example of the changing attitude among Taiwan's engineers. "I graduated from Taiwan National University, got a B.S. in electrical engineering there, and then went to the University of Texas at Austin, where I got my Ph.D. When I finally decided to return here, I found that what I had learned in the U.S. was too theoretical to use in Taiwan. So I literally threw my Ph.D. thesis in the Pacific Ocean and started over again. That was in 1982, and I got interested in the problems of automation and decided the Technological Institute was the place for me to do research in servo technology and teaching. One reason impelling me to return—my mother died. Most Taiwanese would still choose to stay in the U.S., and they need very strong motivation, either family or patriotic, to come back."

But, acknowledges Chang, "things are changing. We have resources and money to develop our own technology, and the future should be very interesting here for young engineers. In our department we do research in the semiconductor area and in electronics materials. Taiwan's National Science Council [NSC] provides about $30,000 a year for research into semiconductor materials to our department."

The total Taiwanese investment in semiconductors is now close to $300 million. It has not, however, been a shotgun approach, simply throwing money at an industry and hoping it will flower. Rather, as far back as 1972, the Ministry of Economic Affairs began to question whether or not ICs could be made economically in Taiwan. Semiconductor assembly had begun as early as 1963, when Philco-Ford set up an assembly plant on the island. Subsequently, that plant was sold to General Instrument Corporation. They were followed by RCA and TI. Soon there were as many as a dozen U.S.-owned assembly plants in Taiwan, staffed largely by Taiwanese engineers under American supervision. A few of those engineers, much like their Silicon Valley counterparts, split off to form their own companies. By 1972, government officials come to the realization that Taiwan had a burgeoning semiconductor industry that was growing wildly, with little if any direction. They also realized that all high-tech industries would flower among the fertile minds and hardworking climate in Taiwan. To oversee that industrial development, they created the Industrial Technology Research Institute (ITRI). Under that umbrella, a second organization, the Electronics Research & Service Organization (ERSO) was set in place in 1974 to develop Taiwan's semiconductor industry. Its head was Professor Ding-Hua Hu, a young professor of electrical engineering at Chiao Tung University. Hu, now the executive vice-president of ITRO, is the son of a Nationalist Chinese air force pilot who brought the family to Taiwan when he was six. He has been with the Taiwan semiconductor industry from the start. He might even be considered one of its founding fathers and is today one of the major planning voices in the Taiwan semiconductor industry.

His years as a professor of engineering left him with the mannerisms of the true pedant, and he illustrated his talk with me on a blackboard. A graduate of National Taiwan University, he has a master's degree from Chiao Tung. In 1967 he went to the States, to the University of Missouri, where he gained his Ph.D. in electrical engineering. After teaching at the University of California at Santa Barbara, he returned to Taiwan in 1970 to teach at Chiao Tung University and head up its semiconductor research center.

"Then, when the government wanted to set up its integrated-circuit project, I joined ITRI," he told me as we sipped tea in his office in the ITRI-built Science Industrial Park, about an hour's ride from Taipei. Approached on wide three-lane highways that cut through rolling hills covered with trees, some farmland, and apartment blocks, past Chiang Kai-shek Airport, the 2,000-acre park might have been lifted bodily from one of the many Silicon Valley complexes and dropped down in Taiwan. Inside the park are several semiconductor companies, the science and engineering faculties of Chiao Tung and Tsing Hua universities, and of course ITRI. Although not quite modeled on the same lines as MITI, ITRI also provides guidance and major industrial planning and funding for Taiwanese companies. ITRI is composed of five main research institutes, Materials Research, Mining and Energy, Energy, Chemistry, Mechanical and Electronics. The Electronic Research and Service Organization under ITRI is instrumental in developing process technology and provides a design service and runs a small fab plant for Taiwanese designers.

"We already had a number of assembly plants in place, and we began to feel we needed to build a wafter fab," Professor Hu patiently explained. "We had in 1964 started a semiconductor laboratory to teach basic techniques at Chiao Tung University. So at an early stage we had both the assembly plants and the wafer-fabrication technical people, mostly the graduates from Chiao Tung Univerity. Another thing we needed was design. At that time, 1975–76, the purpose was to train design engineers and to prove the economics of wafer fabrication.

"We set up a government company, the Electronics Industry Industrial Research Center. That was the basis of ERSO today, the Electronics Research Service Organization."

One of ERSO's first tasks was to build a sample fab line to see if in fact the Taiwanese could turn out high-quality semiconductors and to convert the designs its engineers were creating into silicon reality. "Of course you need the equipment," acknowledged Hu. "You must import it, and chemicals, also import, and other things. Also very important for the wafer fab is the clean-room design and the construction. Equipment was from the U.S. Silicon diffusion plants and clean-room benches, most of the chemicals and wafers, we got from Monsanto and Wacher. We hired a consultant, the R. J. Siegel Company, to do the basic systems design and construction for the clean room. Altogether we spent approximately $9 million. At that time, in 1975, that was a lot of money."

With the research fab up and running in 1978, ITRI decided to go commercial. "After about a year or so," recalls Hu proudly, "we proposed to form the UMC, United Microelectronics Corporation. It was originally government owned and operated, but has since then become a public company."

It has also become a roaring success. From its modest beginnings in 1978 it had by 1986 become the thirty-third largest producer of MOS chips in the world, ahead of such American companies as Rockwell, General Instrument, Honeywell, Fairchild, and others. For 1987, its sales topped $100 million, and it had become the number-one supplier of ICs to the rapidly expanding Taiwan electronics market. One of the major factors in UMC's sales success was the deregulation of the telephone industry in the United States. In 1983, UMC shipped millions of chip dialers to Taiwan and Hong Kong telephone manufacturers, who then flooded the U.S. market with cheap telephones.[10]

Today UMC, which is also in the sprawling Science Industrial Park, is in the process of a major expansion, building a new fab facility and design center that will ultimately cost $200 million. I crossed the park from Professor Hu's office at ITRI headquarters and met with the president of UMC, Bob Tsao. At the age of forty-five, Tsao looks more like a UCLA graduate student than the president of one of his nation's largest and most successful companies. Slim, very youthful looking, he was dressed in a light blue open-collar shirt and Pierre Cardin slacks and loafers. He speaks excellent colloquial English and gives the impression of effortless success as he describes the remarkable achievements of UMC. But Bob Tsao was driven to succeed by a very tough boyhood after his family fled Communist China in 1948, when he was two years old.

"I was born in Manchuria, China. My family were leather traders. My father would travel outside the Great Wall, buy leathers, and sell them in Peking. My father was here taking a tour when the communists took over. My mother took us out in a hurry, and we could take nothing with us. In Taiwan my father taught history and Chinese literature in the high school. He had been educated in China and Japan. But my childhood was very tough; we were very poor."

Unlike most Taiwanese, Tsao did not get much of his education in the States, a point of jocular pride for him. "I'm from MIT," he says, laughing. "Made in Taiwan. I was educated here, got my EE from National Taiwan University in Taipei. After my graduation I joined the electronic IC project in 1971. I was in the first team to be dispatched to RCA in 1976. We sent thirty engineers for ten months to learn the technology."

"We are the only active merchant chip maker in Taiwan. And we do not anticipate becoming a systems producer. That's against my philosophy. We want to concentrate on the processing of ICs. In the next five years this region will still be a major marketplace for us. Of course, we would like to expand our sales to the U.S., Japan, and Europe, but this area will remain our major marketplace."

UMC is a brilliant example of Taiwan's small-enterprise approach to semiconductors. "There is no vertical integration in the IC industry in Taiwan," explains Tsao. "However, there are a number of small, efficient assemblers, design houses, etc., which provide full IC manufacturing capabilities, excellent flexibility, low cost, and full services. With capital investment on the order of $50 million, UMC will generate $100 million in IC sales in 1987. We will produce 20 percent of the total Korean volume with 3.5 percent of their capital investment."

UMC looks to enhance its capabilities and at the same time call on the basic strengths of the other small, specialized IC firms in Taiwan. "In our philosophy we are concentrating on wafer processing and we see no limit in sight, so we will go below 1 micron," says Tsao. "That's why we have to concentrate to the degree that we don't even do our own assembly. Our finished wafers are sent out to Taiwanese subcontractors to be assembled. Of course we then take them all back in house to do the final testing. For us, that's the most efficient way. Our subcontractors have the expertise, and we don't have to carry that overhead. It's worked so well that in 1983 we went to them and told them we would increase the price we paid for assembly. In good times that's the way we share profits with them, and in bad times we are going to reduce the price.

"With the number of startups recently, we now have about forty design companies, and we are going to have at least four new fabs. I think Taiwan will play a major role in supplying ICs to Southeast Asia. In the NMOS IC market in Taiwan, we have 16.7 percent, which is the number-one market share. The rest comes from TI, Toshiba, National Semiconductor, Motorola, and others. We are gradually taking market share away from our competitors in Taiwan. After this year, 1987, Samsung will become an important player. They just started producing DRAMs. I will not go into the memory business unless I have to. In our philosophy, the number-one priority is to get enough loading capacity. In our new fab, we hope to keep capacity at 100 percent. If we can't do that, maybe we will begin producing DRAMs. If we produce memory chips, probably it will not be very competitive if you make memory alone, but if we can fully load the capacity, then the overhead will be shared. Therefore, spread over the entire range of products, the cost of memory can be brought down to competitive levels."

The construction of UMC's new fab points up the battle between American and Japanese semiconductor-equipment makers. For as sensitive as the Taiwanese are to their growing trade surplus with the United States, their number-one priority is to obtain the best equipment at the lowest cost. "The new facility will be designed and constructed by Japanese firms, Mitsui plus Kawasako. They are very experienced at building very clean, automated fab lines," explains Tsao almost apologetically. "Our new fab will be very automated. The first phase will cost us $80 million. In total, we will spend roughly $200 million. The project management will be Japanese, but most of the equipment will be U.S. In photo aligners, the U.S. is becoming inferior to the Japanese. But in a lot of other areas, such as chemicals, deposition, ion implanters, most others will be imported from the U.S., as they still lead in those areas."[11]

Design as well as manufacture is very high on the Taiwanese list of priorities. "We are aiming at regional sales and our own market," explains Professor Hu. "We are promoting the establishment of design houses to support the end products. We'd like to see the fabrication houses' contribution on the island of Taiwan to be 20 or 30 percent of our total consumption. Now our consumption is about 200 million ICs a year. The major items being produced here are personal computers, about $500 million worth, and monitors, terminals, and printers, about $200 million, and other items, like laser printers and Winchester discs, will grow in production. So we are looking to a billion- to a billion-and-a-half-dollar electronics industry in Taiwan within about six years."

Central to achieving that goal was the decision to create a major fab facility capable of converting the ideas of Taiwan's rapidly growing host of design engineers into commercial reality. The small fab line at ERSO is designed primarily to test designs rather than produce volume parts. UMC is already running flat out and has a major expansion under way that will not be completed for another five years. The answer was to build a new state-of-the-art facility to meet the need. But it also had to be in keeping with the basic government philosophy of encouraging private investment. The answer was twofold: first, to find someone to head it up, and second, to find a partner to share the cost and provide the expertise.

The first part was fairly easy. ITRI reached out to Morris Chang, who had left the island thirty years before and had become one of the outstanding executives in the American semiconductor industry. Chang had been a senior executive at TI for twenty-three years before becoming president and CEO of General Instrument Corporation in 1984. Then he heard "the call" and returned to Taiwan and became president of ITRI and chairman of the board of TSMC, Taiwan Semiconductor Corporation, a company that at the time existed only on paper.

Chang then reached out to an old friend in the States to run TSMC, a man named James Dykes. Born and raised in a small town in central Florida, Dykes is a southerner without a southern accent. Tall and tending to overweight, with a thick shock of gray hair, he towers like a redwood in a forest of saplings alongside the diminutive Chinese. But he fits in well with the entrepreneurial spirit that characterizes the Taiwan semiconductor industry. Aticulate and with strongly defined opinions honed by his twenty-five years in the semiconductor industry, he sat in his big, bright office and discussed the Taiwanese strategy that attracted him here.

It could not have been an easy decision to pull up stakes and come with his family to a land so foreign; even the Chinese television shows carry two sets of subtitles in other Chinese and Taiwanese dialects. "I was with GE, corporate vice-president in charge of all the microelectronics of GE and RCA after it had been acquired," he began. "I had graduated from the University of Floria in '62, and I went to work for a little company down in Melville, Florida, called Radiation Incorporated. This was back during the Minuteman missile and the early days of the statellite race, so I hit things at a good time, and I worked on the Telstar satellite; that was my first assignment.

"That thing used some of the early transistors, and I kind of got my feet wet in solid-state technology at that time. Then, in 1967, the Harris Corporation bought Radiation, and the little acquisition ate the parent and changed it, and it metamorphosed into an electronics company. And we went into the semiconductor business initially for our own purposes, to have a captive source for custom devices to use in airborne weapons systems. In 1970 we decided to take this as a commercial venture, and we coined the name Harris Semiconductor. I kind of grew up in Harris Semiconductor. I was in engineering for eight and a half years. Then I went into marketing and was the product marketing manager for three years, doing the product strategies for Semiconductor, and became VP of marketing and then VP of operations. In the last three years I was there, I was general manager of what we call a products division, which was about $120 million.

"Harris at that time was in the top ten among the IC producers in the U.S., largely driven by CMOS technology. Then GE decided to go back into business at that time, and they started talking to me in 1980, and I said, 'GE, no way.' But they kind of got me going, and so I wound up moving to Research Triangle Park in North Carolina and built the facility there for GE in '81. That was the cornerstone that got them back into the semiconductor business. And then the RCA thing came along. That was not a planned acquisition of a semiconductor company; that was a corporate acquisition of another corporation, so we had to put those two businesses together. But GE did not want to be in the semiconductor business, and it's tough managing something that no one wants, so once we got that integrated, I said I wanted to get out."

It was at this point, in 1983, that Morris Chang, who had moved from TI to General Instrument Corporation, another semiconductor maker, came courting Dykes for the first time. "Morris called me and said, 'Look, I'd like to talk to you about taking over the semiconductor business at General Instrument.' I said, 'Morris, I'm weary of old companies and old semiconductors. All I've done at GE is close and rationalize and move, and I don't want to do that anymore. And GI to me is worse off than GE was when I got there. You have too many old facilities,' and so on. He said, 'Fine, but let's keep talking.' So we left it at that, and three weeks later he left GI. Now I didn't have anything to do with him leaving, and obviously it wasn't anything he was thinking about when he talked to me, but suddenly he's gone.

"So time went by, and then he called me again, and he said, 'I really want to talk to you about something else.' And I said, 'What are you doing?' And he said, 'I'm in Taiwan.' They had been after him ever since he left TI to become president of ITRI, and in August of 1985 he finally did it."[12]

Chang also sits on Taiwan's NSC, a panel of five men who evaluate all requests for government support, funding, or shared research in the area of technology. There was a continuing stream of requests to NSC from people to build small wafer fabs costing $10 million and $20 million. Chang thought to pool all those needs and build one facility and have it shared so that the little guys would not be pulled down by the heavy capital demands of constructing a fab facility. Moreover, he believed they could build a fab that would be state of the art to give Taiwan a chance to be competitive. That meant bringing in a partner and a top-flight executive to run it.

The partner was to be Philips, the giant Dutch electronics firm. The top-flight executive had still not been landed.

"So that's when he called me," continued Dykes, "and he said, 'I've got a concept that's not completely solidified yet, but I'm down to talking to the last company, which is Philips, and I'd like to talk to you about coming over to Taiwan and managing the start-up.' And I said, 'Morris, Taiwan isn't in the cards for me. You know, I'm ready to make a move, but my gosh, that's a big jump. And I really can't see why I want to do that. I really can't see it.' "

Chang didn't take no for an answer. "He flew to North Carolina, and we had dinner, and he sent me the shareholder's agreement and a lot of the other things that were drafted at that time. And I started looking at it, and I said, 'First of all, I really believe in this concept.'" It was, in fact, almost a duplicate of an idea Dykes had already spelled out in a five-year plan for GE before leaving the company. "It would have been perfect for GE, a company that did not want to invest in semiconductor production but had to keep it running because it could not afford to alienate its existing customer base. This was a way that would provide any company with production capacity without a major investment. And we said in the plan that our strategy is going to be to use outside foundries for capacity. Now we hadn't validated the assumption that such capacity would be available. And here was the answer to what I was writing into that long-term plan.

"So I talked to Morris some more; then I jumped down on a plane and went to Eindhoven in January, visited Philips, talked to their lab people, talked to their marketing people, talked to management, even talked to the vice-chairman about it, and asked him where this fit into their profile. And they saw the same thing: The semiconductor fabs are becoming increasingly capital intensive; adding increments of capacity means taking $200 million steps. And this is putting the squeeze on a number of smaller companies. A company between $100 million a year and $1 billion a year can't afford now to invest in anything that will put it into an overcapacity situation just to keep it state of the art. So there's a squeeze on, and the foundry concept will feed and supply not only the start-up guys but the people that are caught in the middle."

The concept sold Dykes, and after a family conference, he agreed to move to Taiwan in March 1987, when he became president of Taiwan Semiconductor Manufacturing Corporation (TSMC). The new company is meant to be the linchpin of Taiwan's fast-growing electronics industry. "It's intended to provide three basic functions," Dykes told me. "One, it provides an infrastructure in Taiwan that's needed for competing with the Japanese and Koreans in the electronics industry. The Taiwan market is going the same route that a lot of the other countries have done. As Japan has moved up to a higher level of integration and higher technology, the Koreans have moved in the same direction, and now Taiwan's coming into the consumer and low end of the electronics market. So I place Taiwan about number three in terms of the degree of integration into its own economic infrastructure. But in order to accelerate that and compete effectively with these larger countries, there needed to be a more basic infrastructure here in microelectronics, which is really the key to all of that.

"There are some small companies on the island; one's very successful, UMC, which is across the road here, but nevertheless, they're making standard products that tend to fit the low-end strategy. They are using technology that's 2 and 3 microns, which is essentially yesterday's technology for these low-end products. The government has been putting R and D efforts in here for some time in electronic research and materials technology but never have had the economic scale to make it viable for the country to be an efficient producer and exporter of these type of goods."

TSMC is supposed to address that problem, and the commitment is substantial. "We're funded at NT $5.5 billion," explains Dykes. "At thirty to the U.S. dollar, it's about $180 million, something like that. And that's just pure equity. We're a private corporation now. We have shareholders, and they're looking to borrow even more. So we're looking at this as about a $300 million investment in the future for Taiwan."

Money alone is not, however, all that is required to build a state-of-the-art microelectronics industry. "In order to make it viable, not only did you need to spend the money," notes Dykes, "but you couldn't expect the technology to come up like mushrooms overnight. You had to have a partner that could provide not only a source of technology but a use for this output so that the economy of scale factor would grow rapidly. So there were discussions with U.S. companies, Japanese companies, European companies, and out of all this dialogue, Philips became the partner of choice.

"Philips has a strong presence on Taiwan, has been here for many, many years. Philips saw this as an opportunity to enhance their position on the island and to use this foundry as a Far East source of supply so that they can compete with the Japanese. Because Philips is head-to-head with the Japanese on many of their consumer products. So there was a nice affinity there. Philips took 27.5 percent of TSMC and an option, at least in the future, at a to-be-determined price to take full control of this. For the next three years at least, it's operated and controlled by the government of Taiwan; they're the majority shareholders.

"We have a ten-year technology agreement with Philips, which gives us access to all of their MOS technology, including submicron, for use directly by TSMC. So it's a chance for us to leapfrog and be essentially at the same place Philips and the other large companies are in terms of submicron process technology. That doesn't say that we're gonna be at the forefront of the world or the number-one company, but at least we're competing now, or we're sharing with two companies that are competing as world-class competitors.

"We're chartered then to serve the Taiwan market, to serve the Asia-Pacific market as a part of both Taiwan's export business and part of Philips's export business. And third, to serve those other markets of the world where we can be a viable competitor, and it looks like a large part of it is in the U.S. market. We have a restriction in our charter in that we're not another competitor in the Chip War. We think we're going to wind up being a tool, a partner for companies who are being and have been hurt by the Chip War in terms of their ability to invest. A lot of small companies now are cash short and cannot compete with state-of-the-art fab facilities; they have become too expensive.

"We solve that problem for them. We are a pure manufacturing company. We don't produce any products for sale to the end customer. We don't do chip design, and we don't do chip marketing. We sell wafer fabrications, we sell assembly, we sell testing. So any company that cares to either share their process technology with use or use our existing technology, which is Philips based, can buy wafers produced here and not have to make a capital investment in a wafer fab."

Under this concept, Dykes foresees even the normally non-offshore-producing Japanese becoming customers. "We will not compete with the Japanese for memory products," he points out. "Rather, they could be our customer. Their customers could be our customers, and here again, we're not competing, because what we see is that if a U.S. user comes to us, it's most likely for a design that's already been proven in the U.S."

As a pure foundry that does not compete with its customers, Dykes also feels the concept will offer protection for skittish companies anxious to protect their proprietary designs. "Let me give you some examples of how that works," offers Dykes. "We're working with one company on ASIC [application-specific integrated circuit] chip design. That's the heart of their product, you can think of it as an engine for a graphics processor that does high-speed, real-time, three-dimensional image processing. It's not a commodity item, but they're very concerned about control of the intellectual properties they've invested in that chip. And so it worries them to go to a merchant company that has the potential of perhaps bootlegging that design or learning something they could then use to develop a competitive version of their chip. And so, I think, besides having the foundry operate as a very efficient manufacturing resource, it allows more and more of these specialty companies to build walls around their intellectual properties. And you know from the lawsuits going on today that the proprietary controls are getting stronger and stronger, and even the courts are tougher and tougher, and I think we can help. First of all, our charter is very, very clear. . . . We are not in the business of producing, designing, or marketing chips, so we're never going to compete or have any incentive to steal anybody's design. The second thing is, we sign full nondisclosure agreements in terms of what U.S. companies are used to signing to safeguard their property. We have our own internal controls on customer documents, customer agreements, and our own notebooks. We run this very much like an engineering organization in the U.S. in terms of the control of the intellectual property. And we can have the customer audit us at any time to see how we're protecting his property. We aren't quite as complete, I guess, as the National Security Agency in terms of destroying silicon, but we're accounting for every wafer that's produced under a certain mask set."

Designs are the only capital most of the Taiwanese design houses have, and so protection is as important to them as it is to the much larger American chip developers. "There are over forty design houses in Taiwan now," points out Dykes, "most of which have been spawned since January. And they're starting to use this as a foundry for turning their designs through. And a lot of them are small companies that have no capital other than themselves. They're in everything from video games to computers. Anyway, that's the gist of what we're trying to accomplish, and we think—and Dataquest agrees—that this concept could be the concept of the future with shared manufacturing facilities, because the technology now is becoming portable. The machines are the source of the technology today—the equipment. You can move it around the world, and it would have been very difficult even five years ago to try to set up a state-of-the-art fab in Taiwan.

Dykes inherited an already existing $21 million fab facility that had been built by ERSO in 1985. It was small, capable of only 800 six-inch wafer starts a month. At the time, it was all the small Taiwanese electronics companies needed. "When we took it over," Dykes continued, "we leased that $21 million facility and added $32 million to it to bring it up to a capacity of 10,000 six-inch wafer starts a month. So the total investment here is about $53 million."

A new facility now under construction, called, appropriately if unimaginatively, Fab 2, is capable of 30,000 wafer starts a month. It will also be equipped to run either six-inch wafers or easily convert to the just-now-coming-into-use eight-inch wafer. "We are going to have a very fast cycle time here; we're talking about twenty days, twenty-five days from input to output. That's much faster than the Japanese. The area where the Japanese are worst is cycle time. If they tell you forty days, they're probably running closer to fifty-five. We have a lot of customers who come here on their way back from Japan. The area that they're always amazed at is our faster cycle time. I think that's going to be one of our strong points."

To move chips out as fast as TSMC does requires that a lot of elements be fitted into place. "It's a culture of its own," acknowledges Dykes. "You have to have a mentality in terms of the equipment, setup, inventory placement, and measurement systems in the factory on every shift to do that. So the computer comes in and says, every eight hours, here are the moves we have to make to keep this process line running. There are no 'hot' lines here; all the material moves at the same speed. "We're running three shifts now, six days a week. Our intent is, when we get this new fab up and running, we'll go to seven days, and three shifts, which will give us full utilization."

Maintenance of the equipment is the key to keeping a semiconductor line running efficiently. The Japanese are fanatical about maintenance, the Americans consider it an irksome necessity, and the Europeans fall somewhere in between. The Taiwanese take a uniquely Chinese approach that seems to combine philosophy with the ideograms that are so much a part of education in Asia.

"There was always the feeling in Japan that the operator and the technician were married to the machine, and it was important that it always be up and running, no matter what time it was," explains Dykes. "In the U.S., the maintenance organization was always a separate function. It may have reported to the manufacturing director, but it wasn't measured on the same basis as the output. So I think there is a disparity there. We're somewhere in between that here. We want the operator to be responsible not only for the quality of goods produced at the machine, but we can't expect the high-school-educated, direct-labor operator to know how to repair a $2 million piece of computer gear. But they need to know when that machine is drifting. So statistical control shows us that we're starting to drift. Before the machine goes out of limit, the maintenance operator is called in. He knows that time—we measure lost time as part of the available time—and the up time that he gets measured on is when the machine is in service. Even though it's not being used, if it's in service, it's considered up time. So if you walked out on our line, you'd see a little sticker. It has four faces on it. It can be smiling, which means the machine is in service and in production. It can be asleep, which means the machine is ready for production but nobody's using it, so it sleeps. It can be down for repair, and there it looks hurt. Another one's got its eyebrow all cocked up, which says somethings wrong with the process, the machine's a little skeptical like it's not me, and the engineers are looking at it.

"Those are little psychological things that are in there, and every-body wants to see a smiling face. And they know a sleeping face isn't making any money. If it's down in these other three categories, somebody's on it all the time. You'll see somebody around there until you can move the smiling face back up on the machine. So I don't know whether that's kind of an idiom or not, but it's something that the people thought of themselves; it's all in Chinese, and it really works. It really works, and you know whether the machine's up and running."

Money and technology are not all that is needed to create a first-class semiconductor producer. For Jim Dykes, the work force is equally important. And in the Taiwanese worker hard work, education, and ambition, forged by the Confucian ethic, are an extra advantage. "The technical work force here," says Dykes, "is very well educated. The reason this area exists—Science Park—is there are two major universities here. And one is a materials research center, and one is electronics. So most of the leading faculty and students come from these two unviersities, and their graduates feed directly into TSMC. When we started up, we took 140 people from ERSO with experience and transferred them right into this facility. That's also one of the charters of ERSO, the Electronic Research and Services Organization, to provide students with a job when they get out of school. So they work for ERSO for two, three, or four years, and then they transfer into industry. So we picked up in the beginning an experienced work force. Some of our people have seventeen years in the electronics and microelectronics industry. So that's been a real plus. In addition, we've hired new grads, and we've hired some from other companies around the island, including some computer companies. Over 20 percent of our people have master's degrees or better, 22 percent have had two years or more of college, and 29 percent are high school educated. There are only four people who don't have high school degrees, and their jobs are in security and janitorial services.

"They have good fluency in English—the high school here requires six years in English, and then, of course, most of the textbooks in this field are in English. So that's very good. As to the direct labor force, we only have about seventy-five direct laborers here now, as we're still building up; we'll wind up next year around two hundred. The first thirty of those came from ERSO; they were experienced and no problem. The others are high school educated, have the same six years of English, but their ability's not very good because they haven't used it. We don't try to make them speak in English, but our computer—we run a computer-controlled factory—all of the screens, and all of the inputs are in English. So we've had to go and teach English to these people in terms of training them to use English for the computer-controlled equipment.

"The quality of work is very good—the discipline, time to work, overtime, no problem. Absenteeism is no problem, and I would say that there we're in very good shape. The family culture on the island is very strong. Many of our young workers have their parents with them, so they work hard to take care of their whole family, and that work ethic is helping us in stabilizing the work force."

"The Taiwanese engineers are also a lot younger and have less experience than their American counterparts. The average age of our engineers is thirty years old, and in the U.S. it's much higher. I don't know that age means a lot, but it turns out that we have a lot less experience in the professional side out here than in the American work force. Now that has nothing to do with ability but has a lot to do with how much you can turn loose—that is, let them operate on their own. There's a graying in the U.S. semiconductor companies today; a lot of the people have been there their whole life. My case, I kind of grew up in the business, and I think—I'm not sure what that says for the impetus in the U.S. companies, but a lot of guys who drove it are getting ready to retire, and maybe some of the steam has come out of an industry that has been driven by a remarkable group of leaders.

"So this may not be an inherent disadvantage, because I see that the whole industry is starting to go through a metamorphosis. But we cannot assume here with the work force that they know what to do based on education and experience. We have to go through a long briefing and explaining. But people carry out what you ask them to do very well in comparison to American workers. Maybe if anything, they take direction better, because they have less tendencies to go off and do their own thing, as you have in the U.S., and that means you'd better be giving good direction."

TSMC and other companies are attracting a lot of Chinese back to the island and are also getting a great amount of technological feedback from U.S.-educated Taiwanese. There are more than 130,000 foreign students taking advanced degrees in the U.S., and the largest single group, more than 32,000, are from Taiwan. "So we're seeing that as a resource that I didn't see when I came, and that's very real.

"There's also a Chinese connection between the engineers on this island and their brothers, cousins, and friends in the U.S. who are working for every semiconductor company and computer company in the U.S.," declares Dykes. "And information floats. Now, I don't say this is stealing data, but when these guys want to learn something and they want to find out how things get done, they simply ask, and they get answers. There's a Chinese-American society that meets in Albuquerque, New Mexico, every year. This past year in Albuquerque over three hundred people came together, and they spent several days talking about how they could help Taiwan. So there's a strong nationalistic view coming out of these people, and we're finding even in our business those contacts help us—it's a network, all over. They say, TSMC's in business, we're here to help Taiwan, we need your help. They are always trying to get people to come."

South Korea

Some of the same spirit can be seen in the South Korean semiconductor industry, as Korean engineers and technocrats, trained in U.S. postgraduate schools and honed by ten and fifteen years' experience in

U.S. chip, computer, and electronics companies, head back home. South Korea, however, has entered the Chip War with a very different philosophy and approach than the Taiwanese. For Korea already had in place several large, vertically integrated companies, a structure much closer to that of the Japanese than the free wheeling Taiwanese and American entrepreneurial systems. Indeed, most of the manufacturing—everything from leviathan supertankers to tiny chips—in Korea is done by these huge family-owned conglomerates called *chaebol.*

So dominant are the chaebol that just ten of these firms control two-thirds of the nation's industrial production. In the streets of Seoul there are constant reminders of the great corporations that are the industrial might and muscle of the Korean economy. Hyundai and Daewoo cars, trucks, and buses fill the broad boulevards, giant signs acclaiming Samsung and Goldstar electronics adorn the billboards, while the tall office buildings of the giant companies help make the skyline of Seoul one of the tallest and most modern in Asia. It is a far cry from the desperately poor, mostly agricultural country that had been devastated by World War II and brutalized by thirty-five years of Japanese occupation before that. So undeveloped was the country, there were only forty people trained in science and engineering in all of South Korea.[13]

The Korean War in 1950 further savaged the already impoverished farmers and small shopkeepers, whose average yearly income was less than ninety dollars. But with massive American aid and unbelievably hard work, Korea has made a startling transformation; it has, for example, increased its exports to the United States more than 170 times in the past twenty years. In automobiles, chemicals, shipbuilding, and consumer electronics such as televisions, VCRs, PCs, and microwave ovens, the Koreans have become a major world producer. In 1986, Korea's exports, 40 percent to the United States, grew a remarkable 18 percent, to a record $36 billion. It produced a $3.5 billion surplus, the first in Korea's modern history. In 1987 exports doubled again despite widespread labor unrest, when thousands of workers went on strike in literally hundreds of factories across the nation. Exports of electronic products—everything from VCRs to semiconductors—increased 49.9 percent, to $1.4 billion.[14]

But the cost has been staggering. A $45 billion international debt has been built up to fund Korea's investment in heavy and high-tech industry. More than $2 billion has been invested to create a semiconductor industry.

"Key to the rapid progress and a continuing momentum has been future oriented governmental policies that have been implemented by private initiative," editorialized deputy editor Philip Abelson in the April 1, 1988, issue of *Science* magazine.

Beginning in 1962, a national plan for development was formulated that emphasized import substitution and exports of labor-intensive products. In the plan, the need to increase national capabilities in science and engineering was also recognized. . . . From the mid-1970s, the availability of trained scientists and engineers made feasible full-scale development of the heavy and chemical industries. Much of the technology that was employed came from foreign sources. But the pattern followed was to master the production process and then to move gradually into more sophisticated technology.

Now it is happening to semiconductors.

The government role in guiding Korean semiconductor research and production is not as apparent as it was in the Japanese push to develop its semiconductor industry, but it is there nonetheless. It is also more convoluted. The Korean version of MITI is dubbed MOST, the Ministry of Science and Technology. Within MOST is ETRI, the Electronics and Telecommunications Research Institute, which is charged with the support, guidance, and partial research funding of the Korean semiconductor industry. The majority of ETRI's budget, however, comes from yet another government agency, the Korean Telecommunications Authority, or KTA. So it was to KTA that I went, in the center of Seoul, to see Dr. Jung Uck Seo, the chairman of the project selection board of MOST.

Seo, who is the executive director of quality assurance and the Project Development Center at KTA, is a tall, handsome man with a Ph.D. from Texas A & M and a NASA fellowship among his impressive credentials. His spacious office in the KTA Tower is filled with hand-carved Korean furniture. From it there is a magnificent view of the jagged hills that cup the city. Seo, who points out the window at what is now a high-rise building, tells me that he was born just over there, a few thousand meters from his office.

The view is endlessly reassuring for Seo, who, as a fifteen-year-old, fled with his family when the communists invaded in 1950. After three years as refugees, the family returned to Seoul, and Seo went on to study at Seoul National University. He then spent two and a half years in the Korean air force, serving his compulsory military duty before going to the United States for his advanced degrees. With his NASA fellowship at Texas A & M, he specialized in the measurement of the electron content of the ionosphere. In 1969 he returned to Korea and worked in defense electronics for thirteen years before joining KTA.

"ETRI is spending 40 billion won for research on digital switching systems and semiconductors and other related work," he points out "At the moment, ETRI is acting like a system-management house. They have their own in-house capability, but so far we decied to utilize industrial power to perform most of the actual research. So ETRI is simply orchestrating the research. Samsung and Goldstar are the leading companies, but to do research in semiconductors takes a common effort. You must share the resources and the time. So ETRI assigns projects on the basis of company capabilities."

A major program, funded at 46 billion won, about $53 million, has lumped three of Korea's four major semiconductor makers, Goldstar, Samsung, and Hyundai, together in a joint research project to develop a 4-megabit DRAM. On February 9, 1988, the Associated Press carried a story datelined Seoul announcing that the 4-megabit DRAM had become a reality, with mass production scheduled to begin in 1989. The fruits of the research will be shared among all of Korea's semiconductor manufacturers, just as all of the Japanese companies benefited from the VLSI project. And Seo does not try to hide the obvious MITI-like imitation ETRI is now doing.

"At the moment," he admits, "ETRI is functioning pretty much like MITI in Japan, but they are also building up their own in-house capability for basic research and technology. As far as fabrication and manufacture is concerned, we are going to depend entirely on in-dustry."

But no company in South Korea produces any semiconductor manufacturing equipment. "We do have fine-precision machinery that we make in other areas," adds Seo with pride. "I am the president of the Institute of Korean Electrical Engineers, and we are going to cohost a big technical symposia in conjunction with SEMI. I personally am putting a lot of emphasis on this area. Without that basic infrastructure, you are building your house on top of sand. But we are not going to go big scale like the U.S. or Japan. Rather, it will be limited to the essential areas. That will be my task, to select those areas. If we make an analogy, we are not going to build the Lincoln Continential, but rather the Ford. It is still premature to decide what semiconductors we will concentrate on.

"Up to now we have followed the Japanese pattern, but in future we will be different. Even though the Korean semiconductor industry is growing rapidly, still we import over half these devices from overseas—from the U.S. and Japan."

That dependence disturbs Seo. "So far we have been going outside the country for the technology. Now we are developing our own technology. We are not inventing anything new, just trying to be self-reliant. That's the main thing. We are not reaching the stage of developing manufacturing equipment, but my personal opinion is that we should. Because without going into such a manufacturing equipment stage we are always having to rely on somebody else."

Seo is especially concerned about preparing Korean engineers to fill the needs of their fast-growing semiconductor industry. "We have now invested $2 billion in the semiconductor industry, and we have trained a lot of engineers in it. Most of our engineers are being trained in the U.S. and Japan. When we buy equipment from them, our people go there and are trained in how to use it. Most of our advanced engineering training is received in the U.S. and Europe. But in the last twenty years we have put a great deal of money and effort into upgrading our educational system, but it is not sufficient yet. In terms of quality and quantity, Korea has about 10,000 engineers working in the semiconductor, electronics, computer, and telecommunications areas. ETRI alone has about 1,200.

"We are also getting proposals from the universities to explore materials such as gallium arsenide. Our universities are getting better, but I do not yet have an institute like the Stanford Center for Integrated Systems (see chapter 12). But we do have a similar, functional organization, but the level is not yet at the Stanford level of quality. We are building now an Inter-University Semiconductor Training Center, which is funded by the government. It is at Seoul National University and is basically to train students in semiconductor engineering. We have actually copied the Stanford example. At the beginning of the operation we will have very few students, but as we move up the learning curve, we will of course add many more. The staff will be largely Korean, but we will welcome U.S. professors on sabbatical leave or however we can get them."

Seo feels Korea must be content to play a minor role in the Chip War for the time being. "In the semiconductor field, we must just follow in the footsteps of the U.S. and Japan for five or six years, until we are self-reliant," he declares. "Even then, we may be able to supply all the necessary stuff for consumer electronics, but not for defense; they are so highly sophisticated. So that I won't worry about; if we need that, we go out and buy it.

"That's why, if an American talks about self-reliance, he means something very different than a Korean. Self-reliance for the U.S. means you can live without anybody else's technology if you have to. We cannot. Within the consumer electronics, small computers, and telecommunications fields, we would like to be self-reliant. That's the limit of our capability. If we try to go beyond that, at least in my life-time, we will go bankrupt.

"The Korean approach to the semiconductor industry will be different from the Japanese," declares Seo. "Koreans should be smart enough to take the back-channel approach, but if you repeat every step of the forerunner, you never catch up. If the U.S. and Japan can run at 60 mph, we might do the same, but never go faster. But they might have other interests, whereas we will continue on the same course. We are a small country in terms of size, resources, and technological capabilities. We know we are limited; a small boat cannot catch a big boat. But big countries don't always pay attention to the smaller areas, and we do. So we are challenging the known unknown. You guys are challenging the unknown unknown. There is no risk in what we are doing."[15]

The Korean entry into the Chip War was tentative at the first. In 1974 a small group of Koreans who had been educated in the United States and worked in the U.S. semiconductor industry came back to Korea and founded the Hangook Semiconductor Company. Its primary product was watch chips, and it limped along until 1978, when Samsung bought Hangook.

At about the same time, another company, Daehan Semiconductor, was formed, and it, too, was swallowed up by one of the major chaebol—Lucky Goldstar. Its chairman, In-Hwoi Koo, had owned a drapery shop in 1931. In 1947, following or perhaps blazing the same path as the other chaebol founders, he sold the shop and founded the Lucky Chemical Company, which produced as its first product a cosmetic cream for women. Lucky moved on to make other necessities of life for the then impoverished country—plastic household goods and toothpaste. In 1958 a new subsidiary company was organized, Lucky Goldstar, to manufacture electric appliances. Goldstar then went on to produce the first Korean radio. Lucky Goldstar then expanded into electronics, oil refining, petrochemicals, and most recently, semiconductors. Along the way, Lucky Goldstar has become a $12 billion company with plants not only in Korea but also in the United States and West Germany that produce television sets and VCRs. Goldstar, Samsung, Hyundai, and Daewoo, and a company called Korea Electronics are the five major Korean semiconductor manufacturers.[16]

The Koreans bring to semiconductor manufacture the same qualities the Japanese and other Asian people possess—intelligence and a high regard for education and hard work. "Koreans, until the 1950s, were mainly farmers," Kyu Tae Park, the British-educated professor of electrical engineering at Yonsei University and the president of the Korean Society of Electronics Engineers, told me. I had traveled across Seoul to the campus of Yonsei half fearing a student riot, but all I found were groups of young people with books, tennis racquets, and Walkmans stuck in their ears. Of course, books and tennis racquets can be dropped in an instant and replaced with rocks, but not, thank goodness, on the day of my visit. "Until then," continued Park after he had ordered tea brought into his cluttered office, "maybe more than 95 percent of our people are farmers. Almost no industry at all. Therefore, the people who are working now, and the managers now, are all the farmers' descendants, so they observed then how hard their fathers worked. They wake up at four in the morning, and they have to work until, say, eight o'clock or ten o'clock at night. So that industrious spirit came from the farmers, their work ethic.

"The second thing was from the Confucius teaching. If you don't work, you mustn't eat. So hard working is our spirit. And also it is the Confucius teaching that everybody has to study. Even during the Korean War most of the Korean people went to school, and they were trained theoretically. And they are very good at mathematics, and they are very good at planning. Then, as the economy improved, a lot of the good students go to the United States. And many remained in the United States and Britain and France, and they learned quite well. And they came back and continued to work hard. That's why I think the semiconductor business is very successful—not only semiconductor business but all other electronics as well.

"The semiconductor is a very tedious job from start to finish. You would have to do like caring for children or like a farmer growing rice. If you grow the rice, you must very much till the paddy field. That spirit is also true for the semiconductor business. That's why we are so successful at this work. The Japanese are so all the same. Their industry started before even the Second World War, but you see, they are all children of farmers. They work like hell. Day and night. Like farmers."[17]

The Koreans also have a long and rich cultural heritage and a surprising technological history. They introduced armored warships to Asia in the sixteenth century and invented movable metal type two hundred years before Gutenberg. Like the Japanese, their alphabet is based on Chinese characters that were first brought to Korea in the fifth century. Even after the invention of the Korean alphabet, called Hankul, in 1447, Chinese remained the official script until the late nineteenth century. Professor Park thinks the use of Chinese characters is one reason the Koreans seem to do so well in engineering and math based fields.[18]

"The Korean characters were invented around five hundred years or so ago by King Sejong. It is very much a scientifically combined character," he point out. "It consists of a vowel, consonant, and ending together in the one character. Therefore, this is very much why I think the Korean people are very scientifically well organized people. We are well trained to read and write Korean and Chinese characters. These Chinese characters are like a complete picture; they are ideographs. Therefore, one character gives you a lot of information. Say, if you write all of these Roman letters, other letters, it takes time, but one character gives the ability to capture a lot of information fast."

It must pay off in the teaching of the sciences, which begins in elementary school. In a 1987 survey by the International Association for the Evaluation of Educational Achievement of Science Achievement in seventeen countries, ten-year-old Korean children tied for first place with their Japanese counterparts. U.S. youngsters ranked eighth.[19]

Modern Korea has surged onto the world's economic stage on the wings of what it calls the three blessings—low oil prices, low interest rates, and a high yen. Korea's unit of currency, the won, is linked to the dollar, and as the yen rises against the dollar, it also soars against the won, the Korean currency unit. This makes Korean goods cheaper for Americans than the same Japanese product. The result, especially in consumer electronics, has been dramatic. At the dollar's high of 263 yen, a Japanese-produced television set with an average export price of 40,000 yen was only $152. When the dollar dipped as low as 121 yen in 1987, that same television set would carry an export price of $331 if the manufacturers passed the entire cost on to the consumer, something the Japanese have been reluctant to do. That would make it more then twice the $140 average export price of a South Korean–made color set.

Nor is that the only consumer-electronics area in which the Japanese suddenly find themselves challenged. Eight years ago, the Koreans reverse engineered a Japanese VCR. In 1985, the Koreans sold 1.5 million VCRs. Today they are the second-largest producer of VCRs in the world, with 1987 sales of more than 6 million. By the year 2000, they hope to triple their share of the consumer-electronics market.[20]

The Korean work ethic is similar to that of the Japanese. "It's similar, but not as rigid as the Japanese" is the way Chang Soo Kim, the senior managing director of Goldstar's semiconductor group, described it to me. "We have quality circles. We don't sing company songs in the morning, but we get together before the start of work and try to communicate what has to be done for the next shipment and things like that."[21]

Like their companies, Korean workers are boundlessly ambitious and increasingly restive under the autocratic government and the low wages that first attracted American and other foreign companies to establish plants in Korea. The Korean worker puts in the longest work week in Asia—fifty-five hours for a wage of less than two dollars an hour.[22] Now they want a bigger share of the economic pie, and in 1987 hundreds of strikes disrupted the normally humming Korean work place. Labor unrest was supported and magnified by student demonstrations at campuses across the nation. Much of it was reported on American television, and scenes of riot police firing tear gas at rock-throwing students became commonplace in the summer of 1987.

Despite the television reports, I got the impression that for most

Koreans it was business as usual, the strikes and demonstrations simply another fact of life to be endured as a necessary, if painful, break in the routine, disruptive, but only for brief periods of time. And indeed, after the elections, late in 1987, won by the party in power, the wave of strikes seems to have stopped. Wages, however, will unquestionably rise.

"It will affect our competitive position for the short term," confesses Chang Soo Kim. "We are already trying to overcome that through a lot more automation in the manufacturing area and also trying to come up with a product that will give us more value added so that the labor content, as a portion of the total sales price, is less."

The current wave of labor unrest will undoubtedly result in some erosion of Korea's competitive edge in cheap labor. This will make it even more difficult for the Koreans to climb out of the red-ink bath into which they plunged after making the decision to enter the semiconductor market in a big way. The major Korean investment into semiconductor production was ill timed. The major companies jumped in 1983 and 1984, just in time to add to the glutted world market of 1985 and 1986.[23]

There was nothing subtle about the Korean entry. They bought technology from American semiconductor companies, hired American engineers to staff chip-design companies they either bought or established in the United States, and shipped the prototypes to Korea for mass production. There, working seven days a week, twenty-four hours a day, the Koreans built wafer fabs in only eight months—less than half the time it takes to build a similar factory in the United States. "It took the Japanese twenty years to get where they are. It may take us only fifteen or even ten years to get to the same point," boasts KTA's Jung Uck Seo.

Their marketing strategy was also familiar and based on the Japanese model. "In general, the marketing strategy is similar to the Japanese," says Dr. In Sang Lim, the branch manager of P-E Korea. "We do essentially the same thing in pricing our semiconductor product, making it very competitive. But I think our quality standard is far lower than the Japanese. Our industry in general is not as sophisticated as Japan's. So it will take time. There is the immense concentration on quality, so in the future we hope to achieve parity in quality."[24]

If money can buy quality, the Koreans have taken the right

approach. Samsung, Korea's largest chip producer, plunged $500 million into a pair of fab facilities in 1982 capable of churning out as many as 3 million 64K DRAMs a month. By 1985, however, Samsung was able to sell only 1 million chips a month to a badly depressed world market. Two years later, however, with a world clamoring for memory chips, Samsung was ready. They had mastered the 256K DRAM and were turning out 4 million DRAMs a month. And in what has to be a bitter irony for the American chip industry, Intel, the inventor of and once one of the world's leading producer of DRAMs, before giving up production in 1985, signed an agreement with Samsung in June 1987 to purchase about $25 million worth of 64K, 256K, and 1-megabit DRAMs. Intel will then market the Korean-made chips to its own customers.[25]

Samsung has been relatively successful challenging the Japanese in the jelly-bean competition and now ranks as the world's seventh largest producer of DRAMs. Its 1982 sales, the first year of production, were only $5 million. Sales in 1987 topped $225 million. But Samsung is looking beyond memory chips and has begun a program aimed at diversifying its product line. At the moment, nonmemory chips, such as microprocessors, account for only 10 percent of its production. "Our goal," says Hyeon Gon Kim, managing director of Samsung's semiconductor division, "is to increase our non-memory business to 60 percent of sales within five years."[26]

Goldstar, with very little memory production, is probably the most innovative and impressive of the Korean semiconductor makers. In addition to producing 64K and 256K DRAMs in small amounts, Goldstar is turning out the more complex logic chips and 4- and 8-bit microprocessors. And Goldstar tapped the mother lode by making its semiconductor division a joint venture with AT&T. "They have strong involvement in our operation," declares Goldstar's Chang Soo Kim. We were meeting in his gleaming chrome, steel, and glass office in one of the twin Goldstar towers that sit on an island in the middle of the Hongang River, which runs through the southern edge of Seoul. "We signed a technology transfer agreement with AT&T. The big chunk of the agreement was to get the number-one telephone switching system and manufacture it here and sell it to our Korean government. Part of the agreement gave us the semiconductor technology to produce some discrete and simple digital ICs for the switching equipment."

"That was the start. We sent some engineers over there to be trained, and then from there on we just moved up our operation."

Kim, like virtually all of the South Korean technocrats, is an American-trained engineer who spent most of his adult life in the United States. He speaks excellent colloquial English. He was born and raised in Inchung, about twenty miles southwest of Seoul, where his father was a local government official. After graduating from Seoul National University with a degree in nuclear engineering, he served in the army for two years. A year later, he went to Carnegie Mellon University in Pittsburgh for a master's degree in electrical engineering and then to the University of Florida for a Ph.D. While studying for his doctorate, he met and married a Korean girl who was studying fine art at NYU.

He worked for a number of American companies, including Harris Semiconductor, and then for a small start-up company called Inselect. Unfortunately, Inselect went downhill, and Kim left and joined the RCA Research Laboratories in Princeton. After that, he went to work for Digital Equipment Corporation, and finally, in 1984, after eighteen years, he returned to Korea.

Today Goldstar Semiconductor employs more than two-thousand people, and its 1987 revenues were $100 million. With three fab facilities outside of Seoul, Goldstar is planning to spend the equivalent of those revenues to build a state-of-the-art fab to implement its basic strategy of product diversification. "It's difficult for a very young company to have a very wide product line. On the other hand, even though you're a start-up company in a way, it's very dangerous in this business to be a producer of one kind of semiconductor only. So we tried very hard in our strategic thinking to broaden our product line. We got one chunk in the linear ICs. And that is geared to support our own consumer-electronics equipment. And then we have digital logic—there's a big chunk. And that has some different purposes, essentially to provide some product to fill up the manufacturing capacity, to lower the fixed cost of our product. And then we have a mixture of custom MOS circuits, digital circuits for telecommunications, some memories, microprocessors, and ASICS. It's all mixed up. In the memory area, we stayed as much as we can away from dynamic RAM so far. And we are concentrated on high-speed, so-called niche memory products and also for our own particular situation in

Korea. In this area we needed a high-density MOS to support the local Korean or Chinese characters in the computer.

"As you know, we are producing television sets, VCRs, audio equipment, microwave ovens, refrigerators—everything nowadays needs integrated circuits. All of those we are buying in the past, and we're still buying a great chunk of them from Japan. U.S. manufacturers are out of consumer electronics. The Europeans are stepping out, and so the only source is Japan. And in 1984 and early '85, we had some very difficult times, when there was a shortage of silicon supply, and we couldn't get enough components to make those consumer-electronics equipment.

"Part of the problem, in the immediate business sense, is that we are competing with Japan in the end products and they supply all the components for them. The source of the components dictates how much equipment we can manufacture and sell outside. The next issue is, as you know, that the consumer-electronics-equipment field progresses very rapidly from one year to another year as improvements are made in components and price, attractive features, and so forth. And all of that is determined by the IC component. So if we don't have the capability to develop those ICs on our own, we will be forever behind, one generation behind."

Goldstar is committed to solving that problem no matter the cost. "Our management people are not afraid of spending money," Kim states flatly. "I would say so far, in terms of capital investment alone, we have spent close to $200 million. And the money spent on labor, engineering, technology, and all of that, maybe $300 million easily. And we are not profitable at all today in terms of semiconductors. But we believe that semiconductor technology is the key to make other electronics areas successful.

"So it's a purely defensive mechanism which triggered getting into the semiconductor. Our primary purpose is to supply our own need first. But our system business is so diverse, and there are so many different kinds of semiconductors used, so in terms of volume for each device, it's not that big. So our strategy is to develop as much as possible to supply them, but on the other hand, in order to do that, you have to set up a manufacturing facility, some fixed investment. And it's not enough volume to be cost competitive, so we have to fill up other commodity devices as well to lower the production cost. So the other thing we tried to do is, when we pick out those chips to manufacture, we want to make sure they are something we can use internally as well as something we can sell in Korea and nearby regions like Taiwan, Hong Kong, and Japan. Taiwan, Hong Kong, and Singapore manufacture similar kinds of things as we do, so if we pick the right chip, those are something used in those areas as well."

Nor does Kim fail to include the United States in his market strategy. "We are also looking to the U.S., of course, because the U.S. is the biggest market. But right now our emphasis is more in this area, because first of all, we picked those chips, which are used in the consumer-electronics area, and secondly, our marketing and sales capability is not that strong. We don't have a strong presence in the U.S. Down the road, as we grow, there's no other way but to sell in the biggest market, the United States."

Goldstar has also made a major commitment to semiconductor research. "We have two research facilities," notes Kim. "Each has about 250 people. About half of them are in design. Our biggest weakness is in design, not only chips but the design and understanding of the system in which the chips are used. In the past, we bought the components, we used them, we assembled them, but we never designed or created the system from the beginning.

"One facility is almost 100 percent dedicated to designing our custom linear circuits and custom digital circuits for our end systems. And the other facility is dedicated to designing ASICS and memories. That's for improving two things: One is in our system technology, and the other one is the semiconductor processing control. The memory is a vehicle for that; the ASICS is a vehicle for understanding the system.

"In terms of dollar amount, including the capital spending, annually we spend at least 10 percent; some years we spend as much as 17, even 19, percent of our semiconductor revenue."

Goldstar's semiconductor technology is aided in no small measure by a joint venture with AT&T in telecommunications.

Hyundai, the largest of the Korean chaebols, chose to attack the commodity market, building two fabs outside of Seoul, and invested another $50 million in a Silicon Valley production facility. The results were less than spectacular. The collapsing world-market doldrums, equipment failures, and very poor relations with its American employees delayed the opening of its Silicon Valley plant. Finally, Hyundai decided to just write off the American venture and sold the plant in 1986 at a loss. Its two plants in Seoul are up and running, grinding out limited numbers of RAMs and EPROMs.[27]

Daewoo, the other heavy-industry chaebol to go into the semiconductor business, is not looking to overpower anyone. "We had planned to enter the semiconductor business for a long time, since 1983," Dr. Hong Jo Chang, the executive managing director of Daewoo Telecom, told me. Chang, like most of the semiconductor executives I had met in Korea, had been educated in the States and had spent another fifteen years working there in the semiconductor industry. "I went to Seoul National University, where I got an EE degree. Then I went to the States, to Ohio State, for my Ph.D. I worked for a few companies, including IBM at Poughkeepsie, and returned to Korea in 1983 after fifteen years in the States. I returned because the work I am doing now is more interesting, and here I have a very important position. There I was just another engineer." Chang lived in Seoul throughout the Korean War but fled with his family to Pusan for six months after the communists captured the city. "It was not much of a hardship," he admits. "Other people were gone from their homes for two and three years."[28]

Daewoo's semiconductor facility is located in a working-class district in Seoul. Compared to the sparkling new Goldstar towers, its semiconductor offices and factory that I visited were positively shabby. But in what was for me a now familiar touch, there was a rack of slippers for visitors to exchange for their shoes. I encountered the practice nowhere else in my Korean travels, which were largely limited to factories and offices where American styles and customs have replaced the more traditional ones.

Like the other Korean semiconductor makers, Daewoo's technology was imported from the States, in this case from a semiconductor company called Zymos Corporation. In fact, Daewoo bought a major piece of Zymos. "We have a major holding in that company, and we sent our engineers there to be trained. We made an agreement with them to transfer technology as well. We set up a fab here. At this moment it is producing ICs for use in our consumer-electronics products and telecommunications products. So that is very important, probably one-third of our consumption."

Having achieved a major position in the computer (the company manufactures and sells the phenomenally successful Leading Edge PC, an IBM clone) and consumer-electronics market worldwide, Daewoo found itself in the same position as its Korean counterparts. Beneath the metal-and-plastic covers that bore the Korean nameplates were, for the most part, Japanese chips and components. As the yen skyrocketed against the won-linked dollar, the cost of those components has also risen. Those higher costs will eventually erode much of the price advantage the Koreans have.

In addition, Chang and most Korean industrial leaders believe dependence on others for key components is an inherent competitive weakness. "We realized that without semiconductors we have to rely on others for key parts, and that limits our growth. And we could be controlled by our suppliers. That has happened in the past," admitted Chang ruefully. "If we buy parts from Japan, say, for TV application, then the supplier in Japan sells us only a limited number, not enough, and that limits our production. So we have to have our own sources of supply. I don't know whether that is a result of our competition in TVs and VCRs or not, but somehow we felt that we are not supplied enough. Therefore, the decision arrived at was to manufacture our own semiconductors."

While Daewoo has no desire to become a major merchant player, they still hope to sell a fair number of chips to the rest of the world. "We started this to support our existing systems business. But it is just not enough to survive producing semiconductors for yourself. You must have outside markets. We have made plans for some minor outside sales. There is no reason not to sell other companies the same parts we produce for our own use. We are not interested in producing memories at this point. That is not our business. Our goal is to produce custom ICs to satisfy our internal needs.

"We have a very small fab with a 10,000-wafer-a-month capacity. But it will take at least a year to get up to that capacity. We have just started producing and are not selling outside. In about a year we may be able to support about one-third of our IC need. So we are looking at it almost exclusively as a source of supply for our own use. Later, we may produce 50 percent for sale in Korea and possibly to Southeast Asia.

"For now we buy from domestic manufacturers such as Samsung, and we buy a lot from Japan and also a lot from America. Probably we are buying mostly from America—maybe 50 percent American.

Daewoo's investment is minuscule compared to the other Korean chaebols that have leaped into the semiconductor business. "Including land and everything, we have invested $25 million," says Chang. "That bought us a wafer fab, test facility, and small assembly plant. We also set up a research and design center. At this moment more than half of our expense is for R & D."

Considerable care has also been lavished on attracting and training workers. "Our work force in semiconductors is very small, maybe a hundred people. They are mostly engineers. Recruiting these people is very competitive among Korean companies. To start with, we have recruited some from domestic companies who have some experience in semiconductors, but that's very few. The rest are all fresh graduates, and so we are spending a lot to train them. We also train the assembly-line people, most of whom are high school graduates. They are as motivated as the Japanese. We sing a company song, and we are trying to establish quality control circles. We try to organize not only workers but also engineers into the quality control circles. So in some groups we have a mix of engineers and workers. In some groups, only workers. Sessions are held only after working hours."

Unlike Japan, the workers who participate in the quality control circles do not get paid to attend the meetings.

Daewoo and the other Korean semiconductor maufacturers have adopted a number of other typically Japanese industrial institutions but not high wages. "The salaries we are paying are better than average, but it is not the top," notes Chang. "Top-salary companies are smaller, not big companies like Daewoo. But those people who work here have pride in the company. I try to provide them with good working conditions that make up for lower salaries. We also provide after-work activities, like soccer and other sports, picnics, and parties. The workers all wear uniforms. In the old days the workers needed housing provided by the company, but now they all have their own homes. They prefer their own private places."

Having spent so much time in America, Chang sees the Koreans taking a very different approach to the semiconductor industry. "Americans are very creative, because their education is very different from the Oriental way. "We don't have much strength yet in the Korean semiconductor industry. The only thing we have is, we work very hard, probably twice as hard as the average America. Productivity is much higher than in the States. Of course, our design technology is far behind

America. But I see us working very hard and catching up very quickly, because once we learn something, we use it very efficiently. And even though the labor content in semiconductors is decreasing rapidly still, our engineers and designers work very hard and for much less than Americans, maybe one-tenth as much. Starting engineers here make an average of $6,000 a year. With ten years' experience, they earn maybe $12,000. The average worker, a high school graduate, earns probably $2,000 to $3,000 a year."

Inevitably, wages will rise, but so, too, will Korean productivity, which seems to have the same solid platform from which to surge upward, as does the Japanese. Indeed, there is a great deal of unabashed mimicry of Japanese management style, work ethics, and sales practices in the Korean semiconductor industry. But the Koreans, with a trade imbalance with the United States that could prove to be equally nettlesome, are determined not to repeat that particular Japanese mistake.

"Even though we are talking science and technology, we must always consider the trade balance," says KTA's Dr. Seo emphatically. "If I want to sell something to your country or another country, we must be aware of that. We must have a shopping list for them, too. We don't want to be criticized for overselling. So we must consider what to buy and what to sell."

Yonsei University's Professor Park echoes that idea. "The Korean people are very much grateful to the United States. If we know we'll buy $1 billion worth of goods, then we'll buy the $1 billion from the United States. The Korean people know we sell a lot of things to the United States, and if the United States doesn't exist, we lose a market. So all the foreign ministry people, the industry people, everyone, knows the United States must prosper for us. Then we also are prosperous. The Korean people are blaming the Japanese. These people are misers. They will only sell; they don't buy. They sell to the United States all the time; they don't buy anything. We ask them, open the market. We want to stay open. We want to open the door towards the United States."

Just how far that door will be opened as Korea's own semiconductor industry continues to grow and eventually challenges not only the Japanese is memories but the Americans in the custom chips and microprocessors remains to be seen.

8

The Old World

If you were to draw a line from the Netherlands down through Germany and Italy and then across through France and Great Britain, you would have two sides of a very rough right triangle. Along those two lines you will also find virtually all of the semiconductor companies in Europe. They include Philips in Holland, Siemens in Germany, SGS in Italy, Thomson in France, and Plessy in England.

Only one of the companies, Philips, is counted among the world's top-ten producers. The impact of the European semiconductor companies on the industry has been minimal at best, despite a strong technological base and an early entry into the field.

As early as 1954, Philips and Siemens were successfully producing transistors and between them dominated the European market. With a totally different economic culture, which favored vertically integrated companies and did not encourage the formation of new, young start-ups, few merchant companies gained a foothold in the European semiconductor industry. Those that did, and there were a few, soon were absorbed by the large companies or simply went under.

"Few merchant producers entered the European semiconductor industry during the 1950s," Franco Malerba, a professor of economics at L. Bocconi University in Milan, Italy, explained to me. Malerba has been fascinated by technological innovation and found the evolution of the European semiconductor industry particularly fascinating. With a scholarship to Yale, he made that the subject of his four-year Ph.D. thesis. "The major explanations for this lull include the absence of the spin-off and mobility of personnel among firms which characterized the American industry, the reduced size of the European electronics market during the 1950s, and the concentration of government R & D and production-refinement support in a few established firms."[1]

The larger companies invested enough in research to maintain their technological grip on the leading edge of transistor production. But semiconductors were never seen as having any real value as individual products. Rather, they were viewed in the overall strategy of most of Europe's vertically integrated companies as merely components of larger systems and as a means to increase the competitiveness of their electronic final goods rather than as independent products.

During this time, the 1950s and early 1960s, Philips and Siemens dominated the European semiconductor market. None of the other companies could effectively compete and in fact were kept going largely by their government's insistence on having some representation in the semiconductor industry. Europe was self-sufficient in transistors, and only one American company, TI, had even bothered to invest in European production. But with the introduction of the IC by Fairchild and TI, Europe's technological parity with America was suddenly shattered. "The technological knowledge obtained by European firms about the old technology, the capabilities accumulated in the R & D and production of semiconductor devices, and the commercial success on European markets during the late 1950s and early 1960s was enough to convince these firms that the established technology was the right one," points out Professor Malerba. "The firms concerned decided not to change their efficient production processes and commercially successful products in favor of new, highly uncertain ones. In fact, a high-level Siemens official was quoted as saying, 'There is no real demand for integrated circuits. This can be more easily met through licensing.' Similarly, the management board at Philips entered the 1960s without any decision to begin large-scale production of silicon planar integrated circuits.

"The new integrated-circuit technology was far more complex than the production of transistors, and the Europeans, with their slow start, fell farther and farther behind. Although these firms were ready to move into integrated-circuit technology," explains Malerba, "they did not switch to the new technology as quickly or as radically as American merchant producers. Nor did they commit the same amount of resources as American producers. In addition, because of the limited size of the European integrated-circuit market, they were not able to benefit from either the economies of scale or the dynamic economies in integrated-circuit production from which American producers benefited. Then, when, during the second half of the 1960s, the European integrated-circuit market grew in size, American merchant producers made direct foreign investments into, and dominated, this market."

"The Europeans literally chose to be third in a peculiar way," insists Dr. Sheldon Weining, whose MRC has a plant in France and sells semiconductor manufacturing equipment to all the other major semiconductor producers in Europe. "It's disturbing because if not first, they definitely could have been second. They had established the enormous entities of the Thomsons, the Philips, the Siemens. They were well-established technical companies. They chose literally to not compete because of an inability to commit monies and ideas. . . . In other words, to make a commitment.

"Everything has to start more or less in the sixties. The semiconductor-integrated-circuit breakthroughs occur in the middle sixties. It's about then that all those guys, Siemens, Philips, and so on, should have sat down and said we want to be on the frontier. These are tremendous technological organizations, entrenched in markets and technology; they don't have to be created. Many of the Japanese semiconductor companies had to be created from whole cloth. These existed.

"But, in typical European fashion, we don't see any entrepreneurship; we don't see any new companies. There are no Intels coming up there. So what we have is the staid, tried-and-true European companies sitting still and the Americans coming in like gangbusters. TI, Motorola, IBM—they are all there and start factories. But what the Europeans are doing is studying the shit out of every problem. They must make a decision: Do we go from 4K to 16K? They'll have to think about it; they have to set up a study commission. They didn't have the balls to make a commitment. Let's do it, pow! Which at least the Japanese do. When we come out with a new piece of equipment, the first one is sold to Japan. And everyone has found this to be true. The ninth one, or the ninety-third one, is sold to Europe."[2]

Indeed, in the spring of 1988, one could see in the 1987 rankings just how complete the European decline had been over the past thirty-five years, since they had first begun to manufacture transistors.

EUROPEAN MARKET SHARE ESTIMATES[3]
(in millions of dollars)

Philips	969
SGS-Thomson	835
Texas Instruments	525
Motorola	501
Siemens	446
National Fairchild	382
Intel	295
NEC	258
AMD-MMI	246
ITT	243

So, of the top-ten semiconductor producers in Europe, only three are actually European companies. And the one in second place, SGS-Thomson, achieved that high ranking only as a result of its merger in July 1987. But that merger is indicative of a new spirit of cooperation among the Europeans and perhaps the only hope they have of remaining in the semiconductor business as a significant factor. "In my evaluation, Europe is still lagging behind, but major structural changes in the form of alliances, which are the only way for them to cope, are being made," points out Malerba. "I see these changes as favorable. But the problems of European semiconductor makers are the problem of scale and size. Each player may be too small to compete."

Supporting that viewpoint is Malcolm Penn, a thirty-five-year-old Englishman who is the director of European operations for Dataquest. "I think if they're big enough, the Europeans can compete worldwide. I honestly believe that Philips can still be in the top ten in ten years' time, and that SGS-Thomson stands a pretty good chance of being there with them. The one that's probably got to do something is Siemens, because they're failing to get bigger, and the others will maintain their position of supremacy in the markets they go for, and that could be world-wide."[4]

I had decided to concentrate my research on the three largest players in Europe, Philips, Siemens, and SGS-Thomson, the recently formed Italo-French combination. As background I had the rather academic views of Franco Malerba and Malcolm Penn and the more colorful and certainly more subjective analysis of Shelly Weinig. Now, quite by happenstance, while on my way to an industry conference called SEMICON Europa, I was about to gain another singular and highly personal viewpoint of the three largest and most important European semiconductor companies.

SEMICON Europa began on March 1, 1988, and ran for three days in the sprawling Zuspa Convention Center, not far from the Zurich Airport. It is the premier exhibition for the semiconductor manufacturing equipment makers of the world to display their dazzling wares before the chip makers of Europe. I had included it on my tour of the European chip makers to see firsthand the state-of-the-art manufacturing equipment being offered to Europe's semiconductor producers.

Unfortunately it looked almost from the outset that I was going to arrive late. Leaving the hotel in Munich, I walked out into a furious snowstorm. For a moment I was tempted to return to the hotel and ask them to call the airport and check on the flight. Still, I figured, this was Bavaria, the Alps and all that; surely a little snowstorm would not stop Swiss Air. The taxi driver was equally confident despite the fact that it was hard to see through the swirling snow to the end of the street.

Needless to say, when I arrived at the Swiss Air counter, I was told the flight to Zurich had been canceled. Lufthansa, however, was still flying. "Go to our reservation counter around the corner and they will book you." I soon learned Lufthansa was overbooked and was not even accepting standbys. "But, I insisted, "I must get to Zurich."

"Are you going to Zurich?" asked a short, chunky man with a handsome shock of white hair that belied his forty-odd years. "I, too, must be at the SEMICON conference. Come with me," he continued. "There is a small commuter airline that goes to Basel. My car is there, and we can drive to Zurich."

Gerd Kuhnlein was my rescuer's name, and he is the technical manager of encapsulation systems for Ciba-Geigy, the giant multinational Swiss chemical company that is headquartered in Basel. Gerd was a gold mine of information, and as we bounced along the edge of the Alps after getting above the snowstorm, he filled me in on his view of the European semiconductor industry. After landing, we continued the conversation in his BMW as we drove through Basel to the autobahn that led to Zurich. Of course, we passed right by his office in Ciba-Geigy, which is the largest employer in Basel.

Gerd Kuhnlein joined Ciba-Geigy in 1984. They produce the electronic chemicals that are the base materials for printed wiring boards, laminating epoxy resins, and the materials used for encapsulating semiconductor components. They also manufacture photoresists and recently began to produce silicon for wafers. Born in a small town near Stuttgart, Kuhnlein's father was an optician who owned his own small shop and a very small factory. "It wasn't very profitable, and he said 'It's better you do another thing.' I was interested in chemistry, and so I went to the University of Aachen. Then one day one of my professors who had moved to Stuttgart asked me to join the faculty there, and I did." Subsequently, he went to work for an American chemical company and then on to Ciba-Geigy. His viewpoint, then is as a supplier in a position to compare the strengths and weaknesses of many semiconductor companies.

"Philips is a consumer-electronics-oriented company in a low-price and high-competition business. They are competing with the Far East suppliers, and all their investment is in that sort of production, and there is very little profit. So it is very hard for Philips to make long-term investment commitments, because this is a market-related industry and it is not driven from Europe. It is driven by the Japanese and the other Asian producers. Philips has created, in the last two years, big financial problems. What I hear last week [February 20, 1988] when I am coming from Philips, we have different research and development projects, Ciba-Geigy, with Philips, but they may now have to cut back on the programs. They are also going to let go about 20,000 people. It will be done as early retirement programs. Philips at the moment is not in a very good position."

At just about that time, Philips was reporting a whopping 51 percent drop in fourth-quarter 1987 earnings.[5] "The problem," notes Kuhnlein, "is that Philips, in the last few years, preferred to go its own way, without any joint venture with their Far East suppliers. And they are 100 percent dependent on those Far East suppliers for everything, and that is not a good situation. Philips is a big part of the Far East electronics industry. Philips is the biggest one in Europe, so at the moment the Far East suppliers are supporting Philips, but the risk is Philips will become too dependent on the Far East. All their research and development is done in Eindhoven and Nijmegen [the Netherlands], but very little production. Most of that is done in the Far East. Then Philips waits for the newest technology to come from the U.S., and it takes too long.

"We have so much difficulty just in making contracts with Philips. It is easier for us with Siemens because we have been doing business with them for such a long time. So we can proceed and don't need any contracts until it comes at the end with a patent situation. But with Philips we have to do the contracts first; that is the Philips mentality, and it is very hard and time-consuming.

"Siemens is totally different, a private company that has some research programs financed by the German government. In addition, Siemens is a very rich company, and it is much easier for them to make long-term investments. So Siemens is more flexible in making those decisions. They are in a different business than Philips. They are in the memory business; they have a computer division. It's not so popular, for sure, as IBM or Nixdorf, but it is quite popular in European industry. It's a profitable company.

"Siemens took the 1-megabit license from Toshiba, but it is not the newest technology of Toshiba. The reason why Siemens bought Japanese technology is they realized the grass-roots approach, starting from the beginning, takes too long time and is too expensive. They are the opposite of Philips, who decided to take their own development approach. They believe they are strong enough to develop an equivalent technology in a short time. Siemens didn't believe this. They decided to take the Japanese technology. They said we will spend a lot of money for sure, but of course the government paid part of it. If you don't know what exactly is the newest technology, it is hard to know whether you are buying the very latest from the Japanese. In fact, it is an old technology that Toshiba offered, and Siemens bought it. But it was a starting point, but it is easier then to make the next step. It will cost Siemens a lot of money because it is too expensive for them to produce these chips and sell them competitively at a profit. Siemens will lose money on this project, it is very clear. By the time they are up and running, they will be two generations behind, but they are going ahead with a 4-megabit project. For that there is government funding."

With Kuhnlein's serendipitous input, I began my own research tour of Europe's major semiconductor manufacturers. An hour and a half by train, southeast of Amsterdam, through the flat farmland and canals of the Dutch countryside, is the bustling industrial city of Eindhoven. More properly, the town should be called Philipsburgh, for it is the corporate headquarters of Philips, and much of the town's economy is linked to it. The headquarters of one of the world's largest semiconductor makers is a large compound studded with architecturally uninteresting but highly functional boxlike buildings of two, three, and four stories. Like most high-tech companies I have visited, it was equipped with the almost obligatory gate house, security guards, and visitor's pass. My meeting was with Cees, pronounced Case, Krijgsman, the chairman of Philips's worldwide semiconductor operations. Krijgsman, a courtly white-haired man in his sixties, has been with Philips all of his working life and spends at least 40 percent of his time abroad, most of it at Philips's California subsidiary, Signetics. An engineer by training and an accomplished cellist by preference, his job entails so much travel, he bemoaned, he now has little if any time to play to play the chamber music he loves. Indeed, he confessed, he once played in a Gypsy trio, but now his far-flung duties keep him too busy. But in a sense Krijgsman is living a musical metaphor, orchestrating the far-flung Philips semiconductor operations.

Philips is very much like a Japanese company, a vertically integrated, multibillion-dollar electronics firm with worldwide sales of $27.5 billion and more than a third of a million employees. Its semiconductor division ranks seventh in the world, the only European company in the top ten. Total semiconductor sales in 1986 were $1.6 billion, with about 30 percent of its production used in house.[6]

The nation it calls home, the Netherlands, is very much like Japan, a small country with very few natural resources. Unlike the Japanese, however, as soon as the technology permitted, the Dutch left their tiny land and went exploring, then trading, and then colonizing. Today the world is dotted with present and former Dutch colonies that still pay lip service, if not allegiance, to their former masters. Philips in many ways is in microcosm a duplicate of its national heritage, an international company spread across the world, with its roots firmly anchored in

Holland and strong branches throughout the rest of Western Europe, the New World, and Southeast Asia. And most of its overseas branches, like the Dutch colonies that preceded it, take on the coloration of the native land but retain a Dutch flavor, much like the gabled architecture of Amsterdam one finds echoed in the Caribbean islands of the Netherlands Antilles.

"Well, I suppose there is also a historical background for that," explains Krijgsman. "Most of these companies were acquired—Mallard in England, Signetics in the United States, Valvo in Germany, TSMC in Taiwan, and others. We have coordinated the activities of those companies rather than taking a central role and saying you do this and you do that. We have learned, I think, in the last two decades to make use of the local cultures and to have a regard or understanding, or whatever you want to call it, for the local cultures more than anybody else."

Philips has three core businesses functioning in sixty-four countries around the world. "One is consumer electronics, one is telecommunication and communication technology, and one is components," explains Krijgsman. "Now we at Philips are the biggest component supplier in the world. That ranges from picture tubes to resistors to materials such as ceramics and ferrites to professional tubes, oscilloscopes for medical applications, to transistors and integrated circuits. And I think it is important to know that as a component firm we are approximately a $4 1/2 billion company, apart from the fact that we have 35 percent of MEC Electronics, which is a joint venture of Matsushita and Philips . . . and that's another big component company."

As a producer of semiconductors, Philips is a major force in world markets. "We produce about 1 1/2 billion ICs a year worldwide and assemble them in the Far East in Taiwan, Malaysia, in Bangkok, Thailand, and in Seoul, Korea."

Philips uses its worldwide positioning to improve its manufacturing technology and techniques by comparing the yields of its individual plants in different countries. "More and more we are finding it a useful way of keeping each other on our toes," claims Krijgsman. "We have identical reporting structures in terms of what we produce where, and we compare those statistics. If there's seen an improvement in one of our fabs, in Signetics, compared with what we're doing here, for example, we ship people over to California to learn how they've done it. It's a very good system."[7]

But despite such cross-fertilization of ideas and a major $375-million-a-year research program, Philips, like most European companies, has lagged behind the American and Japanese competition. "There is no question about the fact that from a technology point of view Europe has lagged behind," admits Krijgsman. "One of the main reasons for that, I suppose, is that you found that the semiconductor initiative was very much taken by the data processing industry in the United States, and the data processing industry in Europe hardly existed. And therefore the results of what happened in the United States were utilized here in Europe.

"What has been in my view most interesting is that in 1965—that time frame—the Japanese started to make a real effort in the data processing area. MITI started to put a number of companies together at that time, and that was, of course, the first move. So they sort of grew up, up, up. . . . Grew up a user, all right? And after that they started projects to develop the ICs they needed to support those users.

"Well, we didn't have that stimulus, Europe was consumer driven at that time, primarily. So we didn't have the tremendous amounts of money that went into industry to develop these new technologies."

That is changing, however, and there is a distinct recognition not only that research investments must be made—for Philips and all the major European semiconductor companies have always invested heavily in research—but that it must be done cooperatively. Perhaps the most important of those cooperative ventures is the Mega Project, a half-billion-dollar joint R & D arrangement between Philips and the German semiconductor maker Siemens and the Dutch and German governments.

The Dutch end of the project is based in the town of Nijmegen, an hour from Eindhoven. I was driven there with David Heard, an Englishman with a well-trimmed little beard, who is the head of industrial planning under Krijgsman. A physicist, he has been with Philips for more than twenty years, starting with Mallard, Philips's British semiconductor subsidiary. His decisions determine Philips's future policy in terms of plant investment. "I have a coordinating role on long-term investments," he explained to me on the drive to Nijmegen, "where we put up new fabs, where we close down fabs, how much money we're going to spend, and relate that to our long-term capacity issues, and that's not just fabs; that's assembly facilities and so on. But basically I have to be satisfied that if we're going for a new investment plan, that that's justified in relation to the worldwide business."[8]

So ultimately the money Philips will invest in producing the 1-megabit SRAM (standard random access memory), the end product of the Mega Project, will be in large measure based on David Heard's assessment of the market for it.

In Nijmegen, Philips runs a huge research complex called the Department of Industrial Technology and Mechanization. It is another typical semiconductor complex of low, functional buildings housing 3,000 design and production engineers who work to keep Philips on top of the newest technology. Here, too, is their newest fab, housed in a blue corrugated-steel building, known locally as the "Cathedral," for the giant air ducts that frame its exterior walls look like the vaulting arches of a Gothic cathedral. Inside the Cathedral the newest Philips IC, the 1-megabit SRAM, the end product of the Mega Project, is being run in limited test quantities. In a large office block that overlooks the Cathedral sits Paul De Ruwe, the head of the Mega Project for Philips. De Ruwe is quite tall, at least a couple of inches over six feet, a pleasant man in his early forties with small gold-rimmed granny glasses that give him an aloof look that is belied by his outgoing manner.

"I have two responsibilities," he told me by way of introduction, "being the development head or manager for wafer fab manufacturing. And I am project section leader in the Mega Project. The project goes back to the early eighties, and I missed the real start of it. That took place when the European Economic Community [EEC] was looking for opportunities to strengthen the European semiconductor industry. The Japanese and the Americans were doing much better in solid state than we were in Europe. The European community and large companies like Philips and Siemens were looking for a serious development effort partly subsidized or funded by the government."[9]

"Philips was not in the dynamic RAM or the static RAM business at that stage," Heard chipped in. "There's clearly an advantage in having bulk memory as a technology driver. What we need is to use it for all our technology. We need it for all our activities; we need something that forces our technology and enables us to understand the advanced technologies, and both are ideal for that. So we saw two advantages: (a) the volume of business that is to be had in memory and (b) learning to use the advanced technology in DRAMs."

For Philips to actually come to a decision and move ahead, especially in cooperation with another, rival company was something of a departure from their operating style. "There's an old joke that copper wire was invented by two Philips purchasing agents fighting over a penny," says Dr. Sheldon Weining, laughing. "You have to understand the Dutch. They're capable, they're bright, they're hard-working. But they are trying to run this business as if you were planning a twenty-year cycle of a rubber plantation in the Dutch East Indies. The Dutch do it very well, but somehow, by the time they finish the plan, the train has left the station."

In the case of the 1-megabit chip, however, the Dutch were determined to be on board if not first at least along with everyone else. And in the spirit of cooperation that was sweeping the European community, plus the hefty price tag of more than $1 billion that had been placed on the project, Philips and the Dutch government looked around for a partner. And there, just across the border, was a rich company, Siemens, and an even richer country, West Germany.

It was a natural partnership. "With Siemens there was a working relationship already, long before the project started," explained De Ruwe. "Siemens is a competitor of Philips in the marketplace and in product ranges, but on the technical side, the people involved in technology development and engineering, even in manufacturing itself, knew each other already for ten or fifteen years. When I personally look back on my relations with Siemens, I know those guys from the early seventies onwards. We are almost neighbors; there's only five hundred or six hundred kilometers' distance between Philips and Siemens. And in the technical side of our business we had a lot of things in common. We worked with the same suppliers; we were faced with the same problems in our factories. We met each other in seminars and conferences, and that created at the technical level a spirit of common understanding and, I would almost say, cooperation already before the Mega Project even started."

The two companies and the two governments quickly came up with a working arrangement. "Both companies have almost copied each other's organization on paper for the project," points out De Ruwe.

"Each has a project manager in charge; he has a steering board or strategy board, at both ends the boss of the companies. They both had a project office, with some planning and administrative guys. And on both sides we have cut the project in four parts, the design, technology, CAD development, and manufacturing engineering. And this blueprint existed in both companies."

One of the first problems to be addressed was how to avoid duplications of effort. "In the definition phase we gave a lot of thought as to how to split the task between the two companies," acknowledged De Ruwe. "How to balance the work load, how to avoid double work. And most of that was pretty well defined at the end of '84. We have avoided double work as far as possible with two companies who are finally competitors with each other. In lithography, for instance, there was a split of work made. Siemens was working on "D," line and we were working on "I" line—those are different ultraviolet wavelengths. So we took the 340 nanometer wavelength and Siemens took 560 nanometers. We both did our studies and exchanged the results, and we may both end up with "I" line lithography. It worked that way in everything. After the companies did experiments in all phases, both in their own areas, we compared the results and took the best approach."

Information was constantly exchanged between the two companies. Tasks were split and the results compared, and finally people were exchanged. Siemens people came to work at Nijmegen, and Philips researchers went to Munich, where Siemens has one of its major research laboratories. Indeed, the only major stumbling block the project encountered was at the very outset, when teams from the two companies and the two nations met for the first time to lay out the ground rules for cooperation.

The first session was held at a conference center in Germany, with sixty or seventy representatives from both companies and almost as many bureaucrats from the two countries in attendance. "The first point on the agenda was the language," recalls the De Ruwe. "There was a suggestion made by the two companies to make English the project language. In Philips it is common practice to use English as the working language, and many people involved in solid state at Siemens have no trouble with that. The German government refused to accept English reports, however. They wanted all the project proposals, progress reports, whatever, in German.

"The Dutch government people then said, "If that's the case, we want the reports to be in Dutch." The Germans complained. They said, 'That can never happen, because we cannot then read the Dutch reports, and we want to know what's going on in the Dutch part of the project.' So the Dutch government representative said, 'Well, I can't read German,' but nobody believed him. That was a dispute that lasted for two hours. Finally, it was agreed to make the project bilingual. It would be English for Philips; all our documentation and all our reports and proposals would be in English. The Germans would read it without being translated into German. And all the German reports would be in German and be accepted by the Dutch without translation. It was more of a laugh than a hindrance to the project. I have been in meetings several times where every speaker asked before his presentation what language was wanted and most of the German papers were delivered in English."

Far more difficult was the 1-micron barrier. To produce a 1-megabit chip, wafer steppers capable of printing designs with features of less than 1 micron would be needed. "Both companies had no experience with processes below 1-micron resolution," admitted De Ruwe. "That was the biggest challenge, like breaking the speed of sound in the early fifties."

To develop the new lithographic tools, Philips purchased a 50 percent interest in ASM, a small Dutch company. Siemens turned its development effort over to Canon, and both companies then took the project along separate tracks. "From the start, both companies recognized that we had to build new factories for this product," continued De Ruwe. "We have built several in Nijmegen, and Siemens built in Regensburg. We started out with the idea to keep those two factories, which were built almost in the same time frame, as identical as possible to allow exchange of know-how, and so forth. From the beginning, we recognized that our manufacturing procedures and the computerized factory control systems were totally different in the two companies. Cultural differences, procedural differences—it may well have to do with the German difference in philosophy on how to run a factory. We have struggled with it until today, and we still have problems.

"For instance, Siemens has designed the Regensburg fab to be a memory-production fab. Almost a one-product, one-technology plant. We from the start always said we'll never do so. At Philips we have a culture of having thick catalogs of many products to serve as large a portion of the market as possible. And that's quite a difference in your factory. Memory was selected as a technology carrier, but we had in mind from the start that this technology should serve more products than only memory. From the start, Siemens was somewhat different. They looked upon themselves as a one-product, one-technology manufacturer. You need much less flexibility when you run only one product with one technology. There is much less need for changeover of processes, changeover of recipes, and so on.

"Volume," declared David Heard, "is the key in that school of thought. The more you produce, the lower the cost, and that enables you to try and get a stake in the worldwide market. At least that is the concept. And you could argue that Siemens had that in mind. On the other hand, Philips believes there is a much bigger market than just RAMs. RAMs will always be a dirty business to be in, and we have a responsibility to serve our market on a much broader base. We don't necessarily believe that our future lies in a RAM basket. We want it for its many advantages. And I think that's the different competitive structure. We believe on a return on investment from a broad product mix. Siemens believes in volume as their means of return."

And so, to make essentially the same product, a 1-megabit RAM, Siemens and Philips have built two very different plants. "We have built a factory here that by design and characteristics lends itself very well to diversity of products," explains Heard, "unlike the plant Siemens has built in Regensburg. The original agreement, as far as I understood the intentions, was that Philips and Siemens would not join forces in the marketplace with this project. After the technology development was completed, the idea was that we would split up and each would do business in his own way."

And that, of course, is precisely what happened. But Siemens was not content to stop with their own development of a 1-megabit chip. While Philips limited its development to a 1-megabit static RAM, Siemens is pushing the Mega Project toward development of a 4-megabit dynamic RAM. At the same time, Siemens has purchased from Toshiba its manufacturing technology for the 1-megabit DRAM. All of this scrambling on the part of one of the smaller European semiconductor producers is aimed at independence.

Siemens is a $30 billion company with worldwide sales and

factories in thirty-five countries and, after Philips, is Europe's largest electronics producer. Its semiconductor division accounts for $595 million in sales, putting them well back in the pack, far short of the top ten. Siemens also acquired 11 percent of AMD more than ten years ago, but that is purely an equity arrangement, and little in the way of technological exchange takes place between the two companies.[10]

"The philosophy of Siemens is that we are not a semiconductor house," Dr. Helmuth Murrmann patiently explained to me at dinner in Munich. Murrmann is a tall, slightly stooped physicist who heads Siemens's IC development. Dark and balding, he is a native of Bavaria and grew up in the small town of Starnberg, about twenty-five kilometers south of Munich. His father was the warden of the local jail. "So I really grew up behind iron curtains," he jokes. He joined Siemens in 1964; after earning his doctorate at the University of Munich and teaching there for three years, Murrmann joined Siemens in 1964. "It is a company that serves the total electronic and electrical market," he went on. "Power plants and telecommunications are our main products. The sales of semiconductors and of components, active and passive, are only 5 percent of the total sales of Siemens. So semiconductors are not the main target of Siemens saleswise. But it is a big target for Siemens to have the semiconductors available internally that are necessary for strategic projects and large projects such as telecommunications and computers."

Like many other systems producers who make chips, it is a means of guaranteeing supply. Thus, about 20 percent of Siemen's semiconductor production is used in house. "We do not wish to become dependent upon others for the strategic components," declares Murrmann emphatically. "So if one extrapolates, it becomes not only memory but high-level integration of logic products, especially the CMOS technology. And bipolar technology, as far as computers are concerned. Siemens also produces microprocessors under license, but we are also developing our own designs, so we may become self-sufficient.

"But to do this we must have access to technologies where you can realize such high-level integration chips. And this is CMOS technology, and this is the technology of the 1.5- and 1-micron and submicron chips.

"Our R & D budget is at this time very high, too high, and it's more than 25 percent of total sales. That comes from the fact that Siemens has recognized a time gap, especially in our technology. And Siemens has undertaken a large project called the Mega Project in 1985. It has three target areas: The 1-megabit DRAM, second is the logic associated with CMOS technologies, and third is the 4-megabit DRAM. And the target is to decrease the time gap as compared to the leading chip producers, and these are the Japanese. To close that time gap from maybe two years, when we started, to less than one year.

"You can ask wouldn't it have been wiser for us to try and leapfrog to the 16-megabit device. My experience tells me it is not possible to really leapfrog from one generation of technology and say, 'I am going to go two ahead,' when you don't have the first one fully down. You must go step by step, and if you are behind, you must try to catch up and decrease the time gap from generation to generation. I think success in technology generations is getting more and more difficult. What you could do in the past was to just shrink the dimensions a little bit going from one generation to the other. That has been the history of the seventies and the eighties.

"Now, coming to the age of 1 megabit, 4 megabit, and soon the 16 megabit, it's much more difficult. You have to pay attention to three-dimensional effects, to parasitic effects, to high-electron-velocity effects. And this makes it much more difficult to design a technology.

"This is one side. The other side is that the requirements for the production environment are getting more difficult. It's not only that you are much more sensitive to the defect level; the fact is the fatal defect is getting smaller and smaller.

"Secondly, is the problems in the production environment. There are many more sources of contamination that have to be checked than previously. This means chemicals and gases as well. You have to control contamination levels in the parts-per-billion level, not in the parts per million, as you had to do previously. And third is the material itself. The quality of the silicon and the control of the quality of the silicon are making it more and more difficult to understand and then difficult to use. These are the major reasons why I think it will be impossible to skip one generation, especially now."

To produce its new DRAMs, Siemens has built a new factory at Regensburg, about ninety miles east of Munich. In addition, there are plants in Munich and Valach, Austria, and a small wafer fab in Silicon

Valley. There is also a major assembly plant in Singapore. "The Austrian facility and Regensburg have both front end and back end," explains Murrmann, meaning they do not only wafer fabrication but assembly and testing. "But," he continues, "there is a strong tendency to make most of the back end offshore in the Far East."

One reason for the continued reliance on Far East assembly on the part of virtually all Western producers is labor. "I think what the Europeans have to learn is to take up some attitudes of production culture the Japanese have," concedes Murrmann. "We have done that by running our Regensburg production line seven days a week in continuous production. That's not very easy with all the labor regulations, union regulations, and federal regulations in this country. This took us a year-long battle to get the permission to do that. You need the social environment to be able to do that.

"Regensburg is at the eastern edge of Germany, so you can really get enough workers for seven-day production. We tried to do this in two pilot lines in Munich, but it is very difficult in the environment of Munich to get sufficient workers that will work on seven-day lines. The market for workers is very tight here. It is much overcrowded with foreign workers—Greek, Turkish, Yugoslavian. In our production lines in Munich we have maybe an 80 percent ratio of foreign workers. They work sometimes better than German workers, but they are not a reliable work basis. We still have the problem of language, and you have friction between Greek and Turkish workers. There are so many factors that make it much more difficult in a city like Munich, as compared to Regensburg, or in Austria, where we have a plant.

"So the problems of getting a three-shift production line is, first, federal regulations, and second, the worker's education is not at a very high level. But we make an effort to educate them, to train them. We would hire the average German, but he is not interested to work in semiconductor production. He is doing anything else but production. So production is still looked down upon. And this is also true for engineers. And you can only run a three-shift operation with engineers. If you ask an engineer with a university education where he wants to work, first is the university, second is the research lab, then last is production. Everybody wants to be a designer, a research engineer.

"We are trying for years to change this attitude. We are trying to change that by paying the production engineer a higher salary than a research engineer. We are trying to go that way and transfer people in a sufficiently early stage from research engineering into production.

"To some extent, the old German work ethic is still intact. But what happens is there is an extremely high working attitude at the top level and middle management in Germany, in Europe and the United States, but it decreases steadily as you go down to the worker level. But in Japan just the opposite is true. A top manager in Japan still has time to sit down and think about strategic items. And the farther you go down, the harder they work. And this is another expression of the fact that these guys really concentrate on production and productivity. We are trying, and I mean the U.S. and Europe are trying, to struggle with the management level and are forgetting about productivity. We should try to have skilled people with an attitude for work and backed up by engineers that really don't care that they must do overtime and shift work and then put some reasonable effort into strategic management operations. And this is where I think we have to change attitudes. And the only way to change these attitudes, the only possibility now, is by paying more on the engineering level. If you go back to the university or the secondary-school level, there is still the picture of high-level engineers in research and design, not on the production line. I think this is one of the key issues in determining whether the U.S. and Europe will survive, especially in this crazy field of semiconductors."

Despite his belief that Europe must adopt Japanese production techniques, Murrman does not believe the Europeans are at a disadvantage. "Productivity in Europe has not dropped," he insists, "but I think the understanding of the need for increasing productivity is becoming increasingly important. And we put very much emphasis on increasing productivity and combining it with a technology strategy that is tailored for our needs for the products that the vertically integrated company needs. And all this is combined with the financial power of a vertically integrated company. Siemens and Philips, I think, will be two companies that will play a role not only in Europe but in the semiconductor business worldwide."[11]

A third European company that will make its presence felt in world markets is the newly formed Italian-French combine of SGS-Thomson. In the town of Agrate Brianza, about twenty-five minutes by autostrada from downtown Milan, within sight of the looming Alps, is SGS's corporate headquarters and brand-new Buck Rogers–style VLSI Research Laboratory. And, like Philips and every other semiconductor company, there, outside the flag-draped courtyard that led to the new lab, was the gate house. But this is also Milan, where all the verve and fashion that characterize the city are also on display at SGS. No rent-a-cop security guard, no middle-aged divorcée as receptionist, here. Rather, a young woman who looked as if she were waiting for a casting call from *Cinecitté*, Rome's famed movie studio, took my name but unfortunately showed no interest in my telephone number. I was given a badge and taken in to see Carlo Ottaviani, the director of communications for SGS. Carlo, whose short beard is of the type that has adorned the faces of Lombardy dukes for centuries, is charming, expansive, and informative.

"SGS," he explained, "was created in 1957–58 by two companies, Olivetti and Telettra. Telettra (Telepennia Elettronica Radio S.R.A.), a telecommunications company, is now part of the Fiat group. At that time, those two companies understood the importance of semiconductors and decided to hire some technicians who were working already in a garage in Agrate, where SGS is now located. They played with germanium technology from GE, and then they understood that the solution was silicon. In 1960, Fairchild came in, and after having given some technology, they became a shareholder. So one-third of the company was Fairchild, one-third Olivetti, one-third Telettra. Problems arose in 1968; there was a big fight between the European and American shareholders. The Americans said R & D in America is quite enough; you don't need to do it in Italy. The Europeans said the products needed for America were not suited for the needs of the market in Europe. Those were consumer electronics and telecommunications rather than the American space and missile program.

"SGS at that time was in the process of establishing a real presence all over Europe, with factories, with design centers. They had a factory here, of course, one in Germany, one in France, in England, in Sweden, and that was a big problem. There were too many factories. The idea was to serve each local market in Europe with a local factory and a local design. This was absolutely too much for the size of the European market. Sweden, for example, is a country of 8 million, and although there is Ericcson, with all its military industry, still, it is not enough to support a factory.

"The company always has been a merchant company aiming at the outside market. Olivetti did not use it as a captive source of chips but simply wanted it to have a foothold in the semiconductor industry. They were just customers of SGS like anybody else. We worked in close cooperation with them, and even today Olivetti is linked in a very tight manner with SGS. But this is because they are close; they are Italian. At that time, system feedback from Olivetti and Telettra to SGS was more important than it is today. Everybody was trying to grow up in current, in density, and we knew that anything we produced would be useful to our parent companies.

"In this whole process, the main issue was we need R & D right here; we want it centered here in Agrate. The answer from America was no. At that point they broke apart, but the technological ties and alliances remained intact. Fairchild in essence said okay, you do what you want, but not with our money. In 1969, Telettra was bought by the Fiat group and abandoned its piece of SGS. That left Olivetti as the only shareholder in SGS. At that time, we were in a big expansion. It was the culmination of our construction all over Europe, and we had already built our first assembly plant in Singapore.

"So they spent a lot of money on expansion, with doubtful returns, because the market was not as big as the company thought. Olivetti was having problems then themselves with their big computers and their transition phase between electromechanical typewriters to electronic, which was a painful situation. And there was a break in the semiconductor market in 1970–71. The fact was Olivetti didn't have the money to support SGS."[12]

Fearful of losing a strategic industry, the Italian government intervened. It was not an entirely new role for them. In 1960 the government had purchased a small company called ATES which had been making electron tubes under license from RCA and Marconi. They had switched from tubes to chips and were trying to sell outside of Italy, but without much success, according to Ottaviani. Somehow they had been acquired by the Societa Finanziara Telefonica, a government-owned financial holding company that translates into the Italian acronym STET.

"In the 1930s," explained Ottaviani, "the Italian government created an Agency for Industrial Reconstruction, to save the economy from the depression. The government had taken over a couple of big banks that had gone bankrupt. Around them, after the war, they built up a system designed to help business. It's now very big. I think they hold 30 percent of the shares on the Italian stock exchange. This agency reports to the Ministry for State Participation. It operates like a financial holding company, but it is totally controlled by the ministry. It is further broken down into several agencies, one for mechanics, one for steel, one for banks, and one for telecommunications and electronics, STET."

STET is in essence a government-owned holding company, and its shares are freely traded on the Italian stock exchange. In 1986 its assets were valued at $9.6 billion. Most of its shares are held by yet another superagency, IRI [Institute per la Ruricostucione Industriale], which Ottaviani described to me as a superholding company under the ministry. Although IRI owns the bulk of STET, it shares the ownership with more than 133,000 individual shareholders.

In any event, now that I had the complex financial and ministerial relationships almost sorted out, Ottaviani continued the SGS saga. "STET already owned ATES. When it bought SGS from Olivetti in 1972, they merged the two companies together. A new management team, mainly from ATES, took over. The management of ATES had much smaller ambitions than the management of SGS that they had replaced. And the mission they gave the company was different—much more focused on Italy, less on Europe, not to say the world, because America was too far away. Think small not big was their strategy. Profit was okay but not necessary. They felt that in this industry we would always lose money. That was the mentality. That was not something special for SGS-ATES, it was very common thinking throughout Europe. Every European company felt that semiconductors were something you must do because it represents the future, but you cannot make money.

"As a result of this strategy, the industry focus in Europe was to learn the technology, not necessarily the manufacturing and marketing. Take France; they got out of NATO and developed their own atomic force. They needed the electronics to deliver nuclear weapons. So they absolutely needed the semiconductor technology to support their nuclear force. In that case, who cares whether the transistor used costs $1,000 or 50 cents. You don't need a real manufacturing structure; you needed only to produce the chips you wanted, and the cost was not important.

"The same was true for SGS. Although we had begun with MOS in 1966, we were not able to manufacture it in quantity. But then, when we were able to produce one chip with that technology, our engineers were satisfied. They were always thinking how to upgrade it and make something better. They never thought of how to make money with the generation you had in hand. They never though of producing this in quantity and selling it to the world. Europe had become 9 percent of the world market, and we couldn't produce enough for it. And we were ignoring America, which at that time was 50 percent of the market and is still not far from that.

"So we were restricting our activities to Europe, but mainly to Italy, using very good technology. We invented the isoplanar technology the same time as Philips, we have a lot of patents, we are very good, and technology was never a problem. In one single area, the power ICs, we dominate totally. In 1968 we came out with the first audio amplifier in the world. And in chip technology and packaging of power ICs we were the leaders. But, like many Europeans, we were not able to capitalize on it. Europe has invented a lot of things, from the cassette recorder to the compact disk. Who makes money on them, though, the Europeans or the Japanese?"

SGS was certainly not making money. "Our last year of profitability was 1969. For the next fourteen years, the company lost money and by 1980 had lost 25 percent of its sales. Remember that Italy never had a government plan for funding R & D. In Europe there are huge government contributions to R & D, but not Italy."

At the same time, Italy was being overtaken by political ferment that threatened the nation's continued existence as a democracy. "These were the years of the leftist struggle," recalls Ottaviani. "So the idea was to keep people as calm as possible. Don't make waves or we will have a revolution one of these days. We almost had one in '68. So the idea was to keep up employment, hire people, it doesn't matter whether it's viable or not, do it. So we did it. The losses we suffered were very high. And this went on until 1980.

"You have to see it in the context of the environment. By 1980 we had the kind of economy which was more similar to Yugoslavia than to the U.S. We were not Communist, but of course there was a strong Communist party here and the whole economy was bureaucratic."

Then the president of SGS suddenly died in May 1980. STET

looked for a replacement, and they thought of Pasquale Pistorio. Pistorio was the vice-president of international operations for Motorola, in charge of all their manufacturing and marketing operations outside the United States. He was known throughout the industry as a brilliant executive, and what was more, he was Italian.

Pasquale Pistorio was born in 1936 in a small village in one of the poorest hardscrabble regions of Sicily, the poorest region in Italy. His father, a minor government official, managed an agricultural cooperative. There were two locations where the co-op accepted crops. Every harvest, Pasquale and his brother missed school in order to be on hand at one of the co-op locations in order to accept crops for their father. He still talks about carrying big bags of grain from the wagons to the warehouse. The experience still shows in his massive shoulders and arms.

Still, Pistorio managed to finish high school in Catania. He then went on to study electrical engineering at the Polytechnique of Turin. As in Sicily, he was forced to work while going to school and so took eight years rather than the usual five to get a degree. Pistorio, who has made a career of surprising people, turned down an engineering job at Olivetti to become a salesman for an Italian representative of a Motorola distributor. His reason: The job paid twenty dollars a week more than the Olivetti position. With no money for a car, he made his rounds by bicycle, taking care to remove the clips from his pants before entering a customer's office.

Pistorio was a crack salesman and soon moved to Milan. Then Motorola decided to open its own offices in Italy, and he was hired as a salesman for them. A year and half later, he was named the area manager for Italy and very soon after that was called to Geneva to become the marketing manager for Europe. Then he became the general manager for Europe. In 1977, Pasquale Pistorio was made a corporate vice-president of Motorola, the first non-American to hold that job, and moved to Phoenix as the director of world marketing.

In 1980, he was offered the job as head of SGS-ATES, and he refused. He and his family loved America and the Arizona life-style. His three children were in American public schools and were loathe to leave. But the enormous challenge and the lure of a return to his roots were great attractions. On a trip to Italy he met with the STET people. Pistorio laid down his terms in no uncertain terms. "I come from a purely free-trade company," he said, "and I want to operate the same way. I want full internationalization of the company, manufacturing, marketing, design. If these conditions can be met, I will run the company."[13]

STET agreed, and Pasquale Pistorio, taking a whacking 40 percent pay cut, became the new president of SGS-ATES. Pistorio came to a company that was twenty-third among the world's chip producers. The most glaring problem was losses, at the time equal to 25 percent of sales—close to $30 million a year. Outside Europe SGS sales and manufacturing were almost negligible. "However," points out Otta-viani, "the technology was good; the existence of a production base in Southeast Asia was a strong point. The weak point was a lack of marketing and manufacturing know-how. Pistorio changed it all and turned SGS from a top-heavy bureaucratic organization into an efficient one. He changed one-third of the top managers, replacing six of eighteen. "The population of scientists and managers who knew they were good but were not able to make it work were encouraged," recalls Ottaviani. "Charisma and hard work effected the change. We were waiting for guidance. Italy is a collection of different regions, city-states. This area, which is called Brianza, produces people who are tougher than the Germans and have far more imagination."

Pistorio, a workaholic, came in like an exploding Vesuvius. The SGS managers, used to a more leisurely style where the boss arrived at 9:30 A.M., suddenly found themselves chasing a man who came in at 6:30 in the morning. "He used to read the papers," says Ottaviani, "and wanted the *Wall Street Journal* when he got in, but that was not possible; it didn't arrive in Milan that early. At 7:00 A.M. he had meetings with managers: he had a one-on-one for an hour every day. Then he worked until 10:00 P.M. every day with no lunch.

"Now he's slowed down, arrives at 7:30 and leaves at 8:00 or 9:00 P.M. If he has dinner out, say, in Milan, afterwards he drives back through the plant. When he found somebody not working hard enough, we volunteered to have one manager present every night. So each day one worked from 8:30 A.M. until 6:00 the next morning. Three hours later, you were back at work. We are no longer doing that, because we now have three shifts and many people working late, and so managerial supervision is normally there now.

"A mission was fixed and approved by the shareholders," explained

Ottaviani. "It was to become a broad-based supplier to serve the needs of our microeconomic environment. Second was social, a need to build up strong, productive jobs in each country where we were."

At the time Pistorio took over, the Italian government was using SGS more as a social-pressure release valve than as a profit-making organization, and so there were hundreds of excess workers, many of whom were absent more than half the time. Pistorio promptly fired a dozen of them and cut back the work force by more than a thousand through attrition and early retirements. Most of the lost jobs were in Catania, the capital of Sicily, where passions run deep and custom governs company hiring and firing policies far more than the realities of the economy. There were about eighteen hundred people working in the Catania plant, most of them poorly educated, manual laborers. Pistorio wanted to move virtually all such labor-intensive operations such as assembly to low-wage Singapore, where SGS already had a plant. He also wanted to add night and Saturday shifts in Sicily to meet production schedules.

The immediate reactions were predictable and violent. The unions struck, managers grumbled, but Pistorio held firm. Italy, too, was undergoing a major political catharsis. In October 1980 the innumerable strikes that had plagued the country throughout the seventies erupted in a major strike at the Fiat auto plant in Turin. The strikers occupied the plant and threatened to destroy it. The next day, forty thousand other workers marched through Turin, demanding the strikers leave the plant and that everyone return to work.

"That is recognized officially as the turning point," eclares Ottaviani. "At that point, the attitude of the government also changed. There was always a five-party coalition running the government, and the minister of state participation at that time was Gianni de Michaelis." De Michaelis was young, thirty-five years old, by Italian standards a baby. He flew to Catania, where, to everyone's surprise, he approved Pistorio's plan.

"It said we will stop hiring; we will put some people in *casse integrazioni,* which is the Italian way of not firing people. They are still a member of the company," explains Ottaviani. "Nobody is laid off, but they no longer get a full salary. The other thing they did was to increase the production schedule. We were working, as was all of Europe, only 1,500 hours a year, compared with 2,000 hours in America and 2,200 in Japan and Southeast Asia. We at that time worked only one or two shifts; three shifts you didn't even dream of. Female workers were not allowed to work at night, so how can you compete on the world market?"

The Sicilians and Italians throughout SGS accepted the reasoning. The company was slimmed down, trimming a thousand employees worldwide from the payroll. To make the plants more productive, a night shift was added. SGS was the first company in Italy to employ women at night. It was an enormous cultural change, especially in Catania, where just a few years before only prostitutes worked at night. Pistorio introduced a purely American management style, with slogans and posters on the wall. Everyone at SGS was made aware of quality. Each production worker was issued a plastic card, about the size of a credit card, printed with manufacturing standards that had to be met. Everyone was required to have the card with him at all times. A Japanese approach to quality control was instituted, with quality circles and statistical charts on all machinery. "It was certainly outside European standards and Italian rules," admits Ottaviani. "We don't have a salesman-of-the-week parking place, but it's about the only thing we don't have. We have a company song; all this symbolic management was adopted, and people like it."

When the plan was first drawn up in 1980, STET was hoping to return to profitability after five years. Pistorio said he would do it in four years, and in fact, SGS was profitable within three years of his takeover. "So we were profitable in 1983," says Ottaviani with justifiable pride, "after fourteen years of losses. We were also profitable in 1984, but we lost money, as did everybody else, in '85 and '86. Operating losses in 1985 and '86 were 3 percent of sales, compared with a world industry loss of 9 percent. So we lost one-third less than the average of the industry. Nobody is happy to lose money, of course, but that was quite acceptable.

"The idea was to change the weak points, which we did. We wanted to be broad range, which meant we needed to be big. To be big, you cannot serve only the market where you live, which is Europe; you must sell elsewhere, and that is America. America is the first target we had in mind. It's a free market, where you can succeed, provided you are good. So we planned an attack on the outside market, beginning with America but not neglecting Asia. The idea was first a commercial sales attack, second, the implantation of design centers close to the markets to serve their needs better, and third, manufacturing in those same markets. This is exactly what we did. We started selling more, hiring people in America, to build up a big network; second, we set up a big design center in Phoenix, Arizona; third, we started manufacturing in America. We built a factory in Phoenix, which unfortunately never took off."

It was a leaner, more aggressive SGS that Pistorio had built, but he was not satisfied that it was a company that could continue to compete. Pistorio was convinced that only a large company, one that had achieved a critical mass, could compete on the world scene in the 1990s. To Pistorio that critical mass is $1 billion. "The nineties," he believes "will see the disappearance of all the companies that are less than $1 billion and more than $300–400 million. One billion dollars is the magic barrier, below which you can only be a small specialty house. And by definition, because of our mission, we don't want to be a specialist; we want to be a broad-range supplier. You can probably be more profitable for a short time being a specialist, but you have to shrink to do it. Our object is to grow. The technological objective was to be broad range, because otherwise you don't serve the needs of the shareholders."

While Pistorio was leading SGS through its period of renaissance, the nations of Europe were looking for ways to cooperate with each other in many areas of high technology. In 1984 the EEC came up with a joint research program designed to shrink the technology gap that was growing between Europe and the United States and Japan. Called ESPRIT, for European Strategic Program in Information Technology, it bought together twelve of Europe's largest electronics and semiconductor companies to research subjects considered essential for the survival of Europe's information industries. A year later, in 1985, the EEC developed EUREKA as a means of countering the commercial advantages U.S. companies would gain from their research participation in the American strategic defense initiative, the Star Wars program.

EUREKA is a huge research effort encompassing 165 different projects and funded at $5.8 billion. Fully one-quarter of the budget is aimed at semiconductor and electronics research.[14] It was while participating in these two cooperative projects that Pistorio and his management team concluded that the only way SGS could grow was by cooperating very closely with other companies. Soon the idea of cooperation was translated into merger, and the only problem then was to find a suitable partner. The task was given in part to Dr. Enrico Villa, the director of external technological coordination. Villa's job was to be the coordinator for SGS with the EEC and the several companies SGS worked with under the ESPRIT and EUREKA programs. In addition to his twenty years of experience with SGS, Villa brought a talent for balancing opposing ideas and forces rarely seen in an engineer and doctor of economics. For Enrico Villa is in his third term as the mayor of Vilmacati, a small town of 25,000 people just ten kilometers from Agrate. In Vilmacati, the major employer is IBM, and the coalition government, which Villa, a Christian Democrat, heads, is dominated by the communists.

"The general objective of SGS was to become a large supplier," he pointed out to me, "one that could compete in the big leagues in the market. Up to '85 this kind of strategy could be pursued with our natural growth in sales. But due to the change in the market structure, it is becoming more and more difficult to grow. In those first years from '80 to '85, we grew one position per year in our world standings, but it was not enough. So only natural growth would not keep us in the big league. So then our problem was which is the best kind of partnership to help us grow. You can join with a very advanced company, like a Japanese company, or you can join with a European company with the same problem we had."

There were, in fact, a number of European companies that fit that picture. Plessy in Great Britain was even smaller than SGS and had a smaller range of products. And just across the Italian border was the much larger French electronics firm of Thomson CSF. Like SGS, it was largely government owned, was of a similar size, and had just as many problems.

"Both companies understood that they had little possibility of becoming larger independently. We had been cooperating with France on a number of programs, anyway, such as ESPRIT. We had been cooperating in lithography, in advanced steps of the process, but not in general strategy. So it was a natural partnership."[15]

Alain Gomez, the chairman of Thomson, agreed. "By the first half of the 1990s," he said, explaining his thinking on the merger, "the number of layers in the world electronics industry is going to diminish substantially. Our problem is to remain at the table as one of the players."

"What was also clear," adds Villa, "was that Thomson's strategy had also changed in 1982 and '83 with the arrival of Gomez at the semiconductor division. Their plan was almost identical to SGS's. And the numbers were the same. What was valid for SGS was valid for Thomson. But Thomson was different structurally. It was a collection of many companies, whereas SGS was only two companies. They also had thought of growing by acquisition rather than partnership, but other than that, the strategic thinking was the same.

"Under ordinary circumstances, the decision would be approved automatically by the shareholders. But our shareholders in the STET group wanted to keep SGS strictly Italian to guarantee an Italian source of semiconductors to the electronics companies within the STET group. So the conclusion of the STET group was that the final mission of the company was not only to guarantee a return on investment but also to guarantee a certain strategy to develop the company to remain in electronics. If you merge that strategy with the general picture of the market, it was to find a partner with a similar objective and try to create some synergy between the two. And to be stronger and guarantee both profitability and survival for the future of semiconductors in Italy not only for the shareholders but also for the workers. And the view was that the best partner available was in France, where the problems were quite the same."

The merger almost tripled the product list and enlarged the customer base. "Although some of the main customers are the same," points out Villa, "a lot of different segments are added. We were more strong, for instance, in automotive; we are strong in telecommunications, even in consumer and computers. While Thomson is stronger in industrial and telecommunications, in military it is maybe the third largest producer of military ICs in the world, but that is not part of the deal. In Thomson's strategy, they wanted total control over the military, and there was also the French government concern over foreign interference in their military."

The merger was finalized in April 1987, and Pasquale Pistorio became chairman and chief operating officer of the new company, which was, not unsurprisingly, called SGS-Thomson. Combining SGS and Thomson created the thirteenth-largest semiconductor company in the world, the second largest in Europe. Its combined sales for 1987, the year of the merger, were $800 million. By the end of 1988, with semiconductor sales booming, SGS-Thomson will doubtless go over Pistorio's magical $1 billion target. "We have a very strong presence in Europe in terms of manufacturing and market share," he points out. Thomson brings to the party its manufacturing in the States [Mostek Corporation], and both of us have a good marketing presence. SGS brings its manufacturing base in Asia. This really gives a very interesting combination of complementary manufacturing and marketing, though both of us lack marketing presence in Japan, and this is something we must address. If you want to be a world supplier, you can't ignore any of the three major markets."

Pistorio also believes in the importance of competing in the commodity markets, in DRAMs in particular. "If you are competitive in this particular product," notes Villa, "you can compete in other areas. So it's very important to remain in it not only for the dollars you can generate but for the discipline, attitude, and mentality of the company."

Thomson's Mostek plant in Carrolton, Texas, will be used to turn out DRAMs, especially for specific military uses. The new submicron technologies will be introduced perhaps in 1989 at the now mothballed SGS facility in Phoenix. The new company has aggressive plans. "Next, year," advises Villa, "we expect to produce a 4-megabit EPROM, and we expect to be one of the first to produce the 16-meg EPROM. Today we are doing it in EPROM; tomorrow I hope it will be the DRAM. You cannot be in the big league without those kinds of products. If you want to compete in the world arena, you must be in the advanced state-of-the-art in memories. This is a general condition that applies not only to us but to all companies.

"If you view it as a war, and that is what it is, you can clearly understand the problem is not only the single company; the problem is the system, company plus country. If you are not in a position to manage and to be innovative and to be competitive, you cannot remain at the state of the art."

To gain that sort of capability, SGS-Thomson will maintain a vigorous research program. SGS, in fact, had begun work on a $15 million VLSI research laboratory in Agrate before the merger. It was completed in April 1988, exactly one year after the merger. The VLSI research lab is actually two buildings in one. The central core, which houses pilot fab lines, test centers, and work stations, is surrounded by another building, whose foundations are separate from the central lab core and whose poured concrete walls come within six inches of, but do not touch, the load-bearing walls of the lab. This is to guarantee a vibration-free environment of the inner lab building. Here, where the lines and features of a chip must be etched in lines less than 1 micron in width, the slightest jiggle can ruin the effort. In the outer core are housed the offices of the support staff and engineers. The entry to the building is floored with gleaming marble. Brightly painted yellow walls and red pipe metal stairways lead to the offices on the upper two floors.

The place has a feeling of high-tech elegance, but it is beneath the labs, in the inner buildings, that the truly amazing nature of the building can be fully appreciated. Here, carved out of the rock that underlies the Alps and two stories below the surface, is a new Italian catacombs. One floor below the labs is an amazing world. More than two hundred sheet-metal columns reach up to the floor of the lab. Together they pass 3 million cubic meters of ultraclean, scrubbed, and cooled air to the rooms above. On the floor below, in giant refrigerated rooms, enclosed in stainless-steel walls and doors, are fifty-four giant fans, more than six feet in diameter, that change every cubic foot of air ten times per minute. Atop the labs are the filters, giant gooseneck tubes of sheet metal plugged into the laboratory ceiling. Each tube filters a rectangular space 1.5 meters by .6 meters. Depending on the degree of clean room desired, from class 1,000 to class 1, the tubes are fitted with filters that look exactly like giant slinky toys filled with a special paper. Each filter strains the air of particles down to 0.12 microns, about eight hundred times smaller than the diameter of a human hair. The degree of cleanliness is determined by the number of tubes fitted with filters. For a class 100 condition, say, one in every ten tubes serving a lab space will be fitted with filters; for a class ten, one in every two tubes will have filters, and so on.

To enter the lab, I put on a bunny suit, bootees, and a nylon helmet. Two air showers were required before we could pass into the class 100 area. Although little of the equipment has been installed yet, for the people who work here the rules are strict, but again, this is Italy. While the women are forbidden to wear makesup, the men are allowed to keep their beards, mustaches, and whatever other facial hair may be in style. Smoking breaks are permitted, but no one can return to a clean room for an hour after smoking, and no smoking is permitted in the building.

It was a very impressive tour. I was also impressed by the fact that the building had been designed by an American firm, and virtually all of its semiconductor manufacturing and test equipment was also American. And that speaks directly to what may be the major problem beside size that may always keep the Europeans in third place—an equipment industry.

"Every industry needs a supply business," declares Dr. Sheldon Weinig, the president of MRC. "In the United States the supply business represented by us, MRC, by Perkin Elmer, by Varian, and by a hundred other entrepreneurial firms have all come into existence only since the fifties and sixties. In Japan that supply business didn't get created in the same way; it became part of the zaibatsus. So, NEC owns Anelva, Ulvac is part of Toshiba, and so on. In Europe, it just doesn't happen at all. The entrepreneurs don't exist there any more than they do in Japan. Charlie Brown doesn't get up one morning and say 'Gee I'll start a company.' It doesn't happen that way. The Siemenses, Philipses, the Thomsons do not have that kind of mentality, and they certainly don't see themselves in the equipment business.

"So what they do in typical European style is to take the self-sufficient peasant approach. We'll build a fire, we'll forge the metal and make the machine.' So there's a lot of homemade gear in Europe. And homemade gear is not good enough, because it only has the experience of one customer, the company that built it. And so the first thing we notice is their supply industry doesn't get developed. The entire supply industry of Europe today is either imitation or American in its origin."

It is also growing at a greater pace than ever before. In Zurich, at the SEMICON Europa meeting, where much of the world's semiconductor manufacturing equipment is on display, about half of the six hundred or so exhibitors were European. Not many were in a position to challenge the American or Japanese manufacturers, however, save in a few areas.

"It's a lot like the chicken and the egg," Dataquest's Malcolm Penn told me. "The European semiconductor industry in the seventies was becoming weaker, and therefore it wasn't providing a stimulus for a strong European equipment industry. The equipment industry that was here tended to concentrate on other areas. The Europeans have always been strong in materials, like silicon, for example, and electro-chemicals, gases and things like that. But in terms of actual equipment the European companies were forced to buy American. Now what's happening is because the industry itself is becoming stronger; those same companies want the infrastructure to be there. They want it to be a European source of equipment and products. For example, in lithography equipment, the general consensus is that the Americans are losing the edge to the Japanese. Thus, the Philipses and the Siemenses, the SGS-Thomsons, and so on don't want to be beholden to the Nikons and the other Japanese because they feel that this is a very critical area. Without that equipment you won't have the leadership in the chip design. So Philips, for example, will just deal with ASM as an attempt to build up an indigenous European equipment-manufacturing indus-try, particularly in these critical areas. It doesn't matter quite so much in some other areas."

The Philips investment in ASM Lithography, a manufacturer of photolithographic equipment, has been less than fruitful so far. Faced with fierce competition from American wafer-stepper producers such as GCA, Ultratech Stepper, and Japanese heavyweights such as Canon and Nikon, ASM can't seem to penetrate the market very deeply. In 1987, the company lost $18 million and will probably lose money again in 1988.[16]

But the Europeans are determined to move into the equipment business. "If you look around the show here, you'll see lots of good activity from the European companies now," Penn pointed out. "They're showing leadership position, in fact; in some of the new areas, vapor deposition, for example, is a leadership position, which is a European advance. This hadn't happened in the past. But the industry is strong enough now that it can support this."

Will it be enough to permit Europe to challenge the Japanese and the Americans? It's not likely, but the European semiconductor industry will thrive and carve out a share of the market and will keep them healthy for many years to come. For it is no longer a largely reluctant group disdaining sales and volume production for elegant technology and social welfare.

"I think the Europeans will do very nicely in future for several reasons," asserts Penn. "I think, first of all, that the governments have finally recognized that this is a strategic industry; therefore, they have to support it. The only government that hasn't come to this view is the U.K. But all of the others have realized and recognized the importance of this industry.

"I think that the companies are much more aggressive; they have determined to have leadership position in the electronic equipment, the end-equipment business. They know they must have a strong chip industry to support that. Their eceonomies are the strongest that they've ever been, and I think that from that point of view as well, Europe is going into a new growth phase."

But Europe still faces major catch-up problems. "In my evaluation, Europe is still lagging behind," declares Franco Malerba, "but they are making major structural changes in the form of alliances, which are the only way for them to cope. I see these changes as favorable. But the problems of European semiconductor makers are the problems of scale and size—each player may be too small to compete. Cooperative agreements can help, but they are not the only factor. The second is government policy that must foster inter-European cooperation. ESPRIT and EUREKA helped bring about agreement between SGS and Thomson. It also helped bring together, if not fully integrate, European industry; it made it tighter and better coordinated. It is moving in the right direction despite less than enormous funding. But the governments are still nationalistic and push domestic firms.

"The third factor is international agreements. European industry increased its technological agreements with foreign partners, and that is helping to break their technological isolation. France and Germany always tried to be self-sufficient and so operated for many years in a vacuum. Now a major change has taken place in their attitude toward global competition."

Whether it will be enough to catch up with the enormous lead of the Americans and the Japanese or even to maintain its position ahead of the onrushing Little Dragons remains to be seen.

9

KGB, Chipping Away

The sale to the Soviet Union of giant milling machines by the Toshiba Machine Corporation produced a howl of anguished outrage in the United States. For with the machines, acquired in the early 1980s, the Soviets can manufacture submarine propellers that produce far less noise underwater, making them much more difficult to detect. Eventually, five other companies in Norway, France, Italy, and West Germany were found to have sold to the Russians machinery and computers needed to run the sophisticated milling machines. All of the sales were in violation of U.S. and Western nations export regulations on sales to the Eastern bloc. "Toshiba made $17 million on the sale," bitterly announced Congressman Duncan Hunter, a California Republican, "but it will cost the West $30 billion to regain the superiority that we lost from the sale."[1]

This was the most publicized illegal sale of high technology to the communist bloc but by no means the first. The opening paragraph of a 1985 U.S. Defense Department white paper declared:

In recent years, the United States Government has learned of a massive, well-organized campaign by the Soviet Union to acquire Western technology illegally and legally for its weapons and military equipment projects. Each year Moscow receives thousands of pieces of Western equipment and many tens of thousands of unclassified, classified and proprietary documents as part of this

267

campaign. . . . Targets include defense contractors, manufacturers, foreign trading firms, academic institutions, and electronic data bases.[2]

Announcing release of the white paper at a Madison Avenue–style press conference complete with easel-mounted exhibits and glossy photos, then Secretary of Defense Caspar Weinberger added, "It has become evident that the magnitude of the Soviet's collection effort and their ability to assimilate collected equipment and technology are far greater than was previously believed."[3]

Two years later, the Toshiba scandal underlined just how large an order of magnitude the defense secretary meant. It led his former assistant secretary of defense for international and security policy Richard N. Perle to appear before a congressional committee investigating the Toshiba affair. He told the outraged congressmen:

Far greater than was previously believed is the sort of euphemism to which government officials resort when what they meant to say is this: we had no idea the Soviets were ripping off our technology so skillfully, so comprehensively, so effectively right under our noses. All along, we thought we knew what was going on and then along comes a windfall cache of documents that shows that more than 5,000 Soviet military projects each year have been utilizing our technology. Someone ought to be fired.[4]

Government being what it is, and the U.S. government in particular, no one was fired, of course, but a lot of bleating and vows for revenge, including the absurd sight of half a dozen congressmen smashing Toshiba television sets on the Capital lawn, have followed. It played like a bad television sitcom, but the net effect is that the Soviet military buildup for the past decade or more has actually been subsidized by the United States and its NATO allies. And among the chief and most successful targets for Soviet acquisition are semiconductor and computer technologies.

A 1985 CIA assessment flatly admitted Soviet success, stating:

In certain of the [technical] areas, notably the development of microelectronics, the Soviets would have been incapable of achieving their present technical level without the acquisition of Western technology. . . . This advance comes as a result of over ten years of successful acquisition—through illegal, including clandestine, means—of hundreds of pieces of Western microelectronic equipment worth hundreds of millions of dollars to equip their military-related manufacturing facilities. These acquisitions have permitted the Soviets to systematically build a modern microelectronics industry which will be the critical basis for enhancing the sophistication of future military systems for decades. The acquired equipment and know-how, if combined, could meet 100 percent of the Soviet's high-quality microelectronics needs of military purposes, or 50 percent of all their microelectronic needs.[5]

"There is no question that the Soviet Union and its surrogate intelligence agencies in the bloc do target specific technologies such as semiconductors," I was told by John R. Konfala, director of strategic trade, at the Defense Department's Defense Technology Security Agency (DTSA). One of several organizations charged with keeping America's high-tech secrets out of Soviet hands, DTSA is located in a squat Arlington, Virginia, office building with an unparalleled view of the sprawling south parking lot of the Pentagon across the Potomac.

Konfala, a New Yorker who came to the agency in 1975 after ten years with the CIA, might have studied specifically for the job he now holds. He has a master's degree in Soviet and European area studies and did postgraduate work in political science and Far East–area studies. "We do know that they are actively out there trying to obtain illegally what they cannot obtain legally. There is a long history of this sort of dependency that goes back before the Bolshevik revolution. From the European standpoint there has always existed a symbiotic relationship of selling finished products to Eastern Europe and in return acquiring raw materials. That trade has existed for centuries."[6]

In the late eighteenth century, parliamentary debates in England railed against the transfer of militarily useful technology to Russia. The main focus of concern was the use of coke in iron and steel making, discovered in 1709 by Abraham Darby, a Quaker and pacifist. He believed, correctly, that it would be used for military purposes, and he kept the process a secret for fifty years among a small circle of Quaker ironmongers. After 1750, however, the process spread, and coke became the fuel of choice to make better anchors and naval cannon. The Russians, who used charcoal to make iron, realized that coke was a better process but were unable to master it. To overcome their technology gap, they hired a Scotsman, Charles Gascoigne, to come to Russia with his technicians and coke. He rebuilt the cannon works at Kronstadt to use the newer coke-burning technology.[7]

Now the technology targets for the Russians are far more complex and sophisticated, but so too, are the resources they bring to the problem. "Their resource commitment is enormous by any measure," declares the DOD's 1985 white paper. "It has enabled them in recent years to narrow the Western lead in nearly all key technological areas, particularly microelectronics." The white paper went on to estimate that the Soviets were spending $1.4 billion a year on the acquisition of western technology. A massive bureaucracy involving the KGB, the Russian acronym for the Committee for State Security, the GRU, another Russian acronym for the Intelligence Directorate of the Soviet General Staff, and the Ministry of Trade employs more than five thousand espionage agents, engineers, scientists, and diplomats in the task of gathering by legal and illegal means every piece of Western technology that can be utilized for military purposes.[8]

Each year KGB agents send back to Moscow more than a hundred thousand technological documents and more than ten thousand pieces of equipment from the United States. Soviet technological espionage is a well-organized and well-funded operation that has been so successful that the Russians no longer even attempt to create their own computers. The mainframe workhorse of Soviet science and technology, the RYAD computer, is IBM's old 360 and 370 series, copied right down to the color of their wires.

Many of the newest weapons in the Russian arsenals are virtual duplicates of ours, developed at a cost of billions of U.S. taxpayer dollars. The Soviet version of our AWACS, the radar-bristling detection aircraft, is a prime example. The Boeing short-takeoff and landing (STOL) aircraft, a breakthrough in aerodynamic design made in the early 1980s, joined the Soviet air force just sixteen months later as the AN-72. The missile-guidance system used by the SU-15 fighter plane to shoot down a Korean airliner in 1983 was designed in the United States. The missile itself, called Atoll by NATO, is an exact copy of the American Sidewinder missile.[9]

Soviet and Eastern bloc agents in the United States cluster wherever military secrets or high-tech research is being done. They find especially fruitful picking in the bars, taco joints, and shopping malls of Silicon Valley, where, unfortunately, too many Americans can be bought or compromised into revealing secrets and stealing chips. A 1986 report of the Senate's Select Committee on Intelligence stated:

Soviet trade or scientific representatives travel to California about four times a month in delegations ranging from two to ten people, supplementing the 41-person staff of the Soviet San Francisco Consulate. It is reasonable to assume that, just as 30–40 percent of the personnel in each Soviet establishment are intelligence officers, the same percent of the personnel in a Soviet visiting delegation are intelligence officers and/or co-optees. Thus, the Soviets are able to target more intensively the 1,500 high technology companies in the area known as Silicon Valley. . . .[10]

One security expert, who does not want his name used or his company identified, says flatly that any new computer chip that is produced in volume will be in Russian hands within the first week of production. Even though their movements were sharply limited by the State Department in 1983, Soviet high-tech spies still can virtually stumble over almost everything they want.

At a semiconductor trade show in Palo Alto in 1985, for example, the FBI identified more than fifty foreign espionage agents in attendance.[11] The effort of these and thousands of other spies, diplomats, journalists, scientists, and tourists from all of the Warsaw Pact nations is orchestrated by one of the most powerful of all Soviet agencies, the Military Industrial Commission. Its Russian abbreviation translates as VPK. VPK, says U.S. experts, is the liaison between the Soviet intelligence agencies and the Russian defense industry. When Soviet researchers need a key piece of equipment from the West, VPK tells the KGB or GRU, the Red Army's espionage branch, which plans the theft and establishes a budget, a sort of R & D with cloak and dagger.

Thus, when the Ministry of Aviation wanted a look at the chips and other electronic components used in American cruise missles, they placed their order with VPK. The job was given to the KGB and budgeted at a paltry 170,000 rubles, about $65,000. Test equipment for semiconductor memories, needed by the Ministry of the Electronics Industry, was also assigned to the KGB—budget, 4.5 million rubles, about $1.5 million.

The full extent of Soviet technological spying first came to light in July 1981 at a Summit meeting of the Western nations in Ottawa, Canada. For François Mitterand, the newly elected socialist president of France, it was an anxious time. Ronald Reagan was openly angry at the new French leader's appointment of four communist ministers to his government. But two days after their coldly formal meeting,

Mitterand shocked Mr. Reagan virtually out of his chair with a summary of a supersecret file the French had code-named "Farewell." At the time, only four other people in France knew the file even existed.[12]

A few weeks later, Marcel Chalet, head of the Direction Surveillance de Territoire (DST), the French FBI, came to Washington to brief Vice-President George Bush on the contents of the astonishing file. Collected by the DST, it contained over four thousand Soviet documents stamped *Sovershenno Sekretno,* the Russian equivalent of TOP SECRET. It detailed the almost unbelievable efforts the Soviet Union was making to secure Western technology for its own military machine. It also described in painstaking detail the streamlined bureaucracy—so very different from that which ran the Soviet civilian economy—that had been established to achieve the goal of closing the technology gap.

Considering the almost unbelievable intelligence coup the Farewell file represented, the French came by it in a most mundane fashion; it literally fell into their laps. In the spring of 1981, a French national stopped by 11 Rue des Saussaies, then the headquarters of DST's counterintelligence unit, and dropped off two letters. The first simply explained that the Frenchman was serving as a rather frightened messenger, delivering a letter from a Russian acquaintance of his in Moscow. That letter was written in French by a man who claimed to have been assigned to the Soviet embassy in Paris during the 1960s. He expressed a willingness to serve France. He offered no information to establish his bona fides, but astonishingly, the letter was signed with his full name and patronymic.

The French quickly determined the man had indeed held a diplomatic post in the Soviet embassy and had, as are most junior and middle-level diplomats, been assessed by the DST. Somewhat surprisingly, he had expressed a willingness to work for them. Before anything could be set up, the diplomat had been transferred back to Moscow. Now, almost twenty years later, he was a high-ranking member of the KGB and a member of the Nomenklatura, the Soviet power elite. It seemed most unlikely, but he had dropped a tempting invitation to penetrate Soviet secrets. The DST were faced with a critical decision of whether or not to follow up the offer. In the elaborate chess match that is East-West espionage, double agents, provocateurs, defection, and redefection are part of a very dangerous game.

Even more puzzling, almost unnatural, the Russian wanted nothing for himself save a modest pension should he ever "jump the wall" and settle in France. The French decided to go ahead and set up an elaborate system whereby Farewell would make all the contacts, thus limiting the risk to himself and his DST "handlers." The very first documents stunned the French. In addition to their *Sovershenno Sekretno* stamps, each copy was numbered. Those that came directly from Farewell's office bore the number "1." That was the head of the KGB's Directorate T, the department charged with science and technology espionage. Several files had been annotated and signed by Yuri Andropov, then head of the KGB. One document bore the handwritten notes of Leonid Brezhnev, general secretary of the Communist party and chairman of the Council of Ministers, the all-powerful head of the Soviet Union.

The documents were far more valuable for their information than their autographs. They detailed the two major programs and the apparatus the Soviets used to plunder the West of its technology. The DOD white paper, which is based on the Farewell file, explains:

First Moscow has a program to raise the technical levels of weapons and military equipment as well as to improve the technical levels of manufacturing processes. This program is managed by the most powerful organization in defense production—the Military Industrial Commission (VPK) of the Presidium of the Council of Ministers. . . .

Second, the Ministry of Foreign Trade and Soviet intelligence services administer a trade diversion program to acquire relatively large numbers of dual-use manufacturing and test equipment for direct use in production lines. This program seeks export controlled microelectronics, computer, communications, machining, robotics, diagnostic and other equipment to increase the throughput of weapon-producing industry.

Complicating matters for the Western democracies seeking to protect their technological advantage is that both programs overlap. "Both programs sometimes seek the same products," admonishes the white paper. "Soviet industrial ministries request technology and equipment through both programs. The collection channels overlap and in some cases the same Soviet individuals (intelligence officers and others) are involved in each program."[13]

The Farewell file described in detail the VPK's method of operation. The VPK, according to the file, collects requests from all of the ministries within the Soviet defense establishment and draws up a yearly intelligence plan based on those requests. The plan is then broken down into specific collection requests, a shopping list that is placed with the KGB, the GRU, and the intelligence services of the other Eastern bloc nations. At Directorate T, the shopping list is then sent to the appropriate KGB "residencies" attached to Soviet embassies, consulates, trade missions, and commercial organizations in the West. Here the list, now converted to the size of the Manhattan telephone directory, with pages of double thickness sewn together to prevent removal, is assigned to "X line" agents. These are trained science and technology specialists who, like salesmen, are given quotas to fulfill from the list.

The white paper states:

The VPK program is a Soviet success story. Over 3,500 specific collection requirements for hardware and documents were satisfied for the 12 industrial ministries for just the 10th Five-Year Plan (1976–80). About 50 percent of more than 30,000 pieces of Western one-of-a-kind military and dual-use hardware and about 20 percent of over 400,000 technical documents collected worldwide in response to these requirements were used to improve the technical performance of very large numbers of Soviet military equipment and weapon systems.

Among VPK's earliest successes was the acquisition of the Fairchild Instrument Corporation's Xincom semiconductor memory tester. Bemoans the white paper:

It is a good example of the one-of-a-kind, dual-use product request acquired through the VPK program. Design concepts embodied in the hardware and associated documentation of the tester were copied to develop a Soviet counterpart. The original tester could also be used to help copy or reverse-engineer Western integrated circuits.

Semiconductor technology is one of the key VPK targets, and semiconductors themselves can be picked up by the handful almost anywhere in the Western world. Intel's 8080A 8-bit microprocessor, for example, which was used in many U.S. military weapons systems in the late 1970s was acquired quite easily and copied by the Russians. So diligent was the copying effort, in fact, that the Soviet engineers even retained the Intel part number, assigning their version the number KR5

80 1K 80A. It is this microprocessor that controls the aiming system on the Atoll missile that downed the Korean airliner. Similarly, Soviet ICs known as LOGIKA-2 and series 133/155 were directly copied from the TI 5400/7400 family of microprocessors.

A bizarre item in a 1987 issue of *Electronic News* may be apocryphal, but it does underscore the lengths to which the Russians will go to procure semiconductor technology. "A recent defector, Ion Pacepa, the former head of the Romanian secret police claimed that his agents stole a semiconductor design by secretly photographing diagrams hanging on the office wall of a TI licensee. The theft, according to Pacepa, took place in 1978 and was ordered personally by Leonid Brezhnev."[14]

The white paper explains:

These ICs have been used in Soviet strategic and tactical military systems since the mid-1970s to provide important qualitative improvements. The more advanced Western fabrication equipment acquired by the Soviets in recent years has been used to produce copies of sophisticated Western ICs for their latest generation of weapons.

Nothing is left to chance. The Farewell file documented the rigorous attention to detail the VPK and Trade Ministry apply to their shopping lists. Every request is complete with technical specifications, serial number, firm name, and the locations where the ordered product, process, or material can be acquired. They are as complete as the most demanding purchase order drawn by any Western company. Indeed, much of the information that goes into these "purchase orders" comes directly from the unwilling and often unwitting victims.

The United States, Japan, and the other industrialized nations of the West generate millions of technical documents each year. Virtually all of them are unclassified and readily available in journals, newspapers, press releases, product literature, and dozens of other forms, most of which find their way into electronic data bases. Farewell ascribes the task of acquiring that technical information to the State Committee for Science and Technology, known by its Russian abbreviation GKNT. Here experts translate and assess more than one and a half million scientific and technical papers each year. High on the reading list is *Aviation Week and Space Technology,* a trade journal that often contains important and up-to-the minute news about U.S. defense planning and weapons. To the intelligence shops in Washington, it is known as "Aviation Leak." Each week, several dozen copies are put on the Aeroflot flight to Moscow where every article is translated in flight.

Another prime source of information for the GKNT is the U.S. Department of Commerce's National Technical Information Service. It contains hundreds of unclassified DOD and defense contractor documents.

"There is a tremendous amount of potentially useful information available legally from the NTIS," George Menas, the assistant for strategic trade policy at the Pentagon, told me. Menas, a West Virginian with a Ph.D. in Russian history from Georgetown University, is a policy adviser to the assistant deputy secretary of defense, charged with controlling the transfer of dual-use technology to adversary nations. We started our interview in John Konfala's office and continued it in a taxicab to the State Department, where Menas had a meeting on the subject. "All of it is in the public domain, some of it with direct military use. How to design the F-16, for example, or the B-52, the Sidewinder air-to-air missile, and lots of others. Papers on the use of gallium arsenide in the manufacture of semiconductor devices, for example, is freely available in the NTIS. To the Soviets, this information is worth billions of rubles. It points to directions of research and development that are fruitful and points them away from unproductive areas. What they do is buy up all the technical information, hundreds of thousands of documents, papers and journal articles, that are available in this country and send it back by the planeload.

"The Soviets not only send trucks to the NTIS warehouse in Springfield to buy documents, but they can also access NTIS and other data bases electronically," decries Menas. "That means they can sort through reports rapidly and analyze the data so that even unclassified bits and pieces that are not very meaningful by themselves become very significant when put together. It's a case of the sum being far greater than its parts. A lot of this is specifically related to American military systems. It's not two steps removed; it's actual plans, performance data of weapons and aircraft."

Once the shopping lists are compiled and distributed, the Soviets employ a variety of strategies to obtain the actual hardware they need. Until recently, many items of military value were readily available for export to the Soviet Union through quite legal channels. Prior to 1980, for example, the Russians openly and legally purchased hundreds of tons of silicon from the United States, Japan, and West Germany for use in their semiconductor factories.

"There are a number of Eastern bloc companies licensed to operate in the United States," explains George Menas. "Some East European countries have also bought into American companies. There is nothing illegal about it. There is nothing to stop a Soviet company from buying a supercomputer and setting it up in Kansas or anywhere else in the United States and operating it. They could do any kind of security work on it."

Getting it to Russia, however, is another matter. U.S. export-control laws call for every one of some 200,000 items on a "control list" destined for export to be licensed by the Department of Commerce. "One of the changes in the Export Administration Act," explains Menas, "is that for the first time we require licenses for the direct sale of products or equipment to diplomatic embassies. Now you can't acquire an entire industry through a diplomatic pouch, but you can get examples of technology. You can get a micro VAX or an Apollo work station [both computers used in scientific problem solving as well as IC and weapons-system design], and as long as it is not purchased for sale out of the country, there is no way to legally stop the sale. Of course, these things also require sophisticated technical support, which is theoretically possible, but that does not give them a production capability. They buy a micro VAX or an Apollo work station and then get visas for half a dozen engineers to come over as cooks and chauffeurs, and they are trained on these machines. But it's almost easier to steal what they want and ship it back to Russia. If they are willing to spend the time and money, they can acquire one or two of just about anything they want."

Most of that sort of acquisition is done through third parties, so-called diversion agents. "It's like dealing with the illicit drug trade," notes John Konfala. "There is a lot of money being made and a lot of channels through which one can move equipment. You can buy something on the West Coast, ship it to Panama, send it on to South Africa, then on to Turkey, and finally into Bulgaria. It's laundering equipment instead of money. Very often the customs service of a given country doesn't even look at what is being shipped. It's put into transit zones, where in effect there is no national law governing items that pass through them. It's a no-man's-land, and it just moves out and into the next transit zone in another country.

"Many of the countries have few export-control regulations, and they are encouraging this sort of trade for the hard currency it brings. Countries like Singapore, Taiwan, and South Korea are reluctant to place restrictions on trade. They are not interested in dealing directly with Eastern Europe, but they sell to people who sell to people in Eastern Europe. We are aware of all these problems, and our Customs Service has agreements with other nations trying to alleviate as many of these loop holes as possible."

The Customs Service is the point man of Operation Exodus, a major effort to stop what has been called the "awful hemorrhaging of American technology." Begun in October 1981, the program is coordinated with other government departments such as State, Commerce, Justice, and Defense. To learn just how successful Operation Exodus had been and how it worked, I went to see Steve Walton, who is director of the Strategic Investigations Division of the U.S. Customs Service in Washington. Walton is slender, wears horn-rimmed glasses, and looks more like a college professor than a cop. He has a deep, rich voice, and when I first met him, I almost expected him to be doing voice-overs for television commercials rather than the job he has, which is planning covert operations to catch high-tech smugglers.

"For the diverters, it's a big business," he intoned, "millions of dollars a year. The East bloc will pay 300 percent more than the value of an item. In the semiconductor field, for example, for a piece of semiconductor manufacturing equipment or a dual-use item, they will pay even more, over 400 percent.

"There are middle men involved; there are a lot of Western European businessmen, and their commissions account for most of the markup. They get top dollar, and as we have put more pressure on and enlisted other governments as well as our own, the price goes up. As it does with drugs. And that's one of our tactics, to drive the price up. The East bloc countries have a hard-currency problem; the more it costs them, the more we hurt them.

"We stopped $28 million in one shot. Since 1981 we've made 5,200 seizures worth $585 million. That's high-tech stuff we stopped that was about to leave the United States without at the very least complying with, or for direct, willful circumvention of, U.S. law.

"How much is getting out, nobody knows. We do know we are hurting them, because they are asking neutral countries why they are complying with the U.S. in this regard. A Soviet trade officer in Stockholm complained to a Swedish friend of mine, 'You people are too close to the Americans; you are stopping all their technology from reaching us.' 'We would do the same for you,' my friend replied."

Much of Operation Exodus takes place outside the United States, since it is virtually impossible to stop every piece of control-list items from leaving the country. "Depending upon the level of evidence we have," explains Walton, "and this is all within the context of the law, we can interdict it here, before it leaves the States. We have legal mechanisms in most of the European countries, by virtue of legal assistance agreements, to inderdict it there. We can request our counterpart services to stop something and then return it as evidence to the U.S. There are several different mechanisms we can use. We can make interdictions at virtually any stage until it jumps over the wall."

The route by which illicit equipment ultimately leaves the United States and finally reaches its destination in the Eastern bloc is often as labyrinthine as an ancient Cretan maze. Stationed in Germany before his present assignment, Walton once followed a twisting paper trail from Minneapolis, through England and Germany, and finally to Sofia, Bulgaria. "The trail actually began in England, where we had been keeping tabs on a man named Brian Williamson," said Walton in his rich storyteller voice. "He had been acquiring stuff for the Soviets for years. The U.S. caught on to his act, and the Commerce Department put him on the denial list in 1980. That meant he could no longer receive any U.S. licenses to export technology. He had a firm called Datalac. So he created a German front company called Datagon. The beauty of it was, he had various U.S. suppliers who really didn't know what was going on.

"He stopped applying for licenses; he was on the denial list, so that was to be expected. But suddenly Commerce starts getting license applications from a firm called Datagon in Cologne, Germany. Datagon was headed by another Englishman, named Martin Coyle. He starts entering into business relationships with several U.S. firms selling mostly dual-use computers, DEC, Tektronics, Hewlett-Packard, and IBM. But his major supplier turned out to be a small two-man firm in Minneapolis.

"Now you have to understand the computer business is like McDonald's in this country; computers are a commodity, and these two guys in Minneapolis were just another small hamburger stand. They would get an order from Datagon: 'Get me 3 DEC VAX 1150s for resale in Germany.' The company in Minneapolis would come up with them, even though these big multinational companies try to control their markets. It's like a used-car lot, except the stuff is new.

"The Minneapolis company received a false-end user certificate from Datagon. They used that as the basis for their license application to export the computers to Datagon GmbH Germany for resale in Europe. So far, it's a seemingly legal transaction. But payment was made out of England, and that's what tipped us off. The stuff was actually shipped by air to Cologne, Germany, where it never left the airport. It was then rerouted to England, where Williamson shipped it back to Frankfurt, Germany. Again, it never left the airport, but was transshipped to Sofia, Bulgaria. Datagon Germany provided technicians to the Bulgarians to install the computers. That's always a weak link in the entire mechanism.

"It's one thing to sell a VAX 1185, or some other computer, but it's quite another to service it and provide spare parts for it. Technical agreements can often be part and parcel of the package. The Soviets and the East bloc are no different; they need Western technical assistance. This VAX is classic dual use; it can be used for everything from guiding a missile to running a hydroelectric plant to designing a semiconductor circuit.

"We had an informant in Bonn who made an allegation against Datagon and said they had a business arrangement with a firm in England called Datalac. That, of course, was Williamson's company, and they were on the denial list. Incidentally, a lot of our intelligence comes from informants. They can be business competitors, neighbors, even ex-wives. They make the best informants," says Walton chuckling.

"Datalac was a red flag, and I called Jimmy Macshane, who was my counterpart in London. He and I went to work to follow the paper trail. You don't see much smuggling as such. In the old days, they would mislabel computers as refrigerators or washing machines; now it's much more subtle, and they pretty well accurately describe the stuff.

"At the outset, we went over to Commerce and pulled all the licenses from the United States to Datalac right up to the denial. Then we found the same Minneapolis firm that had been a major Datalac supplier obtaining licenses to ship to Datagon. In Germany I got the air bills and other documentary evidence regarding the shipment from Cologne to London. In England, Jimmy Macshane and the British authorities documented it out. Pretty soon we had it all the way, A to Z. In February 20, 1983, we executed search warrants simultaneously at thirteen locations in three countries. We took everybody down, seized every scrap of paper, and hauled it all away in trucks.

"We were able to document that in a very short time Datagon had sold $4 million worth of computers to the Bulgarians. We identified, interrogated, and used as witnesses technicians who worked for Datagon and had gone to Sofia to install the computers. We arrested everybody in the loop and convicted everyone in all three countries.

The Americans in Minneapolis were not unwitting dupes in the operation. "The Americans knew the stuff was being diverted," Walton snapped angrily. "All we had to prove was that they knew the computers weren't staying where the license said they were to be installed. Moreover, they were paid out of England, and Williamson's name even appeared on some of the checks. A part was also returned from Bulgaria to Minneapolis to be repaired. That's how we finally stopped Williamson.

"Putting him on the denial list wasn't enough. All he did was create another entity. It's like a cancer; it's all got to go. So everybody has to go down. If you leave one peg up, it's a building block."[15]

There are literally hundreds of other Americans, Europeans, and Asians involved in the illicit transfer of semiconductor, computer, and other high technology to the Eastern bloc. The money is simply too tempting for those with little patriotism and even less conscience. They are eager suppliers of the KGB, GRU, or the spy networks of the satellite nations in the Warsaw Pact the Soviets often use to acquire technology for them.

"Bulgaria is probably the leader in terms of acquiring semiconductor manufacturing equipment," notes Walton. "The technology-acquisition effort by the East bloc is not that sophisticated, however, that it assigns technologies by priority to individual countries. But there is a little bit of competition among the bloc countries. The East Germans pride themselves on being a technology-acquisition leader.

The Bulgarians are very good. They have actually acquired whole plants from the cement up."

To prevent the sale of dual-use technology to the Eastern Bloc and the Chinese, North Vietnamese, Cubans, etc., the United States, its NATO allies (excluding Iceland), and France and Japan have joined in a loose agreement called CoCom, or Coordinating Committee on Multilateral Export Controls. "They periodically establish lists of dual-use equipment and their associated technology for which there is a multilateral agreement restricting their sale to Warsaw Pact countries and to a much lesser degree to China," explains the Defense Security Technology Agency's John Konfala. "Therefore, the Japanese, for example, could not sell equipment or materials or products that are encompassed by these lists without the agreement of all the other countries, including the U.S. We have no evidence that any of the CoCom member countries, putting aside illicit sales by individual companies, or third-country sales, have violated the agreement by officially authorizing sales to the Warsaw Pact.

"Nor is there any evidence that they have acted wittingly to circumvent the general agreement on controlling semiconductor manufacturing technology materials or products. These governments have worked with us to investigate companies about whom we have information are engaged in illicit transfers or to assist us in stopping such tranfers. The countries that are members of the CoCom group work closely on a bilateral basis to maintain the control lists, to license for export only those products that are agreed upon by the other members, and to cooperate with the United States to stop illegal transfers."

George Menas points to another facet of the complex problem of preventing the sale and shipment of this high technology to our major adversaries. "The other concern, too, is indirect diversion, through newly industrialized countries, such as Korea, Taiwan, and the other NICs. Recognizing this and recognizing the futility of having even the most sophisticated control system on the front end among the leading Western countries, if we don't close the gaps elsewhere, we have taken steps to gain agreement with those emerging industrial countries and neutrals like Sweden that they will respect the international control list on dual-use items.

"Particularly, when we are talking about U.S. parts, equipment, and components that are embedded in products produced in those countries, we will require a validated export license if it is exported directly from the United States. So in the last three years we have undertaken bilateral and some multilateral efforts to bring neutrals and emerging industrial countries into line with U.S. and CoCom export controls. We've had some success, but there are still a lot of problems, mostly in terms of these countries having the national legislation to put in place a system of export controls. One of the great victories in this area was to bring Spain not only into NATO but into the CoCom agreement. Spain had no export controls, but they are in the process of promulgating the legislation for them."

CoCom is, however, a troubled partnership. In many cases, technologies that Washington bureaucrats consider critical to national security are viewed in other capitals as nonthreatening. "One particular source of tension," declared a National Academy of Sciences panel headed by Lew Allen, director of the California Institute of Technology's Jet Propulsion Laboratory and a former air force chief of staff, "is a practice by which the United States attempts, in effect, to impose its own regulations on foreign companies. The Panel on the Impact of National Security Controls on International Technology Transfer readily admits to the need for controlling dual-use and military technology to the Soviet bloc, but it finds that many of the control attempts are divisive and ineffective.

"Companies that import restricted technologies from the United States, for example, are required to obtain an export license from the U.S. if they subsequently export that technology to a third country. A firm that imports an American-made computer chip then is supposed to apply for a license from the U.S. before it can export any of the products it has made using the chip."

Europeans complain bitterly about requiring a U.S. reexport license for a $20,000 machine simply because it contains a $2 American microchip. "Needless to say, most of our allies resent these restrictions as an attempt to meddle in their affairs," says Allen. "U.S. efforts to extend its national-security-control policies to trade between CoCom countries and third countries have prompted charges of extraterritoriality, infringement of national sovereignty, and possible violation of international law." And these are charges from our friends!

Among the major points of friction is that the U.S. List of

Controlled Items is much longer than CoCom's International List. "As a general policy," the panel said, "the United States should limit coverage of the U.S. Control List and the CoCom International List to those items whose acquisition would significantly enhance Soviet bloc military capabilities. The United States and CoCom countries should also agree on a common approach to the re-export of items originating in CoCom nations."[16]

At the moment, the U.S. list is vast. Some 40 percent—about $62 billion—estimated the panel, of all U.S. exports of nonmilitary manufactured goods in 1985 were shipped under a license requiring prior approval. Perhaps the most absurd example of that sort of licensing was the ruling that kept a TI-manufactured children's game called Speak & Spell, available in virtually every toy shop in America, on the export control list until 1985, because it was driven by a microprocessor.

And the approval process is tortuously slow, an average of fifty-six days to grant an export application from the time it is received. In 1985 there were 122,606 applications, and 5 percent of those took more than a hundred days to process. In Japan, MITI processes about 400,000 export licenses each year within two to three days. The net effect is a sharp reduction in the ability of American companies to compete in the world marketplace.

Consider the U.S. company that applied in March 1983 for a license to export a $450,000 nuclear magnetic resonance (NMR) spectrometer to a medical research institute in Eastern Europe. The application was finally approved 910 days later, in November 1985. In the interim, a German competitor sold several similar systems to Eastern bloc customers. NMR instruments are not on the U.S. Control List, but this sale was subject to export licenses because it incorporated 32-bit microprocessors, components produced in the millions in a number of other countries. Ironically, NMR technology was pioneered in America, but Japanese and German firms now hold two-thirds of the world market for those instruments.[17]

The Defense Department puts much of the blame for delay on the Department of Commerce. "The implication here is that the length of time required to get an export license is caused by review for national-security purposes. And that's not really true," insists the Defense Department's George Menas. "We turn cases around here in less than twelve days. The holdup is with the Department of Commerce, with their bureaucratic ineptitude. We don't review much more than 15 percent of all the licenses issued by the U.S. government. So you have this mass of over 100,000 export licenses that are processed by the Department of Commerce without ever having been reviewed by the Defense Department, and we certainly don't account for the delay."[18]

Nonetheless, American manufacturers who wish to export their products must still wade through an awesome morass of regulations and controls. Those controls, the panel estimates, cost American business $9.3 billion in exports in 1985 alone. The results of a 1987 survey of 170 American semiconductor, electronics, aircraft, instrument, and machine-tool makers showed just how difficult the export controls had made it for them to do business:

52 percent reported lost sales primarily as a consequence of export controls; 26 percent had business deals turned down (in more than 212 separate instances) by Free World customers because of controls; 38 percent had existing customers actually express a preference to shift to non-U.S. sources of supply to avoid entanglement in U.S. controls; and more than half expected the number of such occurrences to increase over the next two years.[19]

The panel recommended several changes in the Export Control Program and in the CoCom system to balance the need for national security with the equally vital need to maintain global competitiveness. These included the elimination of "trade restrictions for all dual-use goods that can be obtained readily in world commerce outside the U.S." and the establishment of "a dialogue between senior administration officials and the private sector on the effects of export policy on economic vitality."[20]

In an open capitalist society any number of sophisticated pieces of technology will inevitably slip through the cracks of control lists and agreements such as CoCom. As a result, some of the Soviet's semiconductor acquisition has been quite legal. Indeed, the Defense Department is still furious over the Soviet purchase of more than a dozen powerful, lightweight satellite receivers, which contain the most advanced gallium arsenide chips ever made. The receivers, produced by several Japanese electronics companies to pick up television programming from satellites, went on sale for the first time in 1986. They no sooner appeared in the windows and stalls of Akihabara than a pair of X line KGB agents from the Soviet embassy in Tokyo showed up and bought out the entire stock. That evening they were on the Aeroflot flight to Moscow. A week later, they were installed at select sites across the Soviet Union, picking up signals from U.S. military satellites. The sensitive gallium arsenide chips, especially designed for signal processing, were easily tuned to frequencies on which the U.S. satellites transmit. The receivers are a perfect example of dual-use technology— designed for home entertainment by the Japanese but turned to military use by the Soviets. And the sale, of course, was perfectly legal.[21]

There is an ongoing tug of war between the Department of Commerce, allied with industry, whose goal is to increase the sale of U.S. semiconductors and semiconductor equipment and other high-tech exports, and the Defense Department, which wants to restrict the export of that technology. "In terms of semiconductor technology," Konfala told me, "our concern is that we know that 90 percent of seimconductor manufacturing in the Warsaw Pact countries goes to the military. Obviously there is very little consumer-electronics industry in Eastern Europe. The military appropriate virtually all of the semiconductor factories. To the extent that there is any production for the civil sector, the best products, the most reliable products, are skimmed off for military use.

"Our concern here is that we don't support Soviet and Warsaw Pact military sector with semiconductor manufactures. The reason for this is that the edge that the NATO and U.S. forces have over the numerically larger Warsaw Pact forces in Europe, in terms of manpower and equipment, is our more sophisticated command-and-control and fire-control capabilities, which are founded on electronic support equipment. Whether it be radars, high-speed digital switching for command and control, or other devices providing infra-red detection capability, we are way out ahead of the Soviet Union and the Warsaw Pact not only in their manufacture but in deployment. And even here our own forces lag behind the civil sector in terms of having fielded current state-of-the-art electronic products. The civil sector, of course, is based upon consumer demand and R & D. It's out in front of the military, particularly in terms of fielded electronic systems.

"So our concern is that under the guise of selling plants, manufacturing equipment, and products for civil applications to the Warsaw Pact nations, we would be in effect transferring to their military-industrial structure the technology that would ultimately upgrade Soviet military systems to a level of electronics technology that is superior to what is presently fielded in U.S. equipment."

In addition to hardware, the Russians eagerly seek scientific and technical knowledge. In 1985, the CIA identified over thirty-five scientific conferences in the late 1970s that had been targeted by Soviet industry as primary sources of military information in areas of "missiles, engines, lasers, computers, marine technology, space, micro-electronics, chemical engineering, radars, armaments, and optical communications." The Soviet Academy of Sciences has also targeted at least sixty American universities for intelligence-gathering operations. Harvard, MIT, Carnegie-Mellon, Michigan, Caltech, Princeton, Stanford, Cornell, Berkeley, Columbia, and NYU are among the top-priority targets. Spying on American campuses may in fact account for as much as one-fifth of the entire Soviet science and technology espionage effort. The CIA also found a "rough correlation" between the number of Soviet military research needs and the number of visits to American universities by Russian and Warsaw Pact scientists with relevant expertise.[22]

One of the primary means of access to U.S. universities is the scientific exchange program, a method we not only condone but actively encourage. "It is hardly surprising that the Soviet Union would turn to its principal scientific organization the Soviet Academy of Sciences to perform a vital intelligence function," acknowledges Richard Perle. "What is surprising is the enthusiasm with which various United States Government agencies should advocate renewing, enlarging, extending and initiating scientific and technical exchanges with an organization known to be part of the Soviet intelligence establishment."[23]

Needless to say, the Defense Department is hardly enamored of the program and tries to limit participation on the part of its researchers. "We have strengthened our controls over what DOD scientists and engineers can release either through public publications and scientific conferences and seminars, particularly where sensitive national defense research is concerned," says John Konfala. "It's a sensitive issue, especially where there are constitutional considerations of freedom of speech and academic freedom. We try to walk a fine line between those basic freedoms and pragmatic national security considerations. We are not in the business of wittingly providing aid and comfort to potential enemies, particularly in the area of high technology. So to some degree it has to be given our open society, largely voluntary, particularly in an academic setting.

"One of the things we look at carefully are Soviet scientists applying for visas to study at U.S. universities. Most are graduate scientists in the physical sciences, and they're coming here to study at our leading universities; they know exactly where they want to go and exactly what they want to study and with whom. We look at it in terms of the potential application to military use, what DOD contracts does the U.S. scientist have, what is his speciality, has he done work for DOD. So-and-so is coming over here for a year to work in your lab, we say. He's going to work on a subject that is closely allied to work you're doing for the Defense Department. How much of that is going to rub off on him. How much of that is he going to carry back with him?

"It's very difficult for people who are working together and socializing as well in the free exchange that goes on in the scientific environment. This fellow is going to ask questions; he's got an inquiring scientific mind. He's also got an agenda from the KGB. And his wife and kids are still in the Soviet Union.

"They send politically savvy, older people. They are not going to be turned around by the glitter of American life. They are in the higher, privileged echelons of their own society; besides, their families are there. And they are handpicked. People just don't apply for study visas to the West or in research areas that are not useful to the state.

"Most Russians come in under the IREPS, the International Research Exchange Program that had been created as part of the scientific and technology bilateral agreements made between the U.S. and the Soviet Union. They also can come under the auspices of a joint program that our National Academy of Sciences has with their NAS," adds George Menas. "There are a hundred or so scientists proposed to come each year under IREPS. Under all programs there are upwards of a thousand Soviet scientists a year who come to this country."

A major priority is scientific computing. "Soviets have great difficulty developing high-power scientific processors. They seem inclined to go toward parallel processing, for one reason, because it requires less sophisticated microprocessor chips. You can use smaller processors to achieve the same results. The leading Soviet scientists in parallel processing are studying at an American university with a scientist who is doing contract work in that field for the Department of Defense.

"Similarly, at a university in the Washington area, another Russian scientist, specializing in computer visual identification, is coming to study with a renowned American scientist who has been working on this problem for the Department of Defense for decades. He wants to spend two years in the American's lab and has been granted a visa. I presume when he leaves in two years he's going to know as much as the American scientist.

"The State Department has the final say but generally follows the recommendations of an intelligence review committee from DOD, State, etc. You can say that Boris is admitted subject to the following caveats. . . . But what happens then is the State Department calls Boris's host at, say, Berkeley and suggests that we will admit him into the country but don't give him access to this or tell him about that, and so on. If the professor wants to say, 'Go fish,' that's it. And what happens more often than not is that the university takes the view that once someone is admitted to the U.S. and the university, he becomes a member of the university community with all the rights and privileges thereof."

The Russians, of course, return the favor, but in not quite the same fashion. "There is a certain reciprocity here," admits Menas, "in numbers if not in kind. The agreement is supposed to be reciprocal, but it's not in practice. The IREP should be a one-to-one basis in fields of study as well. Our people who go to the Soviet Union tend to work in the social sciences. The Soviets who come here work in the hard sciences. They ask to study microprocessor development, fiberoptics, communications networking. . . . They ask to go to Bell Labs, MIT, the submicron lab at Columbia, and our most sophisticated laboratories. In my experience, the Soviets have not treated us reciprocally in terms of giving access to our people to their best labs. Friends of mine who have studied there tell me you are not given access to the stacks in the libraries of the Soviet Union. They would never let him see a book he requested until it had first been reviewed by a Soviet scholar. He was studying the war economy of 1917, but they were afraid he might dig out some piece of information that would be detrimental to them.

"The nature of our society leaves us open to this type of exploitation. But the free exchange of information is one of our strengths. The Soviets, having a closed society, can protect their secrets, but at the same time it limits their ability to change, to adapt, and to innovate."

Such one-sided exchanges have been going on since the era of detente, initiated by President Nixon in the early 1970s, when most political experts believe we virtually gave away the store. "In the detente of the '70s," admits Richard Perle, "the Soviet Union got the lion's share of benefits from exchanges that were supposed to be mutually beneficial. Soviet secrecy prevented us from learning much of interest, while American openness facilitated Soviet acquisition of American technology and know-how."[23]

"Our concern in the Defense Department is not to repeat the mistakes made during detente, when we transferred key technologies to the Soviet Union with our eyes open," says John Konfala testily. "We were not duped into believing that what we were transferring did not have military applications. The decision was made that given the potential political benefits from detente and the aid to their economy from this technology, that would in turn have an influence on Soviet political and military aggressiveness. The Soviets, of course, were selective in what they bought and never subordinated decisions they thought were significant to their national security to whether they got another computer or not.

"Now, of course, we are faced with *Glasnost,* the opening of Russian society under Gorbachev," says George Menas with a smile. "But this opening and closing of the Russian society has been going on for centuries under the czars and the commissars. The Defense Department's concern is that these thaws in our relations with the Soviet Union not be allowed to cloud our common sense where sophisticated, critical high technology with military applications are concerned. Semiconductors, computers, and similar technology will go only to one place—the military. It will not go into the production of VCRs, tape recorders, or other consumer electronics for the general population."

The conflict between national security and competitiveness has paradoxically arrayed the Defense Department against the Commerce Department, and inevitably frictions between them often seem as fierce and grinding as those between the United States and its major trading partners and allies. Now the Russians are again dangling the promise of increased trading as part of Gorbachev's *Glasnost* policy. It is all part of a familiar tactic, according to John Konfala. "The Soviet Union has had for decades an official economic policy where we have seen numerous examples where they have held out bait; they offer to buy hundreds of thousands of products. But invariably, when they sat down to negotiate the details, they would be buying a token number, one or ten rather than the hundred or thousand originally promised. And they would try to acquire the technology involved in the manufacturing process. Their general policy is not to become dependent on the West, particularly in the area of key technologies that have military applications."

The search for technical information is not limited to the Soviet Union and its surrogates or even to the host of other nations that support major espionage establishments. Many major companies, foreign and American, seeking to retain their competitive positions, will pay big money to learn what its competitors are doing. And nowhere is the theft of high-tech secrets a bigger business than in Silicon Valley. Here high-tech crime is committed by junior engineers, senior executives, and just about everyone in between, including the janitor. The main targets are trade secrets, computer crime, and the theft of microchips, according to Santa Clara County municipal judge Douglas Southard, who, until January 1988, was a deputy district attorney in charge of prosecuting high-tech crime.

Hundreds of Silicon Valley people made millions of dollars since the chip became the mainstay of civilized life. With the money has come a very fast life-style that incorporates booze and drugs along with Mercedes cars, mansions, and the other accoutrements of the jet set. But drugs are the great leveler, spilling over the top from the chip and computer millionaires down to the broom pusher. Valley law enforcement officials estimate that 20 percent of the local population is involved in some sort of substance abuse, with cocaine according to Judge Southard, as the drug of of choice. And in the world of high-tech drug dealing, chips are the currency of choice.

"I would say that in just about virtually every case involving chip theft and equipment, cocaine works in somewhere."[24]

For most of the companies involved, it was all part of the Valley life-style. Thus, most treated the problem of chip theft and trade-secret stealing as damaging but a family matter nonetheless. Telling the cops was considered bad for business and just not done. Then, along came Peter Gopal, a Valley electronics dealer. In 1978, Gopal had acquired some of National Semiconductor's most closely guarded trade secrets and tried to peddle them to Intel. Breaking with Valley traditions, Intel blew the whistle, and a major investigation was launched. It turned up Gopal's partner, an Austrian named Rudolph Sacher, who just happened to be working for East German intelligence.

Gopal and Sacher, it was discovered, had already illegally exported semiconductor test equipment and other electronics hardware to Poland. Unfortunately, not enough evidence could be found to prosecute, but they were convicted of theft and the attempted sale of National Semiconductor trade secrets. After four years of appeals, Gopal finally went to prison for a mere sixteen months. Sacher simply fled the country.[25]

The Gopal case revealed the vulnerability of the Valley. It was then underscored dramatically by the 1981 theft of $3.2 million worth of memory chips, one of the largest thefts in industry history. The chips, stolen from Monolithic Memories of Sunnyvale, California, shook the industry out of its closemouthed ways. Sunnyvale, a town of about 115,000, is one of the largest of the communities that make up Silicon Valley. At the time, only seven Sunnyvale detectives were available to investigate the case, and none were specifically trained to deal with high-tech crime. As a result, Santa Clara county, which encompasses much of Silicon Valley, created the nation's only high-tech crime task force. Among its 150 "busts" made since the unit was established in 1981 have been a janitor caught stuffing hundreds of microchips worth thousands of dollars into a cardboard box and executives packing their Gucci attaché cases with chips, computer drives, and trade secrets.[26]

But most of the chip thieves went undetected and even if caught risked little more than being fired. Since then, state and federal officials have realized the danger to America's semiconductor industry and national security, and in addition to California law enforcement agencies, the FBI, the Customs Service, the Commerce Department, and the Drug Enforcement Administration (DEA) have set up shop in the Valley to help local agencies.

The state of California also provided funds for a full-time expert who would not only investigate high-tech crime but also train law enforcement officials throughout the Valley to recognize and properly investigate it. That grant has since run out, according to Assistant District Attorney Ken Rosenblatt, who replaced Doug Southard. The high-tech crime unit now consists of Rosenblatt and one Santa Clara County detective, an expert in high-tech crime who, although not assigned to the unit, is available to "help" when needed.[27]

The boom years of the 1970s allowed dozens of people like Peter Gopal to operate a gray market in stolen chips, computers, electronic components, and chip designs. Gray-market hustlers—known as "schlockers" in the Valley vernacular—can score big hits by delivering scarce chips. A $350 microprocessor in short supply can be sold for ten to twenty times that amount to suppliers desperate to deliver to screaming customers. Deals are made and no questions asked if a part is wanted badly enough. Michael Malone, author of *The Big Score,* a book about Silicon Valley, calls the gray market the "flip side of entrepreneurism." He describes the "schlockers" as opportunists who "just do it when the opportunity presents itself. Most people," he adds, "aren't in the gray market premanently. It's like an underground river of parts moving in and out of Silicon Valley and has many sources and many mouths. It swells and diminishes according to the state of the industry. When things are booming, it's a roaring river, when things are bad, it's just a stream."[28]

While the beefed-up security and increased vigilance have helped cut industrial espionage in the Valley considerably, it cannot be completely stopped. The only hope, then, according to Malone is to "keep pushing the state of the art outward. If we ever falter, we're such a leaky vessel of technological information, the rest of the world will catch up with us almost overnight."

Judge Southard declares:

Silicon Valley needs to protect its intellectual property. If you look at the whole thing from a historical perspective you can see that the capital that's driving the most active part of capitalism in our society—the high tech industry—is no longer the same as the capital of the 19th century. The capital of the 20th century is ideas. Ideas drive companies and are the assets of a high tech company more so than physical plants or raw materials or the labor force.[29]

Unfortunately, ideas are even harder to protect than the chips themselves. Intellectual property, the idea that ideas themselves have an intrinsic value, just as a house or hardware, is largely a Western concept. Patents, copyrights, and trademarks have virtually no meaning in most of the newly industrialized countries where copying of ideas such as chip designs, patented manufacturing processes, and trademarks is virtually endemic. In South Korea, Singapore, Taiwan, Hong Kong, and other countries, for example, it is virtually impossible to gain a patent to protect a manufacturing process or copyright computer software or the intricate microcode that governs the actions of many types of IC. Only recently has it been possible for a foreign company or individual to obtain a patent or copyright for a manufacturing process or IC in Japan. Indeed, it was not until 1984 that U.S. patent law was broadened to provide semiconductor makers with copyright protection for their intricate circuit designs. And still American patent laws offer far less protection to its own patent holders than do the laws of its chief competitors.

American law, for example, does not prohibit the sale of chips, or any other product for that matter, in the United States, even when produced in violation of U.S. patents, so long as the product was manufactured outside the United States. Japan, Germany, Great Britain, and most of the other Western European nations do not permit the sale of products in their countries, no matter where they have been produced, that infringe on their patents. Still, in Japan patent protection is not as shielding as it might be.[30]

Chikara Hayashi, the chairman of Ulvac Corporation, a major Japanese semiconductor equipment maker, told me of his fears at making a patented software program his company had developed available to other Japanese companies to whom he ultimately hoped to sell it. Hayashi, who speaks English fluently, was kind enough to meet me on a Saturday in a crowded Tokyo coffee shop.

"Ulvac has developed a semiconductor plant automation software program called 'The Useful System.' We have never sold it. But the system is working at the NTT lab and was developed with them. It took six years to develop, and it is now working at NTT Atsugi, in their research manufacturing line. It is more difficult to fully automate a commercial semiconductor processing line than a research line since the type of chip being manufactured changes more frequently. Siemens in Munich wanted a presentation of it, but we haven't been able to sell it in Japan. NEC, Mitsubishi, Toshiba—they all have their own and scientific computers that are only now reaching the marketplace. In 1985, however, NEC began selling what it called a "V" series of microprocessors. The V20 and V30 are in fact pin-for-pin copies of the Intel chips. That means that anyone can open his PC, pull out the 8086 or 8088 from its socket, and simply plug V20 or V30 in its stead.

Not only will the PC continue to run; it will even run better, for NEC has made some improvements that have souped up the V20 and V30, making them 10–15 percent faster than the original Intel chips. Such copying, or emulation, as it is euphemistically known in the industry, is accomplished by reverse engineering, literally peeling the chip back layer by layer, as if it were an onion, and noting the exact location of every transistor, interconnect, and circuit on it. And, in the strictest sense, the process is legal.

What is not legal, say Intel's lawyers, is the creation of a "compatible" microcode by NEC, based on the Intel microcode, to run the chips. The case, which has dragged through the courts for almost two years, was further complicated when it was learned that the trial judge, William Ingram, owned eighty dollars' worth of Intel stock. While the case was put on hold to resolve a conflict-of-interest question, a frustrated Intel turned to the Customs Service in August 1987. They asked for a ban on the importation of all NEC V series microprocessors and any equipment that uses them. "We are enforcing the rights given to us by the U.S. government," argued Tom Dunlap, general counsel for Intel. "We have a valid copyright. We have also recorded them with the Customs Service. We've asked them to enforce the copyright."[33]

The Customs Service refused, and the entire case was thrown back to the courts. But Intel will continue to be frustrated. Judge Ingram, even though cleared by a judicial review board of conflict of interest for his eighty dollars' worth of stock, resigned from the case in December 1987. It must now be retried.

While Intel was bringing suit against NEC in defense of its microcode copyrights, it was also suing Hyundai, the giant Korean company, for patent infringement of its EPROMs. Similar complaints by TI to the U.S. International Trade Commission for infringement of its patents on DRAMs were settled early in 1987 with seven Japanese companies. Korean manufacturer Samsung finally bowed to a U.S. District Court order and settled in January 1988, adding perhaps another $650,000 to the $138 million in royalty payments TI received from the Japanese companies during the first half of 1987.[34]

software people, and they would like to have us present
our software to them, but they would then just learn fro
buy it. They want to find whatever nice parts of our system
theirs. I can't say they would steal it—that is too strong a
their interest is more in learning than buying.

"They would be interested in improving their system wi
of ours. There is no way to really protect it. There has been
of difficulty in gaining patent protection in Japan for mici
software. Now we have some protection, but perhaps not
this country the big company is always right."[31]

In the NICs, like South Korea and Taiwan, weak patent l
essence part of national planning. The aim is to ensure access
technology—the key to their economic progress. "Koreans
viewed intellectual discoveries or inventions as the private p
their discoverers or inventors," points out the Korean amba
the United States Kyun Won Kim. "New ideas or technolog
'public goods' for everyone to share freely. Cultural esteem ra
material gain was the incentive for creativity. On a pragmati
asserts Ambassador Kim, "intellectual property protection gi
nopoly power to those who have the technology. Higher price
thereby restricting the ability of developing countries to
technology."[32]

Despite this widely held viewpoint among the NICs, in July
Korea bowed to extreme U.S. pressure and put into effe
intellectual property rights law. Just how much protection
actually offer to patent and copyright holders remains to be seen
is a step toward helping to reduce the more than $1 billion in sale
industry loses each year, according to estimates of the Interna
Intellectual Property Alliance, to patent infringements.

Protecting the rights to intellectual property has become a
tinuing battle on the part of most chip makers and computer
electronic component manufacturers. Perhaps the best example
ongoing series of suits and countersuits between Intel and NEC
microprocessor designs. Intel has over the years produced a serie
microprocessors that are used in a number of applications
especially as the brain of the PC. Intel's 8086 and 8088 were,
example, the basis for IBM's PCs and all their clones. Intel's 80286
new 80386 have made possible a new generation of PC, work statio

"The order," said Richard J. Agnich, vice-president and general counsel for TI, "confirms the value of TI's semiconductor technology. It underscores the role intellectual property can play in providing an adequate return on research-and-development investment to technology innovators."

Toshiba, already under fire for the sale of milling machinery to the Russians, was another target of American patent law. National Semiconductor has accused the Japanese company of violating its copyrights and patents on a chip found in every IBM PC and virtually all its clones. Dubbed the 16450, it is an asynchronous communications device, a chip that allows the PC to communicate with a printer, modem, or other peripheral equipment that may be plugged into the computer. It was a best seller for National; millions of 16450s had been sold for four to twelve dollars apiece. One of National's major customers for the chip had been Toshiba. Then sales to the Japanese company suddenly dropped to virtually zero. Inquiries soon established the reason for the lost sales: Toshiba was making its own version of the 16450.

National bought one of the Toshiba chips and reverse engineered it. It was identical to National's 16450, transistor for transistor, circuit for circuit, and contained exactly the same microcode. National threatened Toshiba with suit and also turned its legal guns on United Microelectronics Corporation of Taiwan, accusing it of infringing its patents on the same chip. Both Toshiba and UMC finally agreed to stop producing their versions of the National chip.[35]

Lest it seem that the Japanese and the NICs are the only companies guilty of patent infringement or other forms of industrial espionage, consider the fact that dozens of American companies are involved in similar legal tangles with each other. In 1984, National Semiconductor paid IBM $3 million to settle an industrial-espionage suit. National and Hitachi have a joint-venture agreement to manufacture IBM-compatible computers. National, although admitting no wrongdoing, had become ensnared in the now famous FBI sting that caught the Japanese giants Hitachi and Mitsubishi with their hands in IBM's cookie jar.[36]

The elaborate sting began with the opening of yet another Silicon Valley consulting firm. In October 1981, Glenmar Associates opened its doors in Santa Clara. Among its offerings was a secret design workbook code-named "Adirondack," for what was then IBM's newest and most powerful computer, the 3081K.

Adirondack and other IBM secrets were dangled before two of Japan's largest computer firms. Hitachi gave Glenmar $622,000 for the information. Mitsubishi paid Glenmar $26,000. A further payment of $1 million was promised, but before it could be made, the FBI moved in. On June 22, 1982, FBI agents "crashed" a meeting just as Hitachi employees were accepting notebooks filled with IBM secrets. Hitachi and Mitsubishi and eighteen of their employees were charged with conspiracy to steal proprietary IBM information. Glenmar had been a classic FBI sting.

The Japanese denied all charges and in fact were furious at having been set up. When the case came to trial, Hitachi plea-bargained an admission of guilt and paid a $10,000 fine. Hitachi subsequently paid IBM $300 million for its part in the affair. In fact, the Japanese claimed to be doing nothing illegal; dozens of consulting firms staffed by ex-IBMers and engineers from other computer and chip makers are available to anyone who will pay their fees. The line between proprietary and public information is sometimes fine and sometimes overlooked. But not this time.

FBI tapes showed Kisaburo Nakazawa, head of the Hitachi plant that develops and manufacturers computers, offering a million dollars for the specifications and actual parts of what were then IBM's latest generation of computers. Other tapes showed senior Hitachi officials ordering information such as manuals and spec sheets and then asking that a receipt and other papers connecting Hitachi with the payoffs be destroyed. Some even joked about how "complicated the spy business is."[37]

It is also almost commonplace and, depending on the viewpoint, an absolute competitive necessity. "I think every offshore subsidiary of a company has very sophisticated intelligence-gathering networks," Sheridan Tatsuno of Dataquest told me. "IBM is probably one of the most sophisticated in the world, and then NEC and Hitachi, in that order, in terms of just gathering information and feeding it back to headquarters. I wouldn't be surprised to find ex–intelligence officers in those companies. Of course, Japen doesn't have anything like a CIA to speak of. But I think a lot of the intelligence gathering is a fairly well developed skill in this country. Major conglomerates use it. The Japanese have developed it primarily for their trading companies; they have ears out there. And the trading companies supply information to their allied companies.

"There's a difference in the focus. American intelligence seems to be geared to geopolitical issues such as the overthrow of a government, say, in the Philippines, and its effect on our semiconductor makers. The Japanese, on the other hand, who are clearly worried about it, still tend to focus more on strategic commercial interests.

"We are talking about a very high stakes, high risk game, with billions of dollars invested. Anybody who thinks they can operate in a foreign country without any kind of intelligence gathering, such as market research or gathering information of strategic interest to their company, is going to fail. No company can succeed offshore without doing that. There are illegal means and legal means, and most companies try to stay within the legal limits. And most of the information is free; by the nature of the technology it's out in the open; it's in the heads of people. You can hire somebody, and you get their technology.

"So you don't have to do anything illegal. American companies raid each other all the time; that's why there are so many lawsuits. If a Japanese company does the same thing, we call it spying. It's a different term for the same phenomenon. It happens; that's the nature of the game. What's important is, are we equally sophisticated in gathering information as they are? The major U.S. guys are. After that, it's miserable, and that's true for the Japanese, too."[38]

While we can beef up our patent and copyright laws, place more and more chips on a control list, follow as many Soviet and foreign company spies as we can, inevitably, America's high-tech genius will be stolen, transferred, and otherwise acquired by the rest of the world. It is the inevitable price we pay for an open and free society. It is also the reason we are capable of achieving the kind of creative, innovative technology that all the world envies and lusts after. And it is this technological genius that is perhaps our best weapon in the Chip War.

10

Fairjitsu—The Monster That Shook the Valley

It was if a central pillar of the American industrial myth were being destroyed. The earth mother of Silicon Valley, Fairchild Semiconductor Corporation, the progenitor of at least a hundred other American semiconductor companies, was being sold to the Japanese. The storm and fury that followed Fairchild president Donald Brooks's announcement in October 1986 that the venerable firm, once one of the most innovative and profitable in the industry, was to be sold to Fujitsu Limited, a giant Japanese computer and semiconductor maker, swept across the Valley and thundered like a tidal wave into the halls of Congress and the White House.

From the floor of the Senate, denunciations rolled with the enthusiasm of pork-barrel appropriation pleadings. Nebraska senator James Exon, a member of the Armed Services Committee, worried that "a major vendor of vital components to the U.S. Defense Department and the sole supplier of certain devices which are vital to important defense programs would be lost to the Japanese."[1]

"The proposed Fujitsu acquisition of Fairchild Semiconductor could permit Fujitsu to dominate the American supercomputer market," cried Ohio's Sen. Howard Metzenbaum. *New York Times* columnist William Safire chipped in with the following news:

Not more than 50 or 60 supercomputers exist in the world today. They are used by governments to make and break codes, predict weather and target nuclear weapons. The few companies that make them already rely on Fujitsu for over half the components and the purchase of Fairchild would be a big step toward monopoly of the components that will go into Star Wars research.[2]

Cray Research, the nation's largest builder of supercomputers and perhaps the world's technological leader in the field, was also worried and let the Congress and anyone else who would listen know about their concerns. Cray has been developing a new family of super-computers code-named Y-MP. Critical to the new machines is a high-density logic chip that does arithmetic calculations at very high speeds. Cray, which used Fairchild logic chips in their current line of supercomputers, was working with them as well as with TI and Motorola on a new generation of the chips. Fujitsu, also a supplier of logic chips to Cray, wanted in on the deal. Cray refused, fearful of sharing their proprietary circuit designs with a company that was not only their supplier of chips but a major competitor in the super-computer market. Now, it seemed, Fujitsu would have direct access to Cray's research by acquiring Fairchild.[3]

The Defense Department, the CIA, and the National Security Agency (NSA), all major Cray customers, also registered their concern. At a White House breakfast meeting in March 1987, then defense secretary Caspar Weinberger and the late Malcolm Baldridge, the commerce secretary, both advised President Reagan to block the deal. "I've come out against it," Secretary Baldridge told reporters after the meeting. "Its really bad policy." Another official, who did not want to put his name to his remarks, declared, "This is a test case. If Japan can come in and buy this company, it can come in and buy them all over the place. We don't want the semiconductor industry under Japanese control."[4]

But for some the question was why on earth did the Japanese want Fairchild in the first place? NYU's Tom Pugel, associate professor of economics and international business at the Graduate School of Business Administration, thought Fairchild was getting far and away the better of the deal. "In the Fujitsu-Fairchild case," he avers, "Fujitsu would have contributed more to Fairchild in terms of technology than vice versa."[5]

What Fairchild would contribute was better access to the U.S. market than Fujitsu already had. "It was pretty clear that Fujitsu was gaining access to a whole set of distribution channels," added Pugel, "that, as I understand it, have so far been closed off to Japanese companies. One could easily argue that U.S. companies are potentially violating the antitrust laws by limiting Japanese firms from accessing U.S. distribution channels. I have heard that U.S. companies have threatened their distributors, both explicitly and implicitly, with loss of their products if they distribute Japanese products. That is potentially actionable under both U.S. and Japanese antitrust laws. It's also the kinds of thing that U.S. companies complain about in Japan. In fact, if I had to guess why U.S. companies were so against the Fujitsu takeover, it was much more that fear than anything else."

A similar view came from Tokyo, from another academician, Professor Gene Gregory of Sophia University. "When weighed in the balance, the principle victim in this affair is Fairchild, not Fujitsu," he wrote in the April 13, 1987, edition of the *Japan Times*.

Had the proposed acquisition been consummated, Fujitsu would have merged its U.S. and European semiconductor operations with those of Fairchild and contributed $400 million in new financing to create the world's third-largest semiconductor maker, with prospects of holding its own in the coming shake-out and consolidation of the industry. . . .

Fujitsu spends approximately four times as much as Fairchild on R & D per annum and has some of the best know-how in the business. Since Fairchild International, as the new company was to be called, would have been Fujitsu's principal international semiconductor interest, it would have become an important conduit of advanced technology from Japan to the United States, of benefit not only to Fairchild but a much larger community as well.[6]

To Fujitsu's management, it was probably more a question of status, of gaining the recognition they felt their $12 billion company deserved on the world scene. "Fujitsu so far doesn't have any alliance with any other foreign company, U.S. or European. We were a frog having its own world but not knowing the outside world," Fujitsu's director of International Semiconductor Operations, Sadao Inoue, explained to me. For our meeting at Fujitsu's headquarters in the high-rise business section of Tokyo called Maronuchi, Inoue, a careful and supremely confident man, insisted I bring an interpreter, and so I did, a matronly lady in her forties. After the introductions, Inoue proceeded to ignore the woman, answering all my questions in accented but excellent English. Only once, when at a loss for a word, did he call upon her for help.

"We have the highest technologies in the world, the highest productivity, but we don't know much about what other people are doing. We have to learn a lot of things from others. It is not true that we would steal technology from the U.S. I think the technology flow in this transaction would have been from the Fujitsu side to Fairchild rather than the reverse. But we needed to know how we could establish a pure U.S.-based company, a U.S. type of business, with distribution and production.

"We have some very lower level of alliance with TI for second sourcing of gate array chips, some with MMI, Siemens. But this is just a business arrangement, just shaking hands, not getting married. But in the case of Fairchild, it was expected to be 80 percent owned by Fujitsu. And this was a U.S.-based company, a place that might be very good for the future of Fujitsu. My opinion was, and still is, that we have to have such a close relationship with some American producers.

"Of course we have a subsidiary business now in the U.S., in Santa Clara, and offices elsewhere in the U.S. and a factory in San Diego, but this is 100 percent owned by Fujitsu. Management people are mostly American, but at the top are Japanese. For instance, Fujitsu Microelectronics Inc., located in Santa Clara—the executive VP is Japanese, and under him are several U.S. managers and directors of the company. And we have now just announced [September 1987] a new factory in Oregon for wafer production. The factories in San Diego were just assembling and testing. The reason why we have given up efforts to buy Fairchild: Last March, our management decided Fujitsu would not be welcome by U.S. industry."[7]

Often lost in what had become an emotionally charged debate over the sale was that Fairchild had already been sold once—and to a foreigner at that. In 1979, a problem-riddled Fairchild was sold to the French-owned Schlumberger Ltd., a giant oil-field-exploration and supply company with worldwide operations. Fairchild had never fully recovered from the defections that had rocked it a decade earlier, when many of its original innovative engineers and technocrats, such as Charlie Sporck, Jean Hoerni, and Bob Noyce, left to form their own companies. To replace them, Sherman Fairchild, the original financial godfather of the company and its primary stockholder, lured C. Lester Hogan, then head of Motorola's semiconductor operations, to Fairchild Semiconductor. In negotiations that sound more like today's free-agent baseball deals than corporate employment contracts, Hogan turned down Fairchild's first offer after Motorola gave him a hefty $10,000 pay raise, boosting his salary to the then princely sum of $90,000 a year. Sherman Fairchild was not to be denied, and with the flair of a George Steinbrenner signing home-run hitters, he brought Hogan to the presidency of Fairchild Semiconductor with an astounding deal. Hogan received a three-year contract that would pay him $120,000 the first year and a total of $1 million over the life of the contract. In addition, he got, as a bonus, an interest-free $5.4 million loan to be used to exercise a 90,000-share stock option at $60 a share. As if that were not enough, Hogan was able to buy another 10,000 shares at the fire-sale price of $10 a share.[8]

It proved to be a shrewd move initially. For, just as the signing of a Reggie Jackson in his prime could double season-ticket sales, so, too, could the addition of a Lester Hogan boost Fairchild's appeal to the investing public. Fairchild stock, which had dropped seven points when Bob Noyce left to form Intel, jumped back up that same seven points when Hogan came on board. At the same time, Motorola dropped eight points. Hogan proved to be an even bigger bargain than even Sherman Fairchild had anticipated, for he brought with him a team of Motorola executives who vowed to turn the company around. Called Hogan's Heroes, after the then popular TV series, they immediately clashed with the remaining Fairchild executives, who had originally built the company. To restore order and establish his iron control in no uncertain terms, Hogan tiffed with Jerry Sanders, the self-styled peddler, Fairchild's chief of marketing. Sanders says he was fired, Hogan says he left of his own accord, but in the final analysis, it made little difference.

Sanders took yet another clutch of Fairchild talent with him to spin off yet another company that would turn around and challenge the Silicon Valley mother of multitudes. Moreover, Sanders's departure came at a particularly bad time, for a year later, 1970, the industry was again struck by one of its periodic recessions, a sort of economic Santa Ana that left everyone feeling woozy and depressed. Fairchild, which had been losing money before Hogan's arrival, was doing it again. When the winds of recession blew themselves out, the two dozen or so companies that Fairchild had spawned, like some giant amoeba splitting off daughter cells, had taken huge bites out of the mother company's markets. In the midst of the recovery, Fairchild was still in the red. And someone had to pay.

In 1974, Hogan, a home-run hitter no longer, was eased out of the lineup. He was replaced by one of his imported heroes, an ex-Motorola executive named Wilfred Corrigan, who became president and ultimately chairman and CEO of Fairchild Semiconductor. Corrigan decided to expand Fairchild's markets and undertook the development of a microprocessor. Sales were good, but Intel, Motorola, and a company called Zilog, had the bulk of the market. Nor was Fairchild ever able to mount a successful challenge in the DRAM business. Still, the balance sheet showed a definite improvement.

Things were looking up, and Fairchild management cast about for new worlds to conquer. To Corrigan, it seemed that the chip-based digital watches and video games that were then dazzling the American consumer might be just the thing to make Fairchild's sales really soar. A consumer products division was created, and the venerable semiconductor company was geared up to do battle with the thousands of Hong Kong grannies who were slapping chips and plastic together, creating watches and video games with blinding speed, and the canny Chinese entrepreneurs who sold them. It was no contest. Fairchild blew $40 million in four years on the fiasco. Still, the company was in the black, thanks in no small measure to the royalties that poured in from all over the world on the patents developed in its early years. Its new chips, however, were less than sensational. The microprocessors and memories it built were good, but not the products of leading-edge technologies that were propelling the sales of chips coming from Intel, Motorola, and others.

Fairchild might have limped along in this fashion for quite some time, but suddenly, financial sharks appeared on the horizon. Gould Inc., a large, cash-laden U.S. electronics systems maker, was anxious to acquire a semiconductor manufacturer. In a hostile takeover bid, they offered seventy dollars a share for the Fairchild stock, then selling at fifty four dollars. The Fairchild board turned noses up at the offer and scrambled around for another, friendlier buyer. They found it in

Schlumberger Ltd., a $3.5 billion French multinational company whose basic business and revenues were based on the services it rendered in the world's oil patches.

In July 1979, Schlumberger officially acquired Fairchild for $425 million, a rather startling $253 million more than the net assets of the company. But Schlumberger chairman Jean Riboud saw this as an opportunity to propel his company into the very heady world of high technology, an arena into which he felt they must move. To replace Will Corrigan as president, Riboud appointed Thomas Roberts, a brilliant, intense man who was then Schlumberger's chief financial officer.

The idea of a foreign company acquiring Fairchild, at the time still a major supplier of chips to the military, did not appear to bother the Defense Department overly much. But Fairchild also provided the CIA and the NSA with vital surveillance and communications devices, and they were concerned. That Roberts, the new president, had been a West Point graduate was a plus. The CIA cleared someone on the board, presumably Roberts, for top secret, thereby creating a sort of corporate cutout for U.S. secrets. He would in essence filter all classified material before it reached the presumably less than security conscious French members of the Fairchild board.

Roberts swept into Fairchild with a clear idea of what needed to be done. He instituted a five-year plan that was to decentralize the company's operations, reduce its staff, and pump big dollars into R & D and upgrade the manufacturing facilities. Over the next four years Schlumberger poured more than half a billion dollars into new plants and new fab lines in existing plants. The Fairchild banner was planted overseas as well in newly built facilities in West Germany and Japan. At the same time, Roberts slashed the work force by two-thirds, reducing the worldwide employee rolls from 30,000 to 10,000.

"This place was fat," Roberts was quoted as saying in 1981. His coldly efficient, military approach to management and his insistence that Fairchild could be run like any other business ran counter to the prevailing Silicon Valley philosophy that the semiconductor industry indeed operated by other rules. It also alienated many of the longtime Fairchild loyalists who had remained with the company through all the rough years. In a déjà vu replay of the late sixties, top management talent fled Fairchild to join other companies and start new ones.

Roberts was left not only with no Indians; he had lost virtually all of his chiefs as well. Riboud advised Roberts to seek help outside of Fairchild. The new savior was to be Donald Brooks, a highly respected executive with twenty-five years' experience at TI. Brooks, with the title of vice-president for North American operations, was to work under Roberts and provide the nuts-and-bolts knowledge of the chip industry that Fairchild had to have at the top.

It was soon apparent to Brooks that he would have to take complete control in order to turn Fairchild around. He finally convinced the Schlumberger management and replaced Roberts as the new president of Fairchild. Brooks was as amiable as his predecessor was cold, but he brought more than a personality change to the company. He brought both new ideas and yet another management team to Fairchild, importing a slew of people from TI. Their task was to move Fairchild out of the murderously competitive commodity-chip business and into the custom area, producing fewer parts that could be sold for higher prices with greater profit. It might have worked, but Schlumberger, which had itself undergone two palace coups in the interim, was tired of the staggering losses, $1.5 billion by the end of 1986. Its own greatly reduced revenues from the main part of its business, the oil industry, made the semiconductor drain just too expensive to continue to support. Schlumberger's new head, D. Euan Baird, a cost conscious Welshman and the first non-Frenchman to become chairman, wanted to unload.

Brooks got the message and cast about for a buyer. And there was mighty Fujitsu, checkbook in hand, prepared to pay $250 million in cash for 80 percent of the company that had only the year before had its book value drop from $800 million to $315 million. Fujitsu also agreed to pump another $400 million into the company, much as Schlumberger had originally done, to once again make it competitive. Schlumberger would retain 20 percent. At that price it seemed a good deal for Fujutsu and to Brooks and Baird; the Japanese must have seemed like on Oriental Santa Claus.[9]

The resulting uproar killed the deal. Fairchild sent a 1,000-page report and more than a hundred boxes of documents to the Justice Department to support their argument that the sale was legal and of great benefit to Fairchild and the nation.[10] Brooks hit the lecture trail, showing up at industry conferences to make speeches about the inevitable globalization of the semiconductor industry. It didn't help. The press, the Congress, the administration, and even the American semiconductor industry were all arrayed in opposition to the deal that would probably have saved Fairchild and made it a viable competitor once again.

Michael Malone, a long-time industry observer, wrote in the *Wall Street Journal*:

Business theorists glibly speak of "The Globalization of Industry" to nodding executives at industry conferences. But the gauntlet run by Fairchild through xenophobia, government bureaucracies, lobbyists and congressmen, a hostile press and industry peer pressure—all in an effort merely to survive—rebukes theory by example. As a window on our society, the Fairchild story can only instill deep doubt about the prospects of a vehemently parochial American industry ever competing with its internationalized competitors in Japan and other Asian countries.[11]

The obituary seemed premature. Brooks and some of his key executives then tried a management buyout, financed in part by Citicorp Venture, Intergraph Corporation, a company using Fairchild's Clipper microprocessor, and that bad penny turning up again, Fujitsu. According to the deal, Brooks would remain firmly in control, and Fujitsu would own 20–30 percent of the company but would have no say in operations. But then MITI, fearful of the growing protectionist sentiment in the U.S. Congress and the Reagan tariffs that had just been imposed at the beginning of 1987, dropped a few words of wisdom in the ears of Fujitsu's management. Fujitsu pulled out of the deal, its long-cherished hopes to align itself with an American semiconductor maker and become a "big frog" dashed again.

Even without Fujitsu, Brooks and his partners had high hopes that Schlumberger would accept their deal. The offer was for a total of $180 million, a good bit short of the Fujitsu offer but a lot better than a continuing sea of red ink that was certain if Schlumberger failed to sell the ailing company. Then, on August 31, came an announcement that stunned Brooks and the entire industry. National Semiconductor, one of the original Fairchild spinoffs and the world's eleventh largest semiconductor company, had agreed to buy Fairchild for $122 million in stock and warrants. Not a nickel in cash would change hands.

"We were stupefied," declared a Brooks investor. "Schlumberger never even came back to negotiate." Why not? "All I can believe is that Schlumberger was scared of being embarrassed if the management group did well," the disgruntled investor told *Business Week*.[12]

"More probably, Schlumberger recoiled at the weight of the debt load the Brooks group's buyout offer entailed," Jay Cooper, the semiconductor analyst at Eberstadt Fleming, advised me. "They were probably concerned as to whether they would ever get paid or not and how long it might take, assuming the management group made it. There are a lot of fixed charges in a leveraged buyout, and this is an extremely cyclical industry, and you might go under trying to pay those fixed charges at some point in the down side of the industry cycle. That's why it's almost a contradiction in terms. You really should have a more stable business for a leveraged buyout."[13]

And so in the end it was one of the original Fairchildren, Charlie Sporck, who brought the almost soap-opera saga of Fairchild Semiconductor to a final conclusion. "The acquisition of Fairchild," he explained to me after a press conference in New York, "was really driven by the fact that we looked at their product line; we found that it was nicely compatible with ours, meaning that it was not a duplication. They built fast circuits; we didn't. They built FACT [Fairchild advanced CMOS technology]; we didn't. They built ECL gate arrays, logic; we didn't, so it fit well. Now there were a few things they built that we also built, and that was a duplication, but they happen to be relatively minor, so it's a nice fit. And we felt frankly we could turn them around without losing a lot of money. With our size, combining Fairchild with us, making the total even larger, we felt that there were economies of scale and opportunities there."[14]

In fact, the combination of National and Fairchild makes the new company the sixth-largest semiconductor maker in the world. But if Sporck has his way, it will be a lean, mean sixth-largest company. "We have significantly reduced the duplication in many areas. There were a number of plants that we did not take with us. Plants that stayed with Schlumberger, which they are selling separately or have already sold. We didn't take them because they were duplications. There was only one product line which we sold off, and that was their microprocessor, called Clipper. We sold that to Intergraph. But we already have a microprocessor, and it didn't make sense for us to have two. We have made a lot of progress with Fairchild, and I think there is no question that we can turn the corner."

So much so that by the end of the first quarter of 1988, Fairchild had just about broken even.

Fairchild was not the first chip maker to attract the covetous eye of Japanese investors. In 1977, Nippon Electric, NEC, purchased Electronic Arrays, a semiconductor manufacturer in Mountain View, California. Now called NEC Electronics USA, the company continues to make chips, but it took five years before it could be turned into a profitable operation. NEC now has nine plants in the United States, and other Japanese companies have followed suit. Toshiba bought Maruman Semiconductor in 1979, and Fujitsu, even though it failed to buy Fairchild, owns half of Amdahl Computer. And despite the Fairchild debacle, Fujitsu, in the fall of 1987, began construction on a wafer fab in Gresham, Oregon, a suburb of Portland. The idea is to complement Fujitsu's existing assembly plant in Santa Clara.

"The press announcement about the opening of our new plant in Oregon was very different from our experience with Fairchild," Sadao Inoue told me with a shake of his head that implied that he would never understand Americans. "We were very welcome in the area. It was a very big difference from Fairchild. Everybody, including the governor of Oregon and the mayor of the city, were attending and welcomed us."

The high yen versus the dollar and fears of increasing American protectionist sentiment in Congress have sent the Japanese on a U.S. buying and building spree. Hitachi, for example, has built a wafer-production facility in Irving, Texas, almost in the shadow of the Dallas Cowboys' football stadium and close by its already existing testing and assembly plant. Beginning in May 1989, Hitachi hopes to produce 256K static RAMS and eventually DRAMs, ASICs, and microprocessors there.[15] Toshiba expanded its Sunnyvale, California, plant to begin assembly on the 1-megabit DRAMs. Also in Sunnyvale, Oki Semiconductor, another major Japanese semiconductor producer, announced its plans in the beginning of 1987 to expand and add fabrication to its existing chip-assembly plant. And Mitsubishi opened a plant to make DRAMs in Durham, North Carolina.[16]

The Japanese are also getting into the venture-capital business, funding U.S. start-ups with an eye to diversifying their own interests. In May 1987, Nippon Steel provided the seed money to four American businessmen and engineers to create a company called Simtek Corporation. Simtek will begin by designing and producing custom memory chips and then logic circuits. Ironically, the four Americans who had been with another semiconductor company called Inmos had become expendable when that company was acquired in 1985 by a British conglomerate.[17]

Operating in America either through wholly owned subsidiaries or acquisitions is, however, never easy for the Japanese. Buying a company in particular is almost antithetical to the Japanese character. Their vision of a company is of a living entity, something much more than a mere collection of equipment and people joined together only by a company logo. "Tradition holds that a company is something another company has no right to buy," declares Kenichi Ohmae, the author of several books on Japanese business. "It's out of our character; it's not in our upbringing," he adds.[18]

Nor are the Japanese the only foreigners buying up U.S. companies. The Koreans, the Dutch, the Germans, the British, and the French have all made substantial investments in the U.S. semiconductor industry. One of the more interesting purchases that had many of the same emotional and historic aspects as the Fairchild imbroglio was the purchase of the Texas-based Mostek Semiconductor Corporation. In 1985, Thomson CSF, the giant French electronics company and defense contractor, bought Mostek Semiconductor. Mostek, once one of the world's largest producers of memory chips, had been a hugely successful spin-off from TI. A couple of bad years in the late 1970s, however, had taken some of the shine off the company, and in 1980 it was acquired by United Technologies, the large U.S. defense contractor. It looked like a smart move, for Mostek finished 1980 with sales in excess of $300 million.[19]

But the bottom was already beginning to drop away. "No one predicted the flattening of total demand for memories," recalled Vin Prothro, who became president of Mostek in 1981. "For the first time in history bit consumption stayed flat for three or four quarters. At the same time, all the Japanese producers started up. There was gross overcapacity. No amount of cost cutting could help you." As Prothro watched, the price of the 16K DRAM, Mostek's most important chip, skidded from six dollars to under one dollar in less than a year.

"The Japanese were quoting ten cents less than us in every account," Prothro declared. "The last part of 1980 was slow, and '81 was terrible." UTC tried to halt the slide by throwing money—$600 million—into Mostek and was still in the red. The onslaught of the Japanese on the memory market, which comprised 60 percent of Mostek's sales, had battered company sales almost to the vanishing point. Finally, in October 1985, Harry Gray, then the chairman of United Technologies, decided to bite the bullet and take a $424 million write-off and shut Mostek down entirely. It was the only way to end the losses, which for the first nine months of that year had reached $328 million. And then, in came Thomson, a blushing suitor so blinded by passion that it failed to see the bride had warts, a wooden leg, and very little in the way of sex appeal.

"Thomson must be crazy to buy that lemon" was the prevailing sentiment. But there was definite method to the seeming madness. "You can't have the ambition of being a world-size semiconductor company without producing and selling in the U.S," argued Thomson chairman Alain Gomez, a Harvard business school–trained Frenchman who took over the largely state owned company in 1982.

Moreover, the biggest part of Mostek that most industry analysts viewed as a disaster area, its memory-chip production capability, was the major attraction to Gomez. He viewed strength in memory chips as essential to his company's very survival, the technology driver every chip maker must have. "The company must keep up in memory-chip technology," he stated flatly, "to fabricate the specialized chips necessary to compete in its core markets. It's vital to the existence of any diversified electronics group."

A second appeal to Gomez was the price. United Technologies was willing to sell Mostek for a mere $71 million. For the money, Thomson was getting four fairly up to date production lines, a research center, and an admittedly dispirited and depleted sales force, but one that was in place in the American market. The deal was struck, and although Mostek's president, Dr. James Fiebiger, a respected former Motorola executive who had been brought in as window dressing for the sale of the company, was retained, most other Mostek employees were not. The company, which had once employed 10,000 people before being briefly shut down by United Technologies, was reopened by Thomson with a work force of 1,200, a number much more in keeping with Mostek's production and sales capabilities. In its first full year after the acquisition, as the industrywide recession began to abate toward the end of 1986, Mostek, with the addition of several Thomson products imported from Europe, just about broke even. And that is just about where the Mostek story ends, for after Thomson merged with SGS, its name was erased from the letterhead; a once proud and successful American company vanished into the maw of a foreign electronics giant.

U.S. electronics defense contractors are much tougher morsels for foreigners to pick up, much less swallow. The Harris Corporation, a Florida-based electronics firm whose products include semiconductors and top-secret communications equipment for the military, was recently the subject of an acquisition attempt by Plessey Company, a British telecommunications-equipment and semiconductor manufacturer. As if trying to prove its evenhandedness, the Defense Department objected strenuously to the bid on the grounds of national security. "It's like the Fujitsu problem," said a Defense Department official who did not want to be quoted by name. In Harris's case the chips in question were not destined for supercomputers, but rather special radiation-hardened parts used in sensitive military gear. "We don't want anyone to control those parts outside the United States," he concluded emphatically.[20]

What raises even more fears in government offices and some academic circles is the joint ventures and technology-licensing agreements and swaps that have seen a vast range of American technological know-how bleed away to hundreds of foreign companies. It is a process that has been going on for a very long time in America and one that is continuing at an increasing rate. "There is hardly any industry where we haven't transferred technology to Japan," angrily declares Clyde V. Prestowitz, one of the Commerce Department's top trade negotiators with Japan from 1981 through 1986. "If we give our technology away, we have nothing to compete with."[21]

For too many American companies, the licensing of technology to foreign companies is an easier way to turn a profit on a research investment than producing and selling the product derived from that research to those same foreigners. Perhaps nowhere has this been more true than among the major companies in the American semiconductor and electronics industries. The case of RCA in this regard is particularly illuminating.

"RCA had a history of transferring technology to Japan," Bernie Vonderschmitt told me. Vonderschmitt, a lean, totally bald man in his early sixties with piercing eyes and an intense interest in the interface between technology and business, is the chairman and CEO of a very successful start-up company in Silicon Valley called Xilinx. We were meeting in his executive offices in San Jose. For many years before, however, Vonderschmitt was with RCA's solid-state division, becoming the general manager in 1972 and finally leaving in 1979. "The management at the time felt the best way to capitalize on their research rather than becoming an international company was to license technology and settle for the royalties. Immediately after the war it was black-and-white television; then, in the middle fifties, it was color television. It was a conscious decision to become a licensing agency instead of an international marketer. Now, it paid well for a long time. RCA for long periods has gotten from $75 million to $125 million a year in royalties. And there was a licensing group within RCA whose job it was to sell the new technologies that came out of the very superb work that was being done at the David Sarnoff Laboratories in Princeton."

Among the technologies RCA licensed during this time was a process called CMOS. "As chips become more complex," explains Vonderschmitt, "the only technology that permits you to go almost indefinitely to increased complexity is complementary MOS, because of its low power consumption. You don't have to worry about thermal management at all."

CMOS was a major breakthrough in chip technology, for it meant that more and more transistors and circuits could be packed onto a chip and that the space between them, previously needed to absorb the heat from the higher voltages required, could now be used to carry working elements. "This was a breakthrough for military and space applications," points out Vonderschmitt, "because suddenly you had a ten-times' or a 20-times' improvement in the density of the chip. Even a two-times' improvement is significant, and ten times was remarkable. We were back at 15–20 microns then, big stuff; you couldn't put much on a chip by today's standards. But nobody really focused on advancing the state of the art, including RCA, to get feature size down.

"We took the CMOS technology and moved it into the product division. As with all things at the beginning, we started with small products. We made a family of standard CMOS devices that were very similar to what was referred to in the industry as TTL, or transistor-

transistor logic, a family of devices which TI had developed and which became an industry standard for random logic. But because it was expensive initially, it wasn't used on complex devices. The only major American companies at that time, now the early seventies, to pick up the CMOS technology were Motorola and National Semiconductor.

"There was never a conscious technology transfer to a U.S. company. Motorola and National Semiconductor looked at what RCA was doing and decided they could do it themselves. Cross-licensing was done on some patents, but I don't think any money actually changed hands."

That was not the case with the Japanese. RCA was more than happy to sell its CMOS technology to companies like Toshiba and Hitachi. "Prices varied," according to Vonderschmitt. "There were usually front-end payments, but they were not very big. Front-end payments were basically requested to make sure there was commitment to development. But the biggest gain was eventually to come out of the royalties, usually 5 percent or less. That can still amount to a fair sum."

The kind of corporate thinking that led to what was in essence a technology giveaway can be seen in the operations of RCA's solid-state or semiconductor division in the 1970s. "At the time I became general manager, that division was about an $80-million-a-year operation," Vonderschmitt told me. "When I left in 1979, it was $350 million. Being general manager of a division like that can be a very frustrating experience. Frustrating from the standpoint that you have a lot of profit centers and product lines and you're not really running any of them. It's doing nothing more than people interface, and you don't have control of your own destiny in that you're part of a large organization. RCA at the time was a very disparate company from the standpoint that in the late sixties and early seventies they decided to become a conglomerate. That was the corporate fad at the time, and Bob Sarnoff did that. The result was that even though in the semiconductor business we were a reasonably good sized company, we were part of an $8 billion corporation.

"Being part of an $8 billion corporation, where the top management has many other things to concern themselves with and where the background of the top management was more from the service side than the high-tech side, it was extremely difficult to get their attention. The company at that time was tied up in a project called video disk.

Video disk siphoned an enormous amount of funds, both capital and discretionary, from the company's coffers. So it was extremely difficult to get capital equipment.

"The misfortune of a company like RCA in that arena was that the visionary who had built RCA, David Sarnoff, was no longer there. He would have recognized the potential of the semiconductor and seen where it would fit in the consumer-electronics business. But if you go back in history and look at the electronic-device business and you look at those in the forefront, you find RCA, Sylvania, Raytheon, and others. All of them in the vacuum-tube business. None of them ever became a factor in the transistor, semiconductor business. And the reason is very simple. The management at the top did not recognize the fact that unless you build new operations to focus on solid state, on transistors, and then integrated circuits, you would never make it. Instead, it was dumped into the old tube operations. The management there had a vested interest in prolonging the life of the tube business and had no interest in developing this new business called semiconductors. The result was it got little attention, was unfocused, and launched very poorly.

"On the other hand, the guys who hadn't been in the tube business, Texas Instruments, Fairchild, and eventually Motorola, became the leaders of the solid-state business. And out of those sprang others like Intel, National Semiconductor, and AMD."

After thirty years with RCA, Bernie Vonderschmitt decided to leave and learn how to run his own business. He began by taking an MBA at Ryder University in New Jersey. "Its very hard to see," he told me as we sat in his executive suite in Silicon Valley in the summer of 1987, "but there is a broad gap between being an engineer and being a businessman. The thinking you do, conceptually the things you do, are quite different. RCA trained the people who ran their business very well. But at RCA you were sensitized to cash flow, and the reason for that was that everything was going into the videodisk project. There were good fundamentals at RCA, but you really didn't have the base to build on. You learned by being thrown into it and figuring out for yourself how do I really do this job.

"When I took the MBA course, I found it to be extremely appropriate. However, if I hadn't been through the mill at RCA, it probably would not have been anywhere near as meaningful to me. I would therefore encourage no one ever to take an MBA without first have rubbed their noses in the business world for at least five to ten years."

Vonderschmitt did not have to scurry about looking for a job upon graduation. A large semiconductor company called Zilog, a subsidiary of the multinational oil company Exxon, had been after him almost from the moment he had left RCA. In 1981 he became general manager of Zilog. "There I met some superb people. One in particular had such an overall understanding from process to systems that it just seemed as though this was one way to leverage an incredible amount of brainpower and ingenuity and innovation capability and develop a business out of it.

"His name was Ross Freeman, a young engineer who covered a broad gamut of information, knowing and understanding the basic process and its limitations all the way to the end use of the product. He was also very adept at implementing ideas, at making a bridge between the process and the systems into which they would be put. And reducing it to products which recognized on the one hand what limitations of the process were and in what direction they were going and on the other hand seeing what the payoff would be. Freeman came up with a basic concept which appeared to solve a problem in the semiconductor industry better than anyone else had been able to do.

"The problem we were addressing," Vonderschmitt went on, "goes way back to 1975, 1976. People were saying we are developing a semiconductor manufacturing process and are able to put many transistors on one piece of silicon and get good yield, yet we don't know how to apply that. Gordon Moore, the chairman of Intel, and the best visionary from the standpoint of process and where it's going of anybody in the industry, bar none, had the concept that every couple of years you can double what you can put on a chip. His constant worry was how do we direct this great process capability we have to end products. How do we challenge the end products?"[22]

That sort of thinking led Intel into the memory business. And the memory chip became a technology engine, driving innovation and ideas to greater densities and more memory. For every computer, from the smallest to the largest, constantly demands more memory. And as the cost of memory goes down, computers become more powerful, and the cost of computing also drops significantly.

"That was the basic Intel concept, and that then led them to develop the microprocessor," continued Vonderschmitt. "At that point Moore said, 'Here we have a product which can be fairly complex and challenges the process capability, yet it's a standard product. We've got memory that stores information and microprocessors which calculate and control its applications. But you need one more thing to tie these two together, to let them talk to each other. And that is logic, random logic. That permits you to build systems with these two building blocks.'

"At the time, in the middle sixties, RCA was making logic chips for what was called the custom market. "It meant that you designed an IC which would be used in a very specific application, such as TV sets, radios, automotive applications, where your volume might be in the millions," explains Vonderschmitt. "RCA built a lot of those. At any one time we had twelve to fifteen custom devices running in production. And we made them by the millions so you could amortize a very significant engineering investment, which was usually a fifteen- to twenty-four-month development cycle. So if you're investing a million dollars to engineer each of these, and some were significantly less than that, and you produced several million of each, there was a substantial profit. So we started saying we've got this great process, CMOS, and we can build very complex chips without ever having to worry about temperature or overheating or dissipating heat. There would be no heat sinks, and packing would therefore be simpler and cheaper.

"It was suggested that what we should do is take those analog or logic functions that we were doing as custom chips and make arrays of these things. That is, put a number of the different logic functions on one chip. Basically, that was the concept of gate arrays. For RCA it was like going from vacuum tubes to transistors. Our division was basically a $250 million company in a lot of other products, and it wasn't focused. The result was it never got enough attention to get it launched properly. Moreover, the computer-aided design tools were not sufficiently advanced at that time."

But other people were also thinking about gate arrays, and in 1980 a company to make them—LSI Logic—was formed. "They did a lot of their own CAD work," continued Vonderschmitt, "but there were some other companies out there, as computational power was becoming cheaper. This minimized the engineering investment and allowed for a large number of designs at moderate cost, and that in effect made gate arrays practical from a design standpoint.

"At Zilog, we looked at this, and we looked at the market that was developing, and we said wow! There's got to be a better way of doing it. The Achilles heel we saw was that if you are making millions of these things it makes a lot of sense. But in the gate-array business they were designing these things for every application no matter what the volume. Payment was for NRE, nonrecurring engineering, charges. Most of their income in the early days, in 1982, was from those nonrecurring engineering charges as opposed to the sale of chips. Therefore, they didn't really care whether the guy bought ten of them or a million.

"The business is kind of the same today, except for the fact that the CAD and CAE [computer-aided engineering] tools are generically available from a lot of difference sources, and any systems manufacturer can go out and buy them and get the layout rules from the IC manufacturer and do his own designs.

"We looked at it in '83 and '84 and concluded that these people were focusing in on the engineering part, and that's a very small part of the total. Usually, when you've finished the engineering, you're 20 to 25 percent home. The rest of it has to do with manufacturing. We saw that there were a proliferation of parts, a nightmare of parts, thousands and thousands of unique custom circuits. The logistics of those, the fact that they could be used in only one application, we believed, would become a nightmare.

"So we looked at the state of the art at the time, and the best that could be done was 3 microns—that is, the feature size that you could individually define in a product. We made some analyses that said that at somewhere between 1.5 and 2 microns there was a better way to do the job. At that point you emulate almost precisely what the microprocessor people had done. You build a standard product, and then you sell that and a development system to a systems manufacturer, who understands his specific problem much better than the chip guy does. Let him do his own custom design with a standard part and, with the development system he's got, determine the sequence of instructions that he wants for his application.

"We said we're going to do the same thing but we're going to apply it to the general logic problem. We'll build a standard part; we'll provide software and development systems and sell that at a fairly inexpensive price to the systems manufacturer and let him do his own design. In the end equipment it looks just like a gate array, but the means of getting there is dramatically different. The situation before we began meant that you had, say, 10,000 customers and those 10,000 might need thirty or forty different custom gate arrays, each one needing a different design. And when it's manufactured it becomes a unique part; it can only be used by that manufacturer in that specific application.

"What we did was to turn it completely around. We made ten or fifteen component parts that addressed all 10,000 customers in all thirty or forty different applications. Instead of making literally tens of thousands of parts, we'll make fifteen standard parts and let the customer customize it. He finds what he wants to do with this standard part and uses the development system to read data into the chip, and it configures itself to do specifically the task he wants it to do. If he wants to change it, it's just a matter of reading new data into it, and it configures itself to do a totally different function.

"That basically is the concept that Ross Freeman came up with. Now, how does all this tie together from a business standpoint? We concluded very early on that at that time 3 microns was the minimum resolution you could achieve, and we felt that for this to be economically viable you had to get down to 2 microns. But as feature sizes get down to 1.5–1 micron, and below it would be terrifically interesting and economical."

"So we concluded here is a market that is not being served and we ought to serve it. The problem was readily apparent to anybody. Here is this board with all these little jelly-bean ICs on it. Why not put all that stuff on one IC. Well, the gate-array people were doing precisely that, except they were making custom parts out of it. So we made the extrapolation. Why not make it a standard part and interconnect it in such a way that you can make it very flexible? That's basically the problem that Ross Freeman and his people solved."

Convincing Zilog that this was a golden market opportunity, however, was to be an exercise in frustration equal to those Vonderschmitt had suffered through at RCA. "There was no way in the world that we could have gotten the attention of Zilog management for this," he recalled somewhat sadly. "I was the general manager, but the purse strings and the shots were being called in New York. Nor was the process capability available in Zilog. And the immobility and the inflexibility of management thinking to even consider going into a development phase where you didn't have the process and had to enlist outside support could never be sold in a hundred years. That's the mental constipation that companies like that have. It was analogous to what I had at RCA, and once you see two in a row, you've got to get out."

Vonderschmitt left Zilog in January 1984, along with Ross Freeman and Jim Burnett, the marketing director for Zilog's best-known product, the Z80 microprocessor. Two months later, Vonderschmitt had raised $4.25 million from some of the Valley's most famous venture capitalists, Hambrecht and Quist, Kleiner Perkins and J. H. Whitney. That was enough to demonstrate feasibility and introduce the first product to the marketplace. But even before he approached the money men, he went to the Japanese, specifically to Seiko, the world's largest watchmaker. In 1972, Seiko had begun to manufacturer semiconductors as a means of guaranteeing their access to state-of-the-art chips for their watches. Five years later, they moved into the merchant market, and it was then, while he was still at RCA, that they first came to Vonderschmitt's attention. "Seiko," he declared emphatically, "is a superbly managed company from a manufacturing standpoint. Part of that is a result of their watchmaking history. And although there's a difference in semiconductor processing, the same excellence and cost effectiveness are applied and show up in the quality of the product. There is enormous attention to detail. The thing we were looking for to make this successful was a tie-in with somebody who was very cost-effective in manufacture. And Seiko was very close to the leaders in feature-size definition and CMOS processing. And they had the financial resources and an interest in staying state of the art."

And here in microcosm may be the frightening future of the American semiconductor industry—ideas made in America, the product made in Japan—yet another example of technology moving offshore. And if that process continues for very much longer, "Made in the USA" will, like so many other bits of America, become a part of our past, a bit of nostalgia to be savored but unfortunately never recaptured. And why does this happen? What were the forces that propelled Bernie Vonderschmitt to look beyond the American semiconductor industry for a company to make his gate arrays?

"The Japanese have the same sort of vertical structure as RCA and Exxon in this case, but the orientation is totally different," explains Vonderschmitt. "The basic differences can be found in the mentality of the Japanese to be willing to accept something on the basis of five or ten years before seeing a profit. In the U.S., with very few exceptions, the mentality of the large corporation to think about payback over that period of time just doesn't occur. Secondly, and this is particularly true of the semiconductor business and I think the basic underlying problem between the U.S. and Japan, is the cost fundamentals. In the U.S., capital is not cheap, and the tax laws no longer favor capital investment, so the mentality is, I'm going to put up a $100 million plant. How in the hell can I get it down to $80 million?

"In Japan the mentality is, if I can improve the performance one little bit, I'm going to spend $125 million. The gross throughput is the same in the two plants, but if your yield is two or three times higher, you're going to have an enormous cost advantage in the better facility. And since the cost of capital and taxes in Japan is roughly half of what it is in the United States, they have a tremendous advantage. And that's basically at the heart of the problem the semiconductor industry faces on an ongoing basis."

There were other advantages besides lower cost that Vonderschmitt saw in manufacturing his gate arrays in Japan. "The way you're successful in getting high yields is to pay enormous attention to detail. And I think that's the other point of difference between us and the Japanese. They take outstanding college graduates, put them in the processing and manufacturing area, and they are willing to stay there. In the U.S. its very difficult to do that. It's not that we don't have people who are just as smart; they're just not willing to stick to that small segment of the knitting."

There was still another compelling reason for Vonderschmitt to deal with the Japanese, the ease of making the deal. "I knew the Seiko people from before, from my days at RCA. The head of their semiconductor operation is a man named Yammamora. We had met through a trading company I had also been associated with through RCA called Okura and Company. They were basically the conduits for us to reestablish that relationship. Yammamora happened to be over here in San Francisco on other business, and my friend at Okura, a man named Itchikawa, put us together.

"Normally, to put a deal like that together, starting from scratch and being unknown, it takes you twelve to eighteen months to get an agreement. He was here in late February, and we went over in April, and we worked out a handshake agreement. We finally signed in March of 1985, and the only reason we actually signed a formal agreement was the venture-capital people wanted it. We had started our second round of funding at that time, and the money people were very nervous about our not having a signed an agreement. To me, it didn't make any difference."

Xilinx is but one of a host of American companies that have made alliances with foreign companies. In 1987, Dataquest Inc. reported over a hundred joint ventures or alliances between Japanese and American semiconductor firms.[23] Nor are such alliances limited to the smaller U.S. semiconductor makers. AMD, one of the nation's largest semiconductor makers, recently agreed to jointly develop a CMOS production process with Sony. The deal gives AMD access to the Japanese market through Sony and reverses the favor for Sony in the United States. The critical item in the arrangement was not so much Sony's market strength in Japan, although very welcome, but the CMOS process. AMD admits it is substantially better than their own.

NMB of Japan and National Semiconductor have entered into a long-term agreement for the joint development and manufacture of ICs. National also struck a deal with SGS-Thomson, the Italian-French venture, whereby each company will serve as an alternate supplier for the other's chips. National also announced a deal with Mitsubishi to market some of their products here and for Mitsubishi to return the compliment to Japan.[24]

The biggest technology-transfer agreement and the one that has raised major concerns has been Motorola's joint venture with Toshiba. Under the terms of the deal, Motorola gained access to Toshiba's DRAM technology and will use it to produce 1-megabit memory chips at three plants in the United States, the U.K., and the Far East. For Toshiba, the acknowledged world leader in 1-megabit DRAMs, it represents the ultimate payoff of the Japanese VLSI program. Motorola had already begun buying 1-megabit chips from Toshiba under an OEM (original equipment manufacturer) arrangement that permitted the Americans to assemble Toshiba's dice into finished chips and put the Motorola name on them.

Toshiba, for its part, will gain access to Motorola's 32-bit microprocessor, a technology Japanese companies have coveted for some time. The pact highlights the increasing trend toward international alliances in the semiconductor industry. For Motorola, long one of the most outspoken critics of Japanese trade practices and of the targeting of the semiconductor industry, this alliance is apparently acknowledgment of the inevitable. "We must get back into DRAM fabrication for the advantages it has as a technology driver for other types of integrated circuits," announced Robert Galvin, chairman of Motorola. The quid pro quo in the deal, the 32-bit microprocessor technology, he viewed as "a complementary sales push to penetrate the Japanese market. No global scale corporation can hope to become a large factor in Japan, or in any other country market, without establishing a major presence there."[25]

Such transfers of technology in the past, such as the DRAM and earlier-generation microprocessors, have eventually resulted in shutting their U.S. producers out of the Japanese market. No sooner have the Japanese makers gained enough confidence to produce their own versions than U.S. sales have plummeted. Even worse, the Japanese have used those products to take markets away from the United States all over the world. But Galvin wasn't worried. "This transfer," he pointed out, "takes place within a joint venture, where we have 50 percent control. I don't think we'll see anything like that happening on the microprocessors to be built by the new venture. It's a good deal for both companies. Toshiba can gain in microprocessors, and Motorola gets fabrication technology for very dense DRAMs, which we can expand into our other integrated circuit areas."

It may be that such agreements are the only recourse left to American semiconductor companies if they are to regain state-of-the-art technology in memory and other areas. But each one of these agreements has raised fears of a further flight of American technology. Ever since the end of World War II, American technology has been given, sold, or bartered away, often for less than a mess of porridge. Originally, the giveaway was the result of a burst of generosity, perhaps overlaid with guilt, as the U.S. government encouraged industry to donate its technology to help restore the war-ravaged economies of Europe and Japan. Arrogance also played a role, as American managers saw no harm in picking up a few bucks by licensing technology to "foreigners who could never possibly compete with us." But suddenly those foreigners who could never compete had driven us out of traditionally American dominated markets, producing superior products using our own technologies that they had improved upon.

"I think the Americans underestimated the ability of the Japanese to pick up the technology and make it successful. After all, the Americans had been licensing and selling semiconductor technology to the Europeans and the Japanese for twenty or twenty-five years, before the mid-seventies, when it suddenly started to bite them," Jack Carsten, now a venture capitalist with U.S. Venture Partners in Menlo Park advised me. For the previous twenty-five years Carsten held a number of high-level executive positions with TI and Intel Corporation. "It was almost a joke among the insiders in the industry that the way to make a fast buck was to license your technology to the Europeans or the Japanese, because they never could make anything out of it. By the way, neither could the vertically integrated U.S. companies. Licenses were granted to a lot of the computer companies, the Honeywells, the Burroughs, the Univacs, and the only one that ever made anything out of it was IBM. Because they had critical mass. So given that perspective, the American companies underestimated what the Japanese could do with it."

In fact, the Japanese have not always been so successful at mastering gift technologies. "The flow of semiconductor technology to Japan went on for a considerable period of time," notes Carsten. "I was involved with it in the late sixties and early seventies, and most of those efforts, I think, were sincere efforts to do licensing and exchange technology; those all failed until the mid- to late seventies. The Japanese were not competitive and were forced to withdraw from the business. That's a part of the story that is not well-known. Fujitsu went into the bipolar digital business in the late sixties and early seventies, during the recession period, and by 1974 or '75 they were forced to withdraw. They went completely out of the business because they were so unsuccessful.

"Then, in the late seventies, the Japanese started importing equipment technology, specifically people like Canon and Nikon. "They started working fairly successfully on steppers, on etch technology, and a number of other manufacturing technologies. In any event, the people who had successful technology transfers, if you want to call it that, tended to be the American equipment manufacturers. There's a whole history of those. But once the Japanese had the domestic-equipment manufacturing capability, particularly in the MOS area, then they began to see success. The early success was in the dynamic RAM, where they could utilize some of their unique skills in process optimization on a given technological base. And they did so very successfully.

"I think the technology transfer that was most damaging was the transfer related to processing equipment. Once the Japanese vertically integrated companies had access to that, and it was in house, and they could do closely coupled R & D for the development of new equipment, that was the part that was damaging to American dominance of the industry."[26]

Whether or not such technologies would have eventually gotten to the Japanese and other competitors is not the question. Technology, like the continually mutating strains of flu virus that sweep the world each year, spreads inexorably. The only question is how quickly it moves from one country to another and what the recipient does with it. Those two factors then determine whether, or how long, the originator can maintain his lead. The Japanese acquisiton of CMOS technology from RCA is a classic case in point.

"My impression," says Professor William Egelhoff of Rensselaer Polytechnic Institute's Center for Science and Technology Policy, "is that their excellence in fine-line CMOS technology today rests on fantastic process-development work, not process research but process development done by the Japanese companies. And that's work that we have not done. Now even the basic transfer of technology to get you going; that could have come not only from the United States but from elsewhere, say, Philips or Siemens in Europe. What that would have done timewise to them, I don't know. They might have had to wait a few years to acquire the technology from elsewhere.

"The point is there are so many routes through which technology diffuses, it's hard to say what happens if you block one route, whether the situation would be substantially different than it actually turned out."[27]

Now the U.S. semiconductor industry is suddenly faced with a dramatic dilemma. On August 14, 1988, Tomihiro Matsumura, the director of NEC's semiconductor division, declared that his company wanted to join SEMATECH. SEMATECH, the American consortium designed to catch and even pass the Japanese in manufacturing technology, had from its conception ruled out any foreign participation. "The SEMATECH project is a U.S. project," admitted Matsumura, "but it is tackling problems that could be solved quickly with multinational cooperation."

Perhaps, but the officials who head SEMATECH want no part of Japanese participation. "SEMATECH is for the benefit of American industry and American taxpayers," snapped Bob Noyce, the recently appointed president of the consortium. Matsumura did not specify the sort of technologies NEC would bring to SEMATECH, and some American executives openly questioned their motives. "NEC," said one industry leader who declined to be named, "would always be viewed as something of a spy no matter how good their intentions."

But even such proponents of keeping American technology safely locked in American hands as Clyde Prestowitz thinks the NEC bid may not be such a bad idea. "At this point," he says, we are playing catch-up ball with them. I doubt there is anything SEMATECH will come up with that the Japanese could not match, if they don't have it already."[28]

The infusion of innovative Japanese technology with American industry would be a definite change in direction. Moreover, if handled properly, technology-transfer agreements and joint ventures might in fact be a way to retain rather than lose competitiveness. "I don't believe the giveaway of U.S. technology is responsible for our loss of competitiveness," George H. Kuper, executive director of the NRC's Manufacturing Studies Board. "It's a piece of the equation, and it's the piece that's most easy to point a finger at, but it's not determinant, I don't think. We're in an era now where it's no longer a technology trickle down; we've got a technology cascade. And it's not just the semiconductor industry, although it's quite obvious there. Without these arrangements, technology is going to move like lightning all over the world. By making technology-swapping arrangements, that should make us understand the strategic advantages and perhaps use them more wisely and apply them more effectively."[29]

Kuper has been an entrepreneur, starting half a dozen small businesses, some of which he admits proudly "are still going." He has also put in time with large corporations such as GE, "then did a lot of consulting, actually odd jobs, but you're supposed to call it management consulting." Kuper went into government work first at the local level in the city of Boston, then ran a small federal agency, the National Center for Productivity. In 1983, he moved to Washington to his present position with the NRC. From these various vantage points he has been in a unique position to see technology flowing from the United States, for the most part with less than the desired results for the Americans.

According to Kuper, seven out of ten joint ventures fail to achieve their aims for the U.S. companies. One reason is that until recently much of American industry has been afflicted with the "not invented here" syndrome. This is the idea that if it didn't originate in the United States it couldn't possibly be of any value. And so in most joint ventures the technology being offered by the foreign partner was too often ignored.

"The problem of joint venturing internationally for us as a country," Kuper told me, "is we have, for one reason or another—hubris, lack of imagination, or whatever else it may be—assumed that we can always do it better than the other guy and take fuller advantage of that joint venture. And that, in fact, is not the case. In fact, we have been very casual about those relationships, and that very casualness has probably cost us something fairly significant."

Kuper can draw from his own experience to illustrate the unbelievable lack of attention displayed by at least one company in its approach to a joint venture. "I was the prime negotiator for a contract between a U.S. firm and a Japanese firm. In one case I was told specifically, 'Don't let this Japanese firm get a hold of X, Y, and Z.' When I met with my Japanese counterpart, to speed up the negotiations, he politely gave me an eleven-page, singled-spaced summary of existing relationships between his company and mine. On that list I saw exchange agreements for the very technologies I was specifically prohibited from disclosing.

"Now I was not told the details of existing relationships, and maybe they weren't spilling any beans that I was not supposed to spill, but I thought it a little strange to see on that summary list things that I was specifically forbidden to talk about.

"So this was a Japanese company carefully orchestrating its relationship with its U.S. counterpart, whereas the U.S. guys were very casual. There was nobody in my company who could have provided me with that same list of existing agreements."

As a result of this attention to detail, this willingness to do one's homework, the Japanese inevitably get the better of the deal. "The Japanese are better able to exploit joint ventures with American companies because they place a higher value on them," notes Kuper. "They go into them with greater calculation; they are after more than just market access or more than just the technology advantage to be gained. It would appear to be an orchestrated effort on their part that is not just a single-purpose effort. When we go into those relationships, it's generally because the marketing guy says there's a market here to be gained, or the finance guy says they've got a better cost of capital and we ought to be piggybacking that. It is not a coordinated, well-thought-out, strategically significant thing for us to do."

Have we learned anything from our technology-transfer experiences with the Japanese?. Probably not. We have now begun the same process on a large scale with the Koreans. "In 1985, Samsung paid Micron Technology, the Idaho-based DRAM house, something like $5 million for their technology," points out Jack Carsten. "Then, after they had developed it, they proceeded to just pound the hell out of the prices. They became a very low cost producer and hurt the Japanese and the Americans significantly. There has also been a technology exchange between Texas Instruments and Hyundai and one between Intel and Samsung on EPROMs. So it seems like history is repeating itself."

When Jerry Sanders, the president and CEO of AMD, was asked about transferring some of their technology to the Korean giant Goldstar, he said, "The way I see it, my enemy's enemy is my friend."[30]

Although technology transfer is seemingly inevitable, the Japanese seem determined not to speed the process where their own special interests lie. "The reason the Koreans are coming to us instead of the Japanese, whom they would probably prefer to go to," explains Carsten, "is that the Japanese won't give it to them. They had a bad experience with that. After they transferred the color TV and VCR technologies to the Koreans, the Koreans have mopped them up."

Yet, in the semiconductor business, especially in Asia, nothing is as it seems. Some Japanese companies are using Korean manufacturers to supply TV sets for resale to their foreign markets as a means of getting around the high yen.

"Every industrial country has gone through this loss of competitiveness after giving away technology," declares George Kuper. "And it's a lack of being able to keep all the balls in the air at the same time. If you look at our economic needs as a hierarchy, the first thing we need is food in our stomachs, a roof over our head, clothes on our back, etc. If you assume the same is true for a corporation, i.e., there is a hierarchy of need, it would appear it gets more and more sophisticated. In other words, management pays more and more attention to more and more things other than making its products. And as that occurs, the corporation's ability to continue the maintenance necessary to the making of its products diminishes. So you have the senior-level management of a corporation today worried about whether it's going to get gobbled up by another corporation, not whether or not it should be investing new resources in a new line or a new product or anything else to do with the marketplace. That's what is happening to us. In a few industries I could point to, since we started playing financial games over the last three or four decades, one of the casualties as a result of those financial games is that risk taking has changed its complexion dramatically. And with the dramatic change in risk taking comes a different product offering, a different rate of growth of products in the marketplace, and a loss of market share to other places that have become more innovative.

"Let me be specific. The U.S. machine-tool industry, where we got to the point where 90 percent of the machine-tool capacity in this country was being made by companies that had absentee capital landlords, conglomerates that did not make the acquisition of that machine-tool company to invest new risk capital, they wanted to generate cash. They bought cash cows. And that's another industry where you bet your company every time you introduce a new product.

"We've got to get hungry enough to see things other people are doing that we want to steal and use ourselves. Once we start doing that, our rate of product-cycle introductions will speed up a great deal. We've got to lose that not-invented-here syndrome."

We have also got to develop a much better means of keeping tabs on what the rest of the world is doing. "The Japanese do their homework," notes Kuper. "They've had 20,000 people tramping through American factories for years. . . . We visit the temples of Kyoto."

The Japanese have in fact developed an exquisite technology whose purpose is to acquire technology in the most efficient manner. "The

Japanese Productivity Center," explains Kuper, "has been bringing Japanese managers over to this country for twenty-five years. They have the best set of industrial espionage files on U.S. factories in the world. All written in Japanese, written by Japanese managers who know what they are looking at. It's fantastic. We have nothing comparable. And that's not dirty pool; it's smart business."

American technology, however, may no longer be the dazzling prize it once was for the Japanese. Indeed, in many aspects of the semiconductor industry it may well be the reverse. "The Japanese are pretty open," says Dataquest's Sheridan Tatsuno. "I go there and speak to them after meetings; they talk about a lot of different emerging technologies and markets. There aren't too many secrets in Japan compared to this country, which is heavily military oriented. Americans, at least from the Japanese perspective, are very secretive people. That is because we spend maybe 70 percent of our research on military projects. Most of that stuff is off limits, top secret, classified. The Japanese are frustrated, especially in leading-edge technologies such as optical computers or bioelectronics, because most of it is being done under the auspices of the Defense Department and it's all top secret. The Japanese routinely complain to me that there are these giant black holes in American research.

"If you compare the U.S. with Europe and Japan, you will see the U.S. clearly behind in certain areas, primarily because its research is all classified. So it never gets into the public press. As a result, Americans falsely conclude that we are behind. Optical computing, for example; Star Wars is pushing billions of dollars into optical electronics, lasers, but none that will ever see the light of day commercially. So we have this situation in the U.S. where 70 percent of our R & D goes into the military. Technically we have it, but we are unable to commercialize it for political reasons or strategic reasons."[31]

As a consequence, we are faced with an ironic paradox: A lopsided emphasis on military research weakens the semiconductor industry's efforts to remain at the leading edge of the commercial marketplace. And this, in turn, reduces its ability to produce the type of chips needed by the military. At the same time, most U.S. merchant chip makers have only reluctantly sought military business and the chip development that accompanies it. The sole exception is TI, which has cornered the military market in advanced weapons such as laser-guided bombs and infrared-detection and ranging equipment.

One company, however, cannot hope to satisfy all of the needs of an increasingly electronic military. For increasingly the weapons of war are chip based, and those chips are in far too many cases produced outside the United States. Thus, America's national security has become a major issue in the Chip War.

11

The Chip at War

They nest in tubes and poke like arrows from the decks of aircraft carriers, destroyers, and virtually all of the navy's surface ships. They are Sparrow antiaircraft missiles, slender, lethal, and guided by microprocessor chips to their targets with unerring accuracy. The Sparrow is but one of a dazzling array of computer-controlled weapons systems that are the new armaments of America's defensive shield. Robot vehicles and weapons, fire-and-forget missiles that recognize friend from foe, spy satellites, night-vision goggles, thermal gun sights, and computer-aided tactics are now in the arsenals and on the drawing boards of many nations.

At the heart of these new weapons and the systems that command and control them is the chip. No missile can be fired, no gun trained, no vehicle or ship moved, without signal-processing chips interpreting data, memory chips storing it, logic chips interpreting it, and microprocessor chips making decisions based on the input from all the other chips. "The most important technology in defense in the next generation, in my opinion, is associated with semiconductor technology and the related fields," declares Norman Augustine, the president of Martin Marietta and the chairman of the Defense Science Board Task Force on Defense Semiconductor Dependency. "That's not to say that aircraft engines are not important and structures are unimportant or aerodynamics is unimportant, but most of those things give you a 10 percent improvement or a 20 percent improvement. None of them give you multiple capabilities, as do the things we build out of semiconductors. If it weren't for the semiconductors on board the F-16, it wouldn't even fly; it is unstable and will tumble through the air. Not only can it not hit the target, not only that it cannot fly at night, not only can one man not fly it; it just flat out isn't flyable. It just can't be done without the chips in its on-board computer and control systems. So it's terribly important that we have these devices."[1]

Augustine offered another example of the revolutionary effect the chip has had on warfare. "Artillery rounds traditionally have not been very accurate. Historically you would never fire a large artillery piece at a tank because you simply could never hit it. Calculations show that to hit a moving tank with an artillery projectile you would have to fire 2,500 rounds to destroy one tank. Now that will take you forever to fire 2,500 rounds. But with the new guided artillery rounds, guided bullets actually, semiconductors provide guidance so accurate we can now hit one tank with every two rounds fired. And so it gives you a whole new capability; artillery can for the first time engage tanks. Take the semiconductors away and you can forget it."

Augustine's task-force report was published in February 1987, and caught headlines across the country. With the signing of the Intermediate Range Nuclear Forces Treaty, the INF, by President Reagan and Chairman Gorbachev in December 1987, the ability of the chip to enhance conventional weapons becomes even more important.

"It's been our strategy since World War II to maintain smaller forces with lesser quantities of equipment than our potential adversaries, particularly the Warsaw Pact," Augustine pointed out. I had gone to see him at Martin Marietta's corporate headquarters in the rolling parkland of suburban Maryland, just off the Beltway that circles Washington. Augustine, an engineer who has spent his entire adult life in the defense industry, is a tall, almost gangling man with a warm manner and an intense concern over the American military's growing dependence on foreign chips. "The advantage they have is not measured in a few percent but by factors of four in the case of tanks, three in the case of artillery. The numerical disadvantage is huge.

"So we have a tough decision. Either we have to have better equipment, or we have got to have more of it. To have more of it is very expensive, and we have chosen not to do that. The last time I checked, the Soviets had about 47,000 tanks, and we have about 12,000. If you were to match the Soviets tank for tank, we must go out and buy another 35,000 tanks. That comes to a great deal of money, which we don't have. Even if someone were to give us those tanks, we couldn't afford to accept. It takes four people to operate a tank. So that means you have to add another 140,000 people to operate the tanks. Then you need additional people to maintain them and still more people to train the tankers, provide security for them, feed them, recruit them, and so on. Then you've got to provide artillery and infantry support to go with the tanks. So you need hundreds of thousands of additional people to match the Soviets tank for tank. In the day of the volunteer army that just isn't going to happen.

"So, like it or not, we are stuck with counting on technological superiority as a major factor in the advantage the U.S. has over the Soviet Union. To date we've had that technological leadership, albeit we are falling behind. The most important technology in defense in the next generation, in my opinion, is associated with semiconductor technology and related fields."

Increasingly, however, that semiconductor technology is coming from abroad. Concerned with the problem, and even more concerned with the fact that its true dimensons are unknown, the Defense Department's Board on Army Science and Technology asked the National Academy of Engineering to form a panel to investigate the situation. The result was the Committee on Electronic Components, a panel including August Witt of MIT, Washington economist and semiconductor-industry consultant William Finan, and other industry and academic experts. Their investigation and that of the Augustine-led task force has generated a growing fear that the nation will become dependent on foreign sources of supply for chips used in weapons, satellites, supercomputers, and other devices vital for national defense.

"The implications of the loss of semiconductor technology and manufacturing expertise for our country in general and our national security in particular are awesome indeed," warns Charles A. Fowler, the chairman of the Defense Science Board.[2]

"This is not another case of citrus, tobacco, or pasta," Timothy L. Stone, then chief of the technology and industrial competitiveness division of the CIA's Office of Global Issues told the Defense Science Board. "The consequences of losing a strong indigenous technology base in this industry could have, in our judgment, widespread effects on this country's ability to maintain technical superiority in advanced weapon systems over the USSR."[3]

The electronic component of those weapons systems accounts for at least 35 percent of the Defense Department's R & D and procurement budget and is growing. It is also the most dependent on foreign sources of supply. "The greatest vulnerability exists for integrated circuits, most of which are assembled and tested overseas at low-cost sites," reported the Committee on Electronic Components. "Of the $16 billion in total sales of U.S.-branded silicon semiconductors in 1984, about one-third of their value was added in such foreign facilities."

"Vulnerability also exists for many semiconductors because of the foreign-supplied content. Most ceramic packages for large-scale and very-large-scale integrated (LSI, VLSI) circuits are Japanese produced or controlled. Silicon substrates of superior quality are supplied in significant and growing measure by Japanese vendors."[4]

The increasing dependence on foreign supplies and the threat of a cutoff during wartime may prove devastating to the military, according to yet another report, this by the NRC. And that dependence is growing.

Of the $1.6 billion the Defense Department spent in 1985 for ICs used in computers, fuses, and aerospace equipment, 44 percent came from foreign manufacturers. The vast majority of liquid crystals and luminous displays used in computer terminals and digital readouts come from overseas makers."[5]

Despite the obvious danger, the council points out that "neither the Defense Department nor any other federal agency has any policy that addresses the overall issue of growing dependency on foreign components for weapons systems."

Indeed, the Defense Department does not even know who makes many of the chips used in its electronic systems. American weapons are built with a large percentage of chips and other electronic components that are made overseas. But no one in DOD is certain of precisely how many chips in just what weapons are "un-American."

"People have a very false sense of what they believe is our level of dependency," declares Bill Finan. "They think that because a missile is made by Hughes Aircraft, for example, there is nothing in there that came from outside the U.S. You go down three or four levels of subcontractors and you find foreign suppliers there. There is a segment of the Defense community that feels that that is absolutely essential to the way the alliance has to work. I think that's true, but again, the question is, are you doing it with some degree of structure and purpose? Or are you just allowing program managers to make decisions based on cost and availability of the parts they need?

"In other words, program managers are given certain objectives, and more and more we want to be able to afford our weapons systems. Low cost becomes a key. There is no incentive in that program to buy American or drive an American alternative into existence."[6]

The result is a set of conflicting policies that pull against one another. The "Buy American Act," for example, is intended to exclude the purchase of weapons systems from foreign sources. But low price is also a critical pressure point on those whose job it is to buy weapons for the Defense Department. And since the Buy American rule applies only to the completed weapons system, DOD program managers, who are responsible for the purchasing of systems and their components, can and often do buy less expensive Japanese, Korean, or Taiwanese chips, which are the brains of the weapons.

In addition, U.S. government policy has long been to encourage the acquisition of weapons systems and components from NATO nations in order to improve its military effectiveness and the strength of alliance members' industry and economy. Thus, DOD Directive 2010.6 states in part:

"Waiver of 'Buy National' restrictions should be sought and applied wherever possible to support this objective."[7] The idea, of course, is to have our allies share not only in the cost of making weapons but in the profit of selling them. Weapons sales are, in fact, major U.S. exports, and we seemingly will sell them to all comers (see, for example, the Iran-Contra hearings). But where the Japanese are concerned, despite our constant bleating and blathering about the trade deficit, we turn a deaf ear to their importunings about weapons purchases.

Early in 1988 the Japanese asked to purchase the Aegis, the U.S. Navy's most sophisticated air-defense system. (A navy investigation of the controversial downing of the Iranian passenger plane by the U.S.S. *Vincennes* in the Persian Gulf in July 1988 blamed human error and not the Aegis system for the tragedy.) It is also the most expensive system at $500 million and would do wonders for our lopsided balance of trade with Japan. But a small group of congressmen tried to block the sale, fearful that the Japanese could not, or would not, keep our technological secrets safe from the Russians. This, of course, is fallout from the Toshiba affair; in fact, however, the Japanese, for the most part, produce many of America's most advanced weapons, such as fighter planes, tanks, and torpedoes, under license locally for their own defense forces.[8]

The Defense Department is unconcerned about the Japanese ability to keep their lips sealed and spent months arm twisting the Japanese to put up the money for the Aegis system. In fact, the United States has spent years trying to get the Japanese to increase their defense spending, especially to expand its protection of Pacific sea lanes out as far as 1,000 miles from its own shores. The Aegis sale is considered, therefore, a major step in that direction. As of this writing, however, the Defense Department has not yet notified Congress of its intent to go ahead with the sale. And so the inevitable debate in Congress over yet another arms sale, this one to an ally we have been pressuring for years to buy more American products, will likely come in 1989. No one I asked in Congress or the Defense Department would speculate on the outcome.

Still another law on the books is the Competition in Contracting Act of 1985, which explicitly requires "full and open competition" in the purchase of weapons. It even calls for each service to have so-called competition advocates, whose job it is to stimulate competition. That competition, however, is not necessarily American.

"Qualified foreign manufacturers will be encouraged to bid for defense business and, if they are the low bidder, they are more likely to be awarded contracts in the future under this new regime," states the 1986 Foreign Production of Electronic Components and Army Systems Vulnerabilities Report by the Board on Army Science and Technology.

As a result of such policies, the Committee on Electronic Components was not too surprised when it tore a number of weapons apart to see just how few of their components were stamped "Made in the USA."

"Every time we opened up a weapon and looked inside, we saw foreign parts," moaned Jacques Gansler, of the Analytic Sciences Corporation, a consulting firm in Arlington, Virginia, and a member of the study team. The navy's Sparrow III was one of the weapons examined. The guidance system was built from ICs and transistors from

Japan, a ferrite phase shifter made in West Germany, a memory chip assembled in Thailand, ball bearings made of raw materials from "various" sources—sixteen alien components in all.[9]

"If shipments of these import-dependent parts were stopped," the study said, "it would be impossible to continue making the missile. If American-made substitute parts were used, the Sparrow could return to production within eighteen months, providing it did not have to be redesigned."

In the brain of the Sparrow can be seen one of the great paradoxes of the Chip War. For the foreign parts of the Sparrow are there not because they are cheaper, although they are, but because they are better! And they are better because the technology used to manufacture them is, for the most part, superior to that used in American semiconductor plants producing the same types of chips.

This simple but chilling fact was spelled out in Madison Avenue–style briefing charts used by the Defense Science Board task force to hammer home their findings to the Defense Department and Congress. There were seven main points, a litany of stark technological truisms upon which, according to the task force, rests America's security.

- U.S. MILITARY FORCES DEPEND HEAVILY ON TECHNOLOGICAL SUPERIORITY TO WIN
- ELECTRONICS IS THE TECHNOLOGY THAT CAN BE LEVERAGED MOST HIGHLY
- SEMICONDUCTORS ARE THE KEY TO LEADERSHIP IN ELECTRONICS
- COMPETITIVE HIGH-VOLUME PRODUCTION IS SUPPORTED BY THE COMMERCIAL MARKET
- LEADERSHIP IN COMMERCIAL VOLUME PRODUCTION IS BEING LOST
- SEMICONDUCTOR TECHNOLOGY LEADERSHIP WILL SOON RESIDE ABROAD.

That chart was then repeated with a bold arrow running from the last point to the first, which was amended to read:

                    WILL   FOREIGN
- U.S. MILITARY FORCES ∧ DEPEND HEAVILY ON ∧ TECHNOLOGICAL SUPERIORITY TO WIN[10]

Needless to say, our chief chip dependency is on the Japanese. "There are some weapons systems that are almost wholly dependent on

Japanese chips," Augustine notes with some dismay. "But I was surprised that the number was actually less than I had thought there would be. I thought, with the Japanese having such a dominant position in the market, that almost every defense system would be dependent upon Japanese semiconductors. It turns out that's not true today. In retrospect, the reason is rather obvious. On the average, it takes eight years to develop a defense system. The average defense system stays in the inventory, if it's an airplane, twenty years; if it's a ship, thirty-three years; if it's a tank, forever. What that says is that the airplanes, tanks, and ships that are in the armed forces today were developed at least eight years ago and were probably started in development at least fifteen years ago. At that time, we had a totally dominant position in semiconductors, and even ten years ago our position was dominant. And so semiconductors designed into the system ten years ago were all from the U.S., and they are the ones that are in the field today. It's only now that the pipeline is becoming more and more dependent on what the Japanese are doing.

"We ran a sample in the case of satellites. Four to 5 percent of the chips came from Japan. That doesn't sound like a great deal, but it only takes one chip, if it's a critical chip, that need be cut off by a foreign supplier and you can't build the system."

That is precisely what happened in early 1987, when a Japanese company called Kyocera suddenly decided to stop making a specialized ceramic package. Electronic eavesdropping devices and military communications utilize those ceramic packages to house and protect customized chips. Customized chips are just that, specially designed to perform specific tasks, and are nested in a protective package, like fragile eggs in a carton. Of the 195 chips involved in the communications devices in question, 171 are housed in packages from Kyocera. At one point, and without warning, Kyocera discontinued the manufacture of the ceramic packaging for one key chip. The NSA, one of the major users of the devices, was sent frantically scrambling to find an American manufacturer to replace the Japanese.

I have been unable to learn from the NSA, one of the most secretive groups in the American defense establishment, whether or not they have been able to replace those packages. It is not an easy task, for American companies that once dominated the ceramic-chip-packaging field have for the most part been driven out by Kyocera. The Japanese company has, through fierce, even rapacious, tactics, managed to take over virtually the entire world market for ceramic-chip packages. "Kyocera came here and destroyed the entire ceramic-packaging industry," angrily says Dr. Sheldon Weinig, president of MRC. "They literally beat it to death. There used to be packaging by Coors, a division of the beer company, called Coors Porcelain. Three M owned a company called American Lava, which made packages in Chattanooga, Tennessee, and about half a dozen others. They all got knocked on their asses. Three M sold to GE, and GE isn't making it and will probably go out of the business. Packaging is a tough business to begin with. Kyocera came in and kicked the struts out from everybody pricewise. They have now run the pricing back up and own all the dies and tools. It's not easy to start over.

"Kyocera made an enormous investment. Kazuo Inomori, the man who runs Kyocera, is a religious fanatic. When a Kyocera employee dies, the family must give half of his ashes to the company, because Inomori says that half of his life was spent with the company. He is a real mystic and was probably destined for this industry.

"Packaging has always been a black art. It's hard to make; it's different from almost any other process in chip manufacture. It's always been a tough game, without big players, because everybody used a different package; everyone was in a different situation. So it's never developed into a rich, fully financed industry. There are a lot of little players. So in comes Kyocera, and he says, 'I would like to make your packages.' And you say, 'Well, I'm buying from Three M or I'm buying from Coors, or XYZ. . . .

" 'I don't charge you for the tooling,' he says. Up to that point all American firms charged for up-front tool and dies, which might be reasonably expected. 'Just give me the drawing,' he says, and there is no charge.

"He takes the drawing back to Japan and makes up the sunnuvabitch and files back with it—not six months later, a couple of weeks later. Then he comes in to show it to you, and it isn't quite right. He picks it up and flies back to Japan with it. Meanwhile, you haven't invested a nickel. Kyocera has picked up all the costs. He plays the game until he's got you by the proverbial. He finally produces a perfect package for you. He then gives you a price which is less than you are paying Coors or Three M or fifteen other American shops. At the beginning, the package was equivalent to the American's in quality and probably better in terms of its reproducibility standards. There's no question he was spending through the nose on this thing.

"You start to buy from him, and you say, 'Screw the Americans. When I buy a thousand from him, there's no deviation. They're all the same; there are no defects.' He does this unrelentingly for years. This Kyoto Ceramic Company, Kyocera as we know it here, is one tough company. It's a billion-dollar corporation; it's no nickel-and-dime operation. It's in many things. Inomori decided he wanted this market, and he went for it. And one day his industry wakes up and says, 'God, there are only two package houses left in the world. One is Kyocera, and the other is IBM, and they do it only for themselves.' In advanced technology today, the only two companies in the world who understand packaging are IBM and Kyocera."

Actually, Kyocera has two competitors, both Japanese companies, and both have small plants in the United States. But for all practical purposes, Kyocera is the dominant figure in the ceramic packaging industry.

"That's an astounding phenomena," marvels Weinig. "Every single merchant in the U.S. has to go to Kyocera to buy his packages. The U.S. government is very upset about this, and Inomori is no fool, and so he opens a plant in San Diego. There's a big play in the papers about his American managers who get up in the morning and exercise and do all these wonderful Japanese things. They all leave, of course, and the next thing you know, he's got all Japanese managers, which is the way he's running it today. Are they good? Yeah. Do they play dirty games? Absolutely."

For example, one of the substrates upon which chips are mounted are ceramic boards, which are covered with thin plastic films and used especially for microwave and hybrid circuits—almost all for defense and space applications.

"Kyocera," points out Dr. Weinig, "has never been in the thin-film-board business, but they're coming in now. They're coming in with prices off the wall. Are the boards being made in San Diego? No, they're being made in Kyoto. And if we don't stop him, he will have this entire industry by the gonads.

"Let me give you an example. There were four quotes on a job, three American companies and a Japanese. We were low bidder, say, $100, and the rest were spread out—$110 to $130. And Kyocera comes in at

$40. When you factor in the price of the yen versus the dollar and the fact that he's in a new technology, there's no way he can produce for that price. He just made a decision to enter this market and capture it. John D. Rockefeller did it in the oil business, dropping the price until he killed off his competitors and then raised the price back up. That's exactly what Inomori is doing."[11]

"When Kyocera was supplying packages for our military, it caused a big stir in Japan," points out Washington semiconductor consultant Bill Finan, a member of the Committee on Electronic Components of the Board on Army Science and Technology. "The Japanese have this mixed sense when it comes to the military, and many felt it violated their constitution. Over here it's forcing the program managers to ask, 'Is there any risk in having a foreign source, is there any cost-effective way to have an alternative? If I had some responsibility to make some capital investments in R & D or facilities, could I wind up with a high-quality, low-cost U.S. source? Or because I'm really not interested in looking long-term, am I simply going to forego that and take whatever is available off the shelf from whomever can give it to me today?' My sense is it's the latter that drives the defense program."

Other chip essentials, even when manufactured by American firms, are increasingly dependent on foreign sources of supply. The Monsanto Chemical Company, for example, is the single most important manufacturer of silicon wafers in the United States. To meet the current and anticipated demand, Monsanto is expanding its plant capacity, but not in the United States. In the fall of 1987, Monsanto opened two new plants in Asia and one in the United Kingdom. As a result, almost all of Monsanto's silicon production is overseas. Only two of the company's plants are in the United States, and one is more than twenty-five years old. Moreover, during the slump years of 1985 and 1986 the electronics materials division lost $80 million, making continued production in the United States questionable.

"Monsanto would prefer to be out of the business entirely," chemical industry analyst George Krug of Eberstadt Fleming told me. "There is no profitibility, it's difficult to compete with the Japanese, and it's just a dog-eat-dog business. Last year [1987] Monsanto sold $185 million of silicon wafers and lost $5 million on it. Monsanto is an $8 billion company. So it's a very small part of their business, and if a buyer came along, they would just as soon sell out."[12]

And that is just what happened. In November 1988, Monsanto

announced the sale of its Monsanto Electronic Materials subsidiary to VEBA AG, a West German conglomerate. The same company had earlier acquired Dynamit Nobel, the only other American-owned merchant silicon wafer producer. "There are several risks in allowing this expertise in materials manufacturing to move overseas," acknowledges August Witt of MIT. "The complication is that if anybody wants to disrupt the technology, it becomes much simpler. And once the skills and equipment depart, they cannot easily be brought back."[13]

Dependency may also become an issue with another semiconductor material that may soon become critical to defense needs—gallium arsenide. A by-product of aluminum mining, gallium arsenide has a number of qualities that make it particularly attractive to the Defense Department. Chips based on gallium arsenide are faster than silicon chips; they can convert electronic signals to laser light and back again. Most importantly, they are radiation hardened. That is particularly important on the battlefield, where radiation fallout from a nuclear blast fries ordinary silicon-chips to useless cinders. And so the military has drawn a virtual shopping list of the chips it would like to see built on gallium arsenide bases. They would then provide a sort of battle armor for such now radiation-vulnerable devices as guidance systems, radar, and communications satellites. They might also make possible devices such as optical computers, "interpreters" that will permit communications between present-day electronic technology and the optical systems of the future.

Gallium arsenide is also a major focus of Star Wars research, according to Sven Roosild, the assistant director for electronic sciences in the Defense Sciences Office of DARPA, the Defense Advanced Research Projects Agency. Roosild, whose family left Estonia when he was six years old, spent nineteen years in the air force as an officer and a civilian at the Hanscom Field Cambridge Research Laboratories. In 1980 he joined DARPA and has been working on the development of gallium arsenide since the strategic defense initiative, or Star Wars, program began. "Our program is funded out of SDIO" [the Strategic Defense Initiative Office], he explained to me, "so we are obviously targeting SDIO applications, primarily signal processing. The first order of business will primarily be sensor data, which will have to be processed. The main role of gallium arsenide in that kind of scenario is that its radiation-hardened properties provide you with a signal processor that will not be affected or destroyed by large doses of ionizing radiation."

Radiation hardness has little value in the commercial marketplace, but gallium arsenide has other advantages over silicon. "The commercial interest in gallium arsenide is speed," Roosild points out. "Its essential properties are speed and radiation hardness. We will use the speed clearly; we may trade it off for lower power, which may be more important for us than trying to reach the ultimate in speed. The technology is no different for commercial applications. The circuit designs may be different, but if you establish manufacturing facilities, they can be used for commercial as well as military applications."[14]

To date, however, the major American chip makers have shown little interest in gallium arsenide as a replacement for silicon. "I do not see gallium arsenide replacing silicon for the following reasons," emphatically declares Intel's Bob Noyce. "One is that gallium is a rare element. Secondly, because it's a binary compound, it's more difficult to work with. You can make silicon in a very pure form. Gallium arsenide has to be exactly balanced, fifty-fifty, and it's hard to make it that good. Silicon is very stable; its oxide is protective. So you've got an environment that has some very natural advantages going for it which do not exist in gallium arsenide. Gallium arsenide will be used in specialized applications where speed is absolutely essential. But what we are talking about in the future, as I see it, is complexity. You can go much much farther at any given time in complexity with silicon than you can with gallium arsenide. And I think that will hold into the future."[15]

"In the commercial world, 99.999999 percent of the demand is for silicon," says Charlie Sporck, the president of National Semiconductor, "and that's what we will continue to use to make our chips."[16]

Roosild sees that as a head-in-the-sand attitude. "The silicon industry would never have gotten off the ground if they had taken the same attitude they have toward gallium arsenide," he points out. "All those silicon guys are uninterested in gallium arsenide, and they cannot therefore see the proper utilization of it. That's typical of the whole industry. You don't have any of the people who were married to vacuum tubes who saw any advantage to semiconductors to begin with, so they all went out of business. People who did alloy transistors saw no future in diffusion, so they went out of business. Everybody thinks that they've got the best thing going and anything else that comes along has got to be bad."

Not that Roosild expects gallium arsenide to replace silicon the way the semiconductor replaced the vacuum tube. "No one should ever expect gallium arsenide to replace silicon," he admits. "That seems to be a continuous theme. On the other hand, it's clear that its going to have its field of application that the silicon people are just not going to be able to get to.

"Basically the materials problems are never going to stop it. I think that's a real red herring that people like to bring out, but they should be reminded that when they started making silicon circuits in the beginning, they didn't have a material that was perfect either, and they made fine circuits. Sure, materials improvements are going to help and make even better devices, but that doesn't mean we've got to sit on our hands and wait around until we get perfect materials.

"If we could find anyone who ran gallium arsenide devices with the same kind of volumes that silicon is being run at, the problems and differences would be minimal. The problem is we don't have the facilities. A gallium arsenide pilot line gets put together with maybe $15 to $20 million worth of equipment and the silicon guys spend $100 million to put a pilot line together. It's the same equipment, it's the same process, just modified."

DOD has managed to get three manufacturers up and running gallium arsenide. "We are constructing pilot manufacturing facilities; we have three of them on line that will produce chips of sufficient complexity to be useful. One is at Rockwell International at Newburgh Park, California. The other is at McDonnell Douglas at Huntington Beach in California, and the third is a new one that just got started at the AT&T facility in Reading, Pennsylvania."

"The devices that AT&T will build for DARPA under the contract," declares Bernard Murphy, director of the Compound Semiconductor Integrated Circuit Laboratory at Bell Labs, "will be faster and more complex than any gallium arsenide chips in production today." To do this, AT&T is using two new technologies considered essential to carry gallium arsenide chips to large-scale integration and beyond. At the LSI level, gallium arsenide is only two generations behind silicon chips but will function at far higher speeds.

Using the new manufacturing technology, experimental circuits made by AT&T have set world records for high-speed performance.

"Until now," says Robert Vehse, AT&T's project manager, "the manufacture of gallium arsenide chips has been a slow, costly process. We will be attempting to jump way ahead in the learning curve associated with gallium arsenide manufacturing by using, whenever possible, processes and methods that we have already perfected for silicon chip manufacture."[17]

Although the United States has an active program to develop gallium arsenide chips, there is no similar program to develop the technology to manufacture the refined crystal from which the chips are made. But the Japanese have. Several years ago, MITI flagged optoelectronics as one of its major development items and pushed a number of companies into intense R & D programs aimed at producing a very high grade of gallium arsenide. The result is that one Japanese company, Sumitomo Electric, now produces 75 percent of the free world's supply of gallium arsenide substrate crystal. Its fully automated plant can produce ten tons of gallium arsenide crystal a year, and the company plans to increase that capacity to fifty tons a year.

No American company is producing at anything resembling that level, nor do they have any plans to come even close to it. "On the contrary," says Richard Brown of Monsanto's electronic materials division. "American companies are taking a wait-and-see approach." For despite all the pump priming by DARPA, the demand for gallium arsenide has not grown as rapidly as hoped. Thus, U.S. companies have not been willing to make the kinds of commitments the Japanese companies do. Monsanto, for example, which had been the chief U.S. producer of gallium arsenide, had abandoned the business in the mid-1970s and sold the rights to its technology to Mitsubishi.

Japanese demand has been growing, and as a result, Mitsubishi has been doing quite well of late selling gallium arsenide. A second result was that Monsanto began to feel a mite foolish about the deal, and so they asked for and got the right to serve as Mitsubishi's marketing agent in America.

But some experts are fearful there may not even be an American supplier of the next generation of high-speed chips made of gallium arsenide. "And," warns the NRC, "a sudden cutoff in the supply of gallium arsenide components due to an economic embargo or other emergency could limit spare parts for old equipment and prevent the production of new systems."[18]

"Most of the top-quality gallium arsenide and indium phosphide, which is another electro-optical material, is made in Japan. We buy a lot from Sumitomo Electric," and American engineer doing research in the field told me. "When we want lower-quality gallium arsenide, we buy from a U.S. company."[19]

Gallium arsenide has other interesting applications that are only just now being explored in this country, areas in which the Japanese have taken a substantial lead. Optical electronics, a technology that enables devices to process information by optical rather than electronic means, for example, is largely based on gallium arsenide technology.

"This field is very much being dominated by Japan," argues Professor Richard Osgood, director of the Microelectronics Sciences Laboratory at Columbia University, "and it's part of our defense posture." Osgood, whose lanky sprawl does not quite hide an almost frenetic energy, heads a team of eight professors and a small legion of graduate students in an $11-million-a-year search for new materials and improved chip fabrication techniques. "Diode lasers," he continued as a small but persistent stream of students kept popping in for a quick word, question, or request, "are used in all optical fiber communications systems. AT&T uses Hitachi diodes in their very long optical fiber links. It's a very important field; it's the telecommunications technology of the future. These electronic-communications laser diodes are integral to that, and the reason this is to some extent scarier than the silicon market is that it's a technology that Japan is beginning to very much initiate research in and dominate the research phase as well as the manufacturing aspect. The United States has just gotten wiped out in that area."[20]

Another area where the American chip industry, almost monolithic in its dominance, is suddenly eating the dust of its Japanese competitors is in the development and production of semiconductor manufacturing equipment. One of the key pieces of equipment, the wafer stepper, which optically prints the complex circuit patterns onto the wafer, is manufactured by, among other companies, a small but once dominant firm called GCA. In February 1987, it appeared as if GCA might go out of business entirely. Only two years before, GCA had 27 percent of the world market, but the slump in semiconductor sales, intense Japanese competition, and apparently poor management pushed GCA to the brink of bankruptcy. Its stock, which traded on Wall Street for more than forty dollars a share in 1984, plunged to two dollars a share, and the New York Stock Exchange threatened to delist the company.

Then, in March 1986, Prince Charming, in the form of Richard Rifenburgh, the head of a Pittsburgh investment firm called the Hallwood Group, came to the rescue. They restructured the company and its debt, fired the hapless management, cut the payroll, and saved the company, at least as of this writing. Ironically, one of the major new investors that helped in the salvage operation was Sumitomo Corporation, a Japanese trading company.

GCA's troubles attracted the concern of the Pentagon and the CIA. For GCA was only one of three major American producers of steppers left in the business. And GCA had definitely lost its cutting-edge leadership. "It's very, very critical," says Donald Latham, an assistant secretary of defense who oversees much of the Pentagon's electronic purchases. "This type of equipment is the key to producing finer- and finer-resolution semiconductor devices. It's simply something we can't lose, or we will find ourselves completely dependent on overseas manufacturers to make our most sensitive stuff."

A CIA report expressed similar concerns. Without such technology, the report concluded, American companies would face extraordinary difficulty in designing submicron circuits—critical to the manufacture of 4- and 16-megabit memory chips and complex microprocessors.

"We are losing some of the base industries that we depend on," moaned Jim Owens, vice-president of technology for National Semiconductor Corporation. "The question we always have in the back of our minds is, are they giving us the best they have, or are they holding back to gain a competitive advantage."[21]

Owens's worries might be well-founded. The experience of Jim Dykes, the American president of TSMC, in this regard is interesting and perhaps significant. Dykes was purchasing semiconductor manufacturing equipment for the new company. The dialogue with the Japanese as he described it to me went like this:

JAPANESE SALESMAN: "We have just begun to export this particular equipment, and we've never sold it before."
DYKES: "Well, why didn't you sell it before?"
JAPANESE SALESMAN: "Because we were using it ourselves, and we sold it to other Japanese companies."

"The more you dig," Dykes explains, "the more you find that they've got a new-generation machine that they're using in Japan; now they're starting to export the old-generation machine. And that's not what we want.

"You don't even know if you're talking about that machine or not. You can't get a direct answer to that question. You would get an answer that would make you think that yes, you're getting the latest machine. What you're actually getting is yes, the latest machine that's for sale.

"There's no question in my mind that that goes on. And I don't know whether that's a controlling situation or whether that's a cultural situation, but they're a very closed society. And even here, among the Chinese, they share the same view, that Japan is a very closed unit. It's very hard to get what you believe to be the ultimate truth out of them."[22]

The pervasiveness of Japanese chips in the American defense establishment goes even beyond weapons systems to the very well-springs of our security. Most weapons today are designed with the aid of supercomputers, which are also essential to weather forecasting, intelligence collecting and analysis, and a host of other military applications. And supercomputers are at the cutting edge of the information revolution.

"The Japanese recognize that whoever controls the information revolution has, in effect, some form of increased geopolitical control," says Professor Michael Dertouzos, director of MIT's Laboratory for Computer Science. Dertouzos is one of the glamour boys of academic computer science, tall, darkly good-looking, and always dressed like a corporate lawyer. His vehemence as he told me about the vital importance of the supercomputer was, however, anything but buttoned down. "This assault is far more serious to our future than the number of automobiles sold by the Japanese or the Americans. The nation that leads the world in supercomputer technology has the possibility of leading the world in the application of all future technology."[23]

The U.S. Defense Department seemingly understood those implications. The biggest user of supercomputers right now, it had, in fact, created the idea of the supercomputer, funded its development, and took delivery of the first one in 1964.

The Japanese started almost fifteen years later and, in the eyes of some American researchers, approached their supercomputer project in much the same way they planned Pearl Harbor. Their announcement in 1981 stunned the American computer community. It is an ambitious project, especially for the Japanese, whose abilities had previously been channeled in the fine art of refining the technology of others. Could the Japanese innovate to such a degree in areas where they had little or no previous experience?

"Japan has come to be considered an economic power by the other countries of the world," MITI's Ryozo Hayashi told me. "Thus, if we consider the direction in which our industries should proceed, it becomes clear that we no longer need chase the more developed countries but instead should begin to set goals of leadership and creativity in research and development and to pioneer the promotion of such a project throughout the world."[24]

Such confidence was based on years of brilliant and solid research done not in Japan but for the most part at Stanford, MIT, and other American universities. "The Japanese," reports Edward Feigenbaum, professor of computer science at Stanford University, "have been sending scientific emissaries to the West for years to study the pioneering artificial intelligence research in the United States, Great Britain, and Europe. The Japanese have grasped the great scientific themes that run through artificial intelligence, and they feel ready to gather up a loosely knit group of ad hoc projects and to consolidate and develop them into what can only be called a momentous national project. Its success—even partial—will vault them into a position of distant leadership in conducting the world's information business."[25]

In October 1981, the Japanese govenment announced that it was going to spend more than $1 billion over the next ten years to develop what they called a fifth-generation computer. Computer eras, like geological ages, are marked by evolutionary changes. In the fifth generation, new arrangements of processing units will allow for sophisticated artificial-intelligence (AI) functions to be built right into the machines. The idea is to produce a machine that will have the power to reason like human beings, that will understand human speech and pictures. To do that, supercomputers will be needed to execute the very advanced programs AI researchers hope to develop.

To further the goal, foreigners were invited to Japan to discuss their research before "interested" Japanese scientists. At the same time, dozens of Japanese scientists scoured the world's major computer research centers picking brains and picking up ideas. Many American scientists came away feeling "ripped off."

"When I looked at their plans," recalled MIT's Professor Dertouzos, "I panicked. I said, 'My God, this is the research charter for my laboratory; these guys have stolen it.'"

In fact, however, Dertouzos and most American computer scientists welcome virtually anyone to their labs, hold conferences that anyone may attend, and publish papers for all the world to read. And the Japanese, once the most secretive of high-technology players, are now doing the same thing. Fifth-generation project conferences in Japan, for example, are open to all. The problem is that the conferences are held in Japanese and the papers are delivered and published in that language. As a result, Americans simply ignore them. At the most recent fifth-generation project held in the summer of 1987, only one American, Mark Eaton, of the Microelectronics and Computer Technology Corp., an industry group that funds semiconductor research in American universities, showed up. Eaton is MCC's designated Japanese technology watcher, and he manages to buttonhole Japanese scientists and collect papers for translation back in the United States.

It's a very expensive undertaking. Translations of Japanese technical reports run more than thirty five dollars per hundred words of English, and many papers can run to ten thousand words and more. Multiply that by the more than fifty papers usually delivered at a conference and the expense can be considerable. Still, MCC has built up a considerable data base of translated Japanese papers on the fifth generation.[25]

Technology transfer, Americans must learn, is a two-way street. With language barriers to be overcome, it is just somewhat harder. But information and innovation alone are not enough to guarantee technological superiority, and the supercomputers Americans build are dependent in large measure on Japanese chips. Intelligence data are especially important in a time when half the industrialized nations of the world possess nuclear weapons.

Norman Augustine considers the collection of intelligence and the ability to break codes among the primary security concerns we face.

"A major concern in the defense area," he argues, "relates to collecting intelligence and using computers to break and to build codes the other guy can't break. It's becoming less and less a case of having very clever people as cryptographers, as it was in World War I and to a lesser extent in World War II, and more and more a case of faster computers. As an example, one of the most powerful supercomputers in the U.S. is the Cray, and it today has 100 percent of its memory built from Japanese chips, and 10 percent of its logic comes from Japanese chips.

"Now Japan is an ally and one can say, 'So what.' The problem one encounters is that firms that build those chips and sell them to Cray also are in the supercomputer business. So when Cray says we want your new generation of chips, the Japanese supplier says sure, but first we want to be sure we give you only the best chips we can make. So the newest chips will be incorporated in our own supercomputers, and we'll test them on the market for a year, and if after that time they are working properly, then we'll sell them to you. Of course, by then the Japanese will have the market. The life span of a supercomputer is a couple of years, and so if you come into the market a year late, you're not there."

The Japanese dispute the very idea of withholding those chips. Sadao Inoue is the general manager of Fujitsu Limited's International Operations Group, which is responsible for the production and sale of chips to the United States, Europe, and Asia.

"I have heard stories that Fujitsu will not sell the most advanced chips for supercomputers to U.S. supercomputer makers until after they have been installed in our own supercomputers and have established them in the marketplace," he admitted. "But those American companies such as Cray and ETA Systems [both supercomputer makers] are some of my biggest customers. Some of the higher upshots of those companies told me of those stories, but I told them what I tell you: It is not true!"

Inoue, who chain-smoked through the entire interview, emphasized the point with a slashing movement of his hand, spilling ashes down the front of his dark silk suit. By the time I left, he was covered with ashes from his frequent, emphatic hand chops.

"We have never refused to sell any chip to an American company. Also, a couple of years ago, we developed a few samples for a new generation of high-speed memory chips. At the same time we introduced them to our systems group, a couple of those samples were given to Cray engineers. This was at a very early stage. But I know that Cray, Control Data Corporation, and other American supercomputer makers have such concerns. But we have never refused to supply anyone with even the newest samples of our latest designs. Also, every time I sign a nondisclosure agreement with those companies, it limits Fujitsu to just a few engineers in the semiconductor group who can touch on this information."[27]

Not so according to Lloyd Thorndyke, the president of ETA Systems, who flatly states that the Japanese are not supplying their best chips to American supercomputer companies.[28]

That may be just good, hard business practice, but to some industry observers it poses a very real threat to the viability of the American supercomputer industry. "So the U.S. supercomputer industry dries up, and we are then dependent upon Japan to break codes, build weapons, and for other aspects of our national security" is one bleak scenario offered by Norman Augustine. "And the Japanese can sell those computers to the Russians if they choose, or the Russians can steal them from the Japanese or from us, as they have done in the past. So what it all boils down to is, if you really are dependent upon an ally for your own national defense, that strikes some people, including myself, as not a very good circumstance for a world power."

The Japanese are apparently determined to ensure their access to supercomputer technology and to overtake the U.S. lead in the field. Robert Price, the chairman of Control Data Corporation, met with the Joint Congressional Economic Leadership coalition at a luncheon in February 1988 and warned that "the U.S. is losing momentum in the supercomputer race against the Japanese and is failing to build the necessary infrastructure to maintain its lead.

"The Japanese not only are building supercomputers for sale," added Price. "They are embedding them into their infrastructure. By placing supercomputers in both their universities and workplaces, the Japanese are gearing up their industries to utilize this very powerful tool." Price noted that Japan has set up more of its own supercomputers in their universities than U.S. universities have installed American-made supercomputers. Indeed, just a few months before Price issued his warning, government pressure had been applied to MIT to prevent what is probably the single most important technology institution in the country from buying a supercomputer from Fujitsu rather than from an American maker.

The problem, as Price sees it, is that U.S. government policy has been to put most of the supercomputer risk and expense on industry.

The government, he pointed out, did not buy advanced military aircraft in this fashion. "If the government told industry to spend $200 million to develop a new airplane and if we like it we might buy some," he argued, "no company would build an airplane.

"The risk in the supercomputer industry could be reduced in the same way. Government would guarantee the purchase of a number of supercomputers as an investment in national defense. What's really needed is a proactive, affirmative U.S. government policy supporting technological excellence in supercomputing. There also needs to be an ongoing rigorous dialogue with Japan to achieve a level playing field in supercomputer trade."[29]

Asking the competition to play less rough will not work for very long or very well. Nor is American industry totally on its own in the development of supercomputers. The Defense Department is well aware of the disadvantages of dependency and has mounted two major programs to guarantee its supercomputing capability and its chip supply. Supercomputing is being addressed for the most part by a program called "Strategic Computing." Starting in 1983, DARPA began the task of creating a new generation of "intelligent machines." The idea was to combine AI with supercomputers fast enough to process millions of bits of information virtually instantaneously. The ultimate result will be weapons systems that seemingly owe more to H.G. Welles and Jules Verne than to Gattling and Colt. Already constructed and being tested are an autonomous land vehicle (ALV) for the army, a battle-management program for the navy, and a pilot's associate for the air force.

The ALV is an unmanned armored vehicle that literally drives itself. In a 1987 test, and ALV successfully drove itself down a 4.2-kilometer winding stretch of Rocky Mountain road at 20 kilometers per hour and avoided obstacles by slowing to 3 kilometers per hour. A video system allows the vehicle to discover its location on the road, negotiate sharp turns, and find its way through road intersections. The ALV uses a superconductor to understand the images it sees and apply them to performing its mission. With multiple sensors on board and its superbrain, the ALV will be able to choose the appropriate direction at a complicated intersection and select an alternate route when the one originally planned is blocked. In its combat form, the ALVs will reduce and in some cases eliminate entirely the crews needed to man tanks, trucks, and other military vehicles.

The battle-management program uses computers to make instant assessments of tactical and strategic battle situations and provide commanders with the implications of complex combinations of events and decisions. It in effect provides commanders with a series of "what ifs," predicting the consequences of a host of alternative actions.

The pilot's associate is just that, a computer co-pilot that gives the pilot of a single-seat fighter plane a "crew" of experts to plan missions and tactics, monitor all of the aircraft's systems, and assess all of those situations during every moment of the flight.

The heart of all three programs are supercomputers based on a concept known as parallel processing. Ordinary computers can only process information serially; that is, the bits are all lined up and must pass through the computer's processing unit or chip one bit at a time. This becomes a bottleneck that limits the speed at which information can be processed. The idea behind parallel processing is to have not one but dozens, even hundreds, of chips acting as computer processing units each processing a small piece of information, but all at the same time, greatly increasing the speed. DARPA has harnessed more than a hundred universities, industrial research centers, government national laboratories, and the Defense Department's own research establishment to the task of enhancing America's supercomputing capability.[30]

The other part of the Defense Department's concern, getting the specialized chips it needs, is being addressed by the VHSIC program. Fueled by a $1 billion Defense Department program, IBM, TRW, and a dozen other American companies are trying to develop VHSIC as a superchip with the computing power of a supercomputer. With the VHSIC program, the Defense Department hopes to put military systems at the leading edge of IC applications.

The key words are "leading edge" and "applications." Since the 1960s, the DOD has been in a follower position relative to IC technology. Commercial applications, not military, drove the technology to higher levels of integration, and the IC devices developed were optimized for commercial applications, primarily memory and simple microprocessors. ICs produced for the commercial markets have not met important military requirements such as temperature, radiation hardening, and reliability assurance.

The technological advantage that VHSIC devices will provide is greatly increased on-board signal processing and data-processing capability. At the same time, VHSIC chips will require less power, be lighter in weight, and take up less space. Even more important, they will be more reliable that "civilian" chips. Military threats and requirements for the 1990s and beyond require orders-of-magnitude improvements in electronic warfare, acoustic signal processing, radar/infrared target search, classification, acquisition and track, and adaptable communications. Only smaller and faster ICs, tailored to military requirements, can provide this capability. At the same time, the smaller circuits reduce the numbers of chips and modules in a weapon system, thereby reducing the amount of interconnections, which in turn reduce the number of points where failures can occur.

A key goal of VHSIC is to develop a chip with feature sizes that are in the 0.5 micron range. That is half again as small as the most advanced chips now being produced. There are, for example, more than 25,000 microns in an inch. To squeeze the number of circuits onto a chip requires that the traditional means of etching each layer of transistors and circuits onto the chip be replaced by an advanced method using electron beams.[31]

Under a VHSIC contract, P-E has developed a lithographic technique that uses electron beams rather than light rays to "write" the chip design directly onto the mask. Called AEBLE, the system can draw lines as small as the 0.5 micron VHSIC chips demand on both silicon and gallium arsenide. With its capabilities proven in the VHSIC program, P-E was then able to develop and sell a civilian version of the AEBLE machine to commercial semiconductor makers—a spinoff with positive economic benefits.

VHSIC chips, however, are another story. Given its experiences over the past decade with the legal and illegal transfer of technology to the Soviet bloc, the Pentagon is understandably reluctant to see VHSIC technology join that eastward flow. Thus, VHSIC chips will not be available to commercial customers. But some of its producers are already applying the designs and manufacturing techniques to civilian versions. Many chips based on VHSIC technology should therefore reach the commercial marketplace by 1990. With its tremendous computing speeds and vast memories, VHSIC-type chips may form the base for PCs that will be completely interactive; that is, you talk to it, it talks back. Home robots and a host of "smart" household machines may well be the new products of the 1990s that will enable U.S. producers to regain lost markets and give American chip makers a leg up in the Chip War.

"The new game will favor companies that can churn out good designs fast, not the low-ball manufacturer," says Dan Hutcheson, president of VLSI Research, a San Jose company that studies the semiconductor industry.[32]

Even with the success of the VHSIC program, which will keep design and production of militarily vital chips in exclusively American hands, the Defense Department is a long way from solving its dependency problems. All of the problems however, can be solved according to the Committee on Electronic Components. The most urgent need is to identify the foreign components now in American weapons systems.

"The original sources of all electronic components and the materials they contain (if they are foreign sourced) should be identified. Of critical importance is whether there is an equivalent domestic source. If not, then the part becomes identified as a critical item," notes the committee.

Stockpiling of critical parts, creating a domestic or standby capacity, substituting and redesigning of the weapon to eliminate dependency, are all suggested short-term corrective actions recommended by the committee. The third recommendation was to take action to reduce future dependency on components in the next-generation weapons systems. "In the future," said the report with an echo of horror, "whole classes of components might be made in foreign countries only. In fact, a whole generation of technology might be implemented exclusively abroad, and thus the United States could be totally dependent for a major share of its weapon systems. Hence, current R & D on systems, subsystems and components, and broad long-term trends in industrial components production and materials should all be analyzed.

"In this area, one DOD research office should be made fully responsible for determining what actions are necessary to reduce or eliminate all future dependency."[33]

While these recommendations, if followed, will give the Defense Department much tighter control over the chips that are so vital to its weapons, in the final analysis it is the civilian, commercial development of semiconductors on which the national security must ultimately rely. "One can say our problem in this country is with the Japanese, not the Soviet Union," muses Norman Augustine, "and in fact the Russians are much worse off than we. The difficulty is, we are counting on our technology not to give us a 10 percent edge over the Soviets but a factor of several, and to do that with technology is very, very difficult, and as technology leaks from Japan, that will help the Soviets move ahead more quickly. Furthermore, the technology of semiconductors is driven principally by the commercial marketplace, not by the defense marketplace. And that's important, because as the U.S. commerical market falls behind, then U.S. defense manufacturers will also fall behind in the technology of semiconductors. It's not like rocket engines where the technology and state of the art in the U.S. is related to what defense companies do. The state of the art of night-vision systems is determined by the defense manufacturers. The state of the art in semiconductors is determined by the commercial marketplace, by the consumer market, in fact.

"High-volume production is the key, we believe, to leadership in semiconductors," continues Augustine. "This is true for two reasons: One, the need to drive down prices requires high volume. Secondly, it's that volume that provides the financial base to companies to do R & D for the next generation. The new generation is only about two and a half years away normally."

Those high volumes, however, can only be achieved in the open commercial marketplace. "High-volume production is supported by the commercial market, not the defense market," says Augustine. "The DOD used to represent about 70 percent of U.S. semiconductor end use. Today it's less than 10 percent in terms of sales; in terms of number of chips, it's about four percent. . . . Leadership in commercial volume production, which underpins all things, the evidence would suggest, is being lost. If one looks at the various companies that have had lead positions in terms of ranking and sales in '75, '80, and '86, you can see the U.S. firms moving steadily toward the bottom of that list and dropping off of the list. If you look at how many U.S. firms are building the most advanced version of semiconductor in any given generation, you can see we've gone from where we used to have fourteen U.S. firms building the then most advanced generation. Today there are three, and only one of those sells on the merchant market available to Defense Department contractors as a whole. The other two build chips only for themselves. I'm referring to IBM and AT&T in the latter regard. In Japan there are seven firms building megabit chips, and that number is increasing."

"Chips which seem to be only silicon and very dense integrated circuits," declares Colombia's Dr. Richard Osgood, "are also interrelated very closely with other things like infrared detectors and other electronic instrumentation which are not as well known commercially. For example, to make a satellite receiver is a very commercial, very mainstream type of development effort. Whereas to make an infrared detector is much more specialized. Nonetheless, there is an interrelation. Both use semiconductors; both are concerned with very small features. It turns out the U.S. is not only losing the battle over the big, standard silicon integrated circuits, but I think we are beginning to lose a lot of our edge on these other specialized electronic components. Like mercury cadmium telluride detectors, which are chip culture. That is to say, the people who are doing research in those areas, which may not be directly chip related, are part of the same culture; the problems that must be solved are the same. My understanding is in the area of mercury cadmium telluride, which is an infrared detector, the United States was historically a leader. Now Japan is beginning to do very, very well in that field and is assuming an important, if not dominant, role.

"Another area is charged coupled devices, which is not only very useful in defense but also the base of the consumer camera market. That's a hybrid between silicon and other materials. The Japanese have become expert in that and are producing and selling millions of camcorders. So we are not only losing the battle in the standard silicon integrated circuits, but also in these very specialized ones which play a very important role in defense. And that's a big thing to be concerned about."

"There is a symmetry here," adds Norman Augustine, the task force head, "that says the defense market isn't terribly important to the worldwide semiconductor market, but the U.S. semiconductor industry is critically important to the Defense Department to provide for the nation's defense. When you add it all up, it's not an overstatement to say that our ability to provide for the defense of the nation is dependent in no small part ten years or twenty years from now on how the semiconductor war comes out."

Unfortunately, that symmetry has been badly skewed, and as the U.S. manufacturer's hold on the commercial marketplace has slipped, so, too, has our ability to control our own national security. "Our options vary all the way from accepting our dependence upon Japan," points out Martin Marietta's Augustine, "for the technology we need to provide us with a military advantage to getting our industry on its technological feet so it again can provide the leading edge in semiconductors. We make very few semiconductors for our own systems, but we are basically dependent upon the merchant market for most of our semiconductors. From a business standpoint it's not terribly important to us as long as we can get the chips from somewhere. But from the standpoint of our obligation to the country it's terribly important that we get the most advanced chips possible from American makers."

That may not always be possible given the present state of the American semiconductor industry. The Defense Science Board's task force came to the following gloomy assessment:

If the established trend in the critically important semiconductor manufacturing technologies is allowed to persist, it appears likely that in the 1990's U.S. military system designers will be faced with a choice of but two alternatives. The first of these alternatives is to buy foreign semiconductors and accept the implications of technological and material dependence attendant therewith. The second is to settle for "second-best" semiconductor devices and the systems they support. In terms of implications for the overall U.S. economy semiconductors truly are "the industrial rice" of the information age and, as the information industry becomes a growing element of the world economy, it would appear critically important for the U.S. to regain and maintain a strong competitive position in this field.

How? The task force was not hesitant about its recommendations.

The principal and most crucial recommendation was for the establishment of an Institute by a consortium of U.S. firms, somewhat along the lines already practiced in Japan, to jointly advance the state-of-the-art in generic semiconductor manufacturing technology. . . .

That recommendation has already been followed with Congress passing the legislation that has allowed SEMATECH to come into being. On it rests much of the hope of both the Defense Department and American industry of winning the Chip War.

12

Winning the War

Cooperation

Charlie Sporck was on a roll. The tall president and CEO of National Semiconductor with the drooping mustaches of a Mexican bandido and the vision of an American robber baron had, in September, bought Fairchild Semiconductor in a stunning coup and now, on January 26, 1988, stood before a group of reporters in the rooftop ballroom of New York's St. Regis Hotel. For Sporck was also the chairman of SEMATECH, which is supposed to restore competitiveness to the American semiconductor industry. And he was about to announce SEMATECH's first major undertaking. Beside him on the platform sat Jack Kuehler, the executive vice-president of IBM and William J. Warwick, president of AT&T Microelectronics. These two men represented the two largest captive chip makers in the world, and they were about to make a pair of major contributions to SEMATECH.

Seated in the first row among the experts "available to answer the press's questions" was IBM's Sandy Kane. It was almost two years to the day that he had shown the IBM management committee the numbers he believed indicated the precarious position of the American semiconductor industry. His belief had been shared and he was sent on a mission, a modern-day Paul Revere warning the countryside of the approach of its enemies. Yet the obvious place to go, the ones for whom

365

the warning was most urgent, the members of the SIA, had always turned a deaf ear to such dire pronouncements.

Nonetheless, in February 1986, Kane began traveling the country, buttonholing everyone who would sit still long enough to listen. "I started to take that pitch around to the industry, and the feeling was that if we went directly to the SIA and said, 'Hey, we want to show you this presentation,' we would have had a great deal of difficulty. They may not have wanted to listen, and if they were willing to listen, they would have sat there, pointing fingers at one another. We decided that the best way to do it was to hit these guys privately, one-on-one, and show them the pitch. I showed it over the next four or five months about a dozen times to most of the key executives of the major semiconductor companies.

"I showed it to Bob Noyce [Intel], I showed it to Bill Sick [TI], to Jim Norling [Motorola]. I showed it to Charlie Sporck, and I showed it to some people in the Digital Equipment Corporation. I also showed it to some of the equipment manufacturers: Perkin-Elmer saw it, GenRad, Ken Levy at KLA saw it, and a few others."

A number of agencies and departments within the federal government had begun expressing concern, and Kane also went to them. "They tended to be various kinds of people. I showed it to a group in the CIA. I showed it to a group in the Commerce Department, some people from the State Department, some people from the White House, the Science Advisory Committee, that kind of thing. Small groups and stuff. A couple of things came out that were interesting.

"The first thing that surprised me and concerned me was that in spite of the fact that nothing in that presentation had been developed from special sources—it was all publicly available information—not a soul that we showed it to was familiar with the information. No one!

"There were pieces they had seen before, but no one had taken the trouble to put that story together. I mean, they just hadn't seen the totality of that story. And, without exception, no one argued with the information, and they were all quite stunned by what they saw, some to the extent that they really didn't even know what to say. They sat there quietly and hardly ever asked a question. At the end they didn't even say, 'What do we do?' Most were able to have a reasonable discussion, and we got into what needed to be done. The net effect was that the presentation became the catalyst that led to SEMATECH."[1]

At the next SIA board meeting in June 1986, in Santa Clara, the heart of Silicon Valley, the industry as a whole discussed the numbers in Sandy Kane's presentation and their implications. It was not even on the agenda, but virtually all of the SIA representatives, some forty people from the top semiconductor companies in the country, had seen the presentation and began talking about it. After two hours of discussion, Charlie Sporck volunteered to pursue the matter further.

"Charlie was going to go and kind of test the tea leaves," recalls Kane, "and talk to a whole bunch of people during the next few months, in the industry and in government and universities, to get a feel for what was in the realm of the possible. Charlie did that over that summer of '86 and reported back to the board when we met in Santa Clara in September. The consensus he brought back was that, in fact, something had to happen and that something could happen, but they weren't sure what."

It was enough of a start; the industry was aroused and ready to take action. "It was from that feeling that something had to be done," Kane told me, "that we formed a task force that first met in October of 1986. I was a member of that group, and we tried to put some form and structure on the solution, and it was from that that SEMATECH eventually came into being."

"We're not dealing with a research project," Sporck told all who would listen to the grand design that was evolving for SEMATECH. "The American industry has fallen behind Japan not so much in research but in the manufacturing of products at low cost with high quality. SEMATECH will be aimed at improving manufacturing ability."[2]

The original thought had been to appoint a workshop group to select a test chip for which production equipment would then be chosen. But the startling and totally unexpected offers from IBM and AT&T promised to get SEMATECH off to a much faster start. Both companies were to contribute the designs, masks, test data bases, and a shopping list of all the equipment required to manufacture their most advanced chips. From IBM it was a 4-megabit DRAM, from AT&T a 64-kilobit static RAM. The two chips represent next-generation technologies not yet in commercial production. The offer, which meant that both companies were prepared to give up some of their technological advantage, clearly surprised many of their fellow consortium members. It also underscored their deep commitment to SEMATECH.

"We offered what we consider our most advanced state-of-the-art technology," declared IBM's Kuehler. "We're very pleased with the opportunity this decision presents us to demonstrate the strength of our commitment to SEMATECH. We believe the successful fulfillment of SEMATECH goals is important to U.S. industry and the U.S. economy."[3]

Engineering teams from IBM and AT&T then flew to Austin, Texas, to begin supervising the installation of duplicates of the same chip-making machinery both companies will use in their own commercial production of the two chips. But once installed as a pilot line in Austin, the equipment will be subjected to the most rigorous testing and tinkering in the hopes of producing enhancements and innovations that will generate an entirely new generation of manufacturing technology.

"Companies will work on process improvements, with some SEMA-TECH money and some of their own," declared AT&T's William Warwick. "SEMATECH members will have right of first refusal on any equipment suppliers develop specifically for SEMATECH."[4]

Manufacturing technology, the ability to produce high volumes of high-quality chips at the lowest possible cost, is where the Japanese have excelled. It is the area where American industry has fallen behind and is one of the keys to winning the Chip War. By developing the manufacturing expertise at the leading edge of semiconductor technology, SEMATECH will then distribute its knowledge to all its members.

"That's why the SEMATECH project is of so much interest," declares the Commerce Department's Jack Clifford. "We are hoping it will provide the manufacturing technology that's necessary to keep the U.S. competitive in the high-volume areas. The industry itself comments that the high-volume areas are where they get the funding from to do the research and development for the next high-volume product and the microprocessors and the ASICs and custom products. I don't think it makes sense for our industry to hope to compete by backing itself into specialized devices. The markets are going to be relatively small. The Japanese can push far enough and fast enough so that even that is not a safe area to be in. We have real concerns about that."[5]

With the 4-megabit and 64K chips, SEMATECH was off to a running start. "These vehicles will allow us to demonstrate a baseline against which we can measure progress into phase II of our plans targeted at half micron geometries," enthused Charlie Sporck.

Each new generation of chip packs more and more transistors and circuits into ever smaller spaces. This can be accomplished only by reducing the size of the elements, the so-called geometries of the chips. The 4-megabit DRAM, for example, a half-inch- by one-quarter-inch rectangle of silicon, contains more than 4 million transistors and other components and circuit lines that measure 0.7 micron. AT&T's contribution, the 64K SRAM, also features geometries 0.7 micron. A micron equals one millionth of a meter. Approximately 25,400 microns equal one inch. Such incredibly tiny dimensions push the tools of semiconductor manufacturing to their very limits. In addition, although both chips use different process sequences in their construction, the same processes can be used to make logic chips, microprocessors, and ASICs.

With the 4-megabit chip, the American semiconductor industry might even regain some of the memory markets lost to the Japanese. "We are starting essentially even with the Japanese on developing manufacturing technology for 4-megabit DRAMs," says Bob Noyce. "They aren't in the lead, as they were in 256-K and 1-megabit DRAMs. The major confrontation with Japan is even more likely in the 16-megabit area."[6]

The Japanese experienced more than a little difficulty in moving from 256K to 1-megabit chips. "Initial yields were a problem for every producer," complained NEC's Tomihiro Matsumura. The low yields have also slowed the payback of the enormous investment required to go from 256K to 1 megabit. Thus, the Japanese would prefer a much longer sales life for the latest-generation chip not merely to recoup their investment but to turn a profit for all their effort. The development of 4-megabit manufacturing technology by SEMATECH may not give Japanese producers that luxury. If Americans start up 4-megabit lines, Japanese semiconductor makers will have to follow suit much sooner than they would like. For not only will they lose the opportunity to recoup their huge R & D and capital investments in the 1-megabit chip; they will have to plunge even deeper to finance the next-generation production.

And the 4-megabit chip promises to be even more difficult to produce. "I estimate there are fifteen hundred manufacturing processes alone subject to minute amounts of contamination that can severely impact yields," says Tsuyoshi Kawanishi, the executive director of semiconductor operations for Toshiba.[7]

With the 4-megabit chip and subsequent generations, the American semiconductor industry might well regain some of the memory markets it lost in the 1980s, but much will depend on a change in what has been more often than not an arms-length relationship between equipment makers and chip makers. Unfortunately, the two groups have often functioned in isolation, and the lack of cooperation has meant long delays between the design of new chips and their high-volume production. The Japanese have solved this problem by having both equipment manufacturers and semiconductor makers as part of the same vertically integrated company. Cooperation is in that instance close and continuous. SEMATECH hopes to draw together both design and equipment makers in a cooperative web that will to an extent emulate those same close ties the Japanese have established.

SEMATECH is designed to produce the sort of innovations that will lead to enhanced manufacturing equipment. And it will inevitably be selecting one manufacturer over another, conferring advantages to those chosen to produce a given piece of equipment over their competitors who have not been so knighted. "It's not immediately obvious how we deal with the advantages some companies will gain over others in developing equipment for SEMATECH," admits Charlie Sporck. "And obviously we are not interested in playing God. At the same time there is a fundamental problem in that there are too many equipment manufacturers. There's over a thousand of them, or some enormous number. How does that all work out? We are not only going to want to work with them; we are going to want to finance them. I don't want to describe SEMATECH as having all its problems and issues resolved; we don't. Nobody's done this before, and we are walking in a swamp right now. Obviously, we are going to run into problems and will have to work our way through."

SEMATECH also represents a basic change in American corporate philosophy, which for the past twenty years has been actively shifting from smokestack industries to service. With its emphasis on developing new manufacturing technologies, it will also help to bring manufacturing back as a respectable, indeed vital, sector of the economy. "We are pointing out some very simple truths in that the service industries are very closely associated with manufacturing," Bob Noyce, SEMATECH's president, told me during our wide-ranging discussion. "If manufacturing disappears, the service industry disappears. You can't create wealth by taking in each other's laundry. You've got to make a new shirt now and then. I think the general realization of that problem is coming around."[8]

SEMATECH will also foster a sense of cooperation within the industry that has for the most part been lacking. Sandy Kane thinks it can accomplish even more. "I think that if we put our emphasis on the right kinds of projects and not just on manufacturing, SEMATECH can show the world that that kind of concept can succeed in the United States. I believe that on the development and research side of the equation we can do the same kinds of things, pool some of the resources that already exist in our companies, national laboratories, universities, and so on. Then I believe we can either regain or hold our own with regard to technological leadership."

Congress has already reacted to this sort of thinking, including in its 1988 trade bill an amendment calling for the establishment of a National Advisory Committee on Semiconductors. Its chief function would be to recommend allocation of R & D resources throughout the United States. The aim would be to prevent duplication of effort in federal laboratories and academic institutions and the apportionment of funds among consortia such as SEMATECH and other industry/government groups.

"Assuming success," continues Kane, "we need to use the example of SEMATECH and start spreading that gospel to other areas, to find different and increasing ways in which we can cooperate. And the combination of industry, universities, and government is essential in that process. That message needs to get across."

R & D is conducted on three basic levels in the United States: by individual companies, in universities, and in national government laboratories. It has long been one of the major competitive advantages this country has had. Rarely have the research aims of those disparate organizations been coordinated to achieve a specific goal. Now, the NRC has proposed a radical change in the traditional strategy that combines the short-term focus of industrial laboratories with the long-range perspective of universities. The new American approach will have to be a long-term one, with far less emphasis on fiscal-year, bottom-line profits. Instead, like the Japanese, we will have to make commitments to projects that may not reach the marketplace for ten years or more. These and other strategies, which will invest more in basic research and at the same time couple semiconductor research with the development of equipment and new products to use the new chips, will not only ensure the ongoing health of the semiconductor industry but help create the new products that are the lifeblood of any industry.

Companies must also cooperate with each other. Beyond the mergers that have already taken place and will continue as American companies seek to increase their size, joint ventures, and investments in individual firms that are important suppliers must be made by the wealthier companies. A classic example was the 1983 infusion of $350 million cash by IBM into Intel. With the money, Intel was able to launch a $100 million research project that led to the development and production of its phenomenally successful 80386 microprocessor. For IBM it was a means of guaranteeing the viability of one of its major chip suppliers. In 1988, Intel forged an alliance with Micron Technology, the small Idaho chip maker, aside from TI, the only American producer of DRAMS. Under the deal, Intel committed itself to long-term purchases of chips from Micron, along with $11.6 million for a 2 percent piece of Micron. For Micron it means a guaranteed supply of cash to fund a major new memory fab it is building in Boise. For Intel it means a guaranteed supply of memory chips, which it stopped building in 1985. It will enable Intel to sell and have access, for its own use, to memory chips far more inexpensively and quickly than if it were to build its own memory-production facilities.

It will also "help decrease American equipment firms' dependency and vulnerability to Japanese DRAM domination," declared Bob Noyce. Clearly, mergers and cooperative ventures between semiconductor companies, systems equipment makers, and systems manufacturers are essential to improving the competitive position of many companies. The vertically integrated companies of Japan and Korea and Europe have far more resources than their American competitors. NEC, Hitachi, Toshiba, Samsung, Goldstar, Philips, Siemens, are all $10–$15 billion plus companies. By contrast, the two largest U.S. merchant companies, Motorola and TI, had sales of $6.7 and $5.6 billion, respectively. Intel, AMD, and National Semiconductor, even after mergers and acquisitions, are in the $1–$3 billion range.[9] In the cyclical semiconductor industry, most American chip makers are simply too small and underfunded to withstand the bust part of the cycle. Nor can they make the capital investments every two or three years that are required to maintain leading-edge technologies.

Clearly, the structure of the industry itself must be changed. Merchant chip companies must merge with or otherwise become intimately tied to systems and equipment makers. In that way, a relatively constant demand for chips can be guaranteed, and revenues from other sources—the sale of computers, electronic systems, and chip-making equipment—can be used to subsidize chip-making operations through boom and bust cycles. Profits generated in those other areas can offset the inevitable losses that occur early in a new chip's product life and especially in the production of those commodity chips that are the technology drivers of the industry.

Unfortunately, antitrust laws in America prevent many of these types of mergers. Originally designed to prevent a monopoly from being created in a specific industry or geographic region, antitrust legislation must now consider not merely a U.S. market but a global market. The world is indeed now a global marketplace, and in that arena monopolies are created by the very policies that seek to curtail them. In other words, preventing the creation of a possible monopoly in the United States virtually guarantees its creation in the form of much larger and more powerful foreign companies with whom the much smaller U.S. firms cannot compete.

Capital

The semiconductor industry has become enormously capital intensive. Whereas once deep pockets were needed to build a new fab plant, now bottomless pockets would be a more appropriate fashion. "To stay in this business you require enormous amounts of capital," declares Michael Attardo, the president of IBM's General Technology Division, the unit responsible for making the more than $3 billion worth of chips they produce each year. "To put in place a state-of-the-art fabricator that's going to make a 4-megabit device, for example, or leading-edge logic, you are talking about half a billion dollars in capital equipment and building."[10]

Admittedly there are less expensive ways to build and/or equip a plant, but even at half the price, the numbers for most American semiconductor companies are discouraging. So, too, are their efforts to raise money to finance expansion, new equipment, R & D, and the multitude of other demands made on the financial resources of a business today.

The difficulty of raising capital is one reason companies like Xilinx have looked to Japan for production of their designs. "The insidious thing about the semiconductor business is that if the cost of capital is cheap and you don't need instant payback and if the tax laws favor you for capital investment, what will you do?" asks Bernie Vonderschmitt, president of Xilinx. "In the U.S. capital is not cheap, and the tax laws no longer favor capital investment, so the mentality is, I'm going to put up a $100 million plant. How in the hell can I get it down to $80 million?

"In Japan the mentality is, if I can improve the performance one little bit, I'm going to spend $125 million. The gross throughput is the same in the two plants, but if your yield is two or three times higher, you're going to have an enormous cost advantage in the better facility. And since the cost of capital and taxes in Japan is roughly half of what it is in the United States, they have a tremendous advantage. And that's basically at the heart of the problem the semiconductor industry faces on an ongoing basis."[11]

The relative ease with which Japanese semiconductor companies can obtain capital and its lower cost has given them an enormous advantage. Virtually everyone I talked to about the increasing capital-intensive requirements of the semiconductor industry pointed to the disparity between the cost of capital in the United States and Japan. "There's a huge debate now going on in American financial circles about structural impediments to competitiveness," Roger Altman, a managing director of Shearson Lehman Brothers, said to me. Altman, a former assistant secretary of the treasury and one of the chief architects of the Chrysler bailout, spoke with me in his richly appointed office just off Wall Street, with its spectacular view of New York Harbor. Filled with Washington mementos—photos of Robert Kennedy and Jimmy Carter, in whose administration he served—the office was a reminder of the different ways in which the Japanese and American semiconductor companies raise the huge amounts of capital they need to stay competitive. This was no understated banker's domain, but a more obvious citadel of power, one of several investment-banking and brokerage houses that put together the stock and bond deals that provide the money for American business to operate, expand, or merely survive.

Altman, a cigar-smoking, lean man in his early forties, rolled his words over carefully before speaking. "Access to and cost of capital must be improved. In Japan, because of a variety of factors, semiconductor producers are better financed and have more stable financing, that is, in terms of the amounts they get and the stability of their sources. They are thus able to run their business without nearly as much concern for the vagaries of their capital base and their capital sources as U.S. companies. In contrast, in our country there is a boom-bust cycle. When our capital markets, the stock markets, are doing well, the semiconductor industry is able to raise capital; not as much as it should, but some capital. When those markets are going badly, cycling down, that industry is the first to go; nobody wants to give them a nickel. So they go through periods when they can raise money at relatively good terms and then long periods when they can't raise any capital at all. So what do they do? They worry about it. They manage the company from a short-term or medium-term perspective. One of the immediate constraints is, how is it going to look to the markets, how will it affect our ability to finance?"[12]

Just how closely the industry watches Wall Street can be seen in one of the first efforts made by the Commerce Department to warn of increasing foreign competition. Jack Clifford, director of the Department's Office of Microelectronics and Instrumentation, told me of the attempt by his office to document their growing concern as far back as 1974. "When we started to work on that report, we saw how fast the Japanese semiconductor industry was developing, the areas in which it was developing, and our concerns about what that competition would do to the U.S. semiconductor industry. We talked about the production base of consumer-electronic products and how it shifted to Japan so that it gave them a huge market in their own country to operate from, a protected market because there were no U.S. products competing against their products. There was little competition by U.S. semiconductors at that time because the Japanese had vertically integrated companies that were buying only what they had to from the U.S. and supplying the rest themselves. All these factors said there is going to be a problem from that Japanese semiconductor industry in 1973 and '74. We did a preliminary report, and it sailed up through a few of our offices, and we decided to have a meeting with the semiconductor industry to lay this out for them.

"There was no SIA then, just WEMA, the Western Electronics

Manufacturers Association, and the Electronics Industry Association. We had representatives from those two groups come in, and we showed them what was a preliminary report, a draft form. One of the problems we saw at that time was a rather significant falloff in semiconductor R & D spending. We said with what's going on in Japan, if you continue down this path and not stay ahead in both the manufacturing technology and the product technology, the situation is going to get worse. The Japanese are going to be a major competitor. At that point, the Japanese had about 2 percent of our market. Those people looked at us and said, 'You're crazy as hell. We don't want that report out because if that hits Wall Street our stock prices will drop way down.'

"It really surprised me, but their biggest concern was what it would do on Wall Street. If the government came out with a report, a stamped government report, that said U.S. industry was not spending enough on R & D and they were going to have problems with the Japanese in the future, they didn't want that out. And they said, 'We don't believe those R & D figures.'

"We said, 'You build a model, and we realize you can't get all the R & D figures you would like, but we've got some good ones, and those R & D figures at least point in the direction in which you're going.' Qualitatively they say what we think is going on. We've had a lot of other people look at those numbers, and they agree with us."

Despite the numbers, the industry view prevailed. "Our bosses said we don't want any problems; don't issue the report." Clifford admitted. "A year or so later there was some more yelling and screaming, and we were told to pull out the report and revise it, bring it up-to-date. More pressure came on us, and again we couldn't publish. The government doesn't want to lock horns with industry if they can help it. There was also a reorganization of the Commerce Department at that time which tended to work against publishing the report. At one time we were all pulled off into one big room upstairs and told, 'You are no longer going to communicate with industry. You're going to sit in your little ivory towers and do research work, but don't communicate with industry. You're creating too many problems.'

"In the last five years that's all turned around. The SIA came back in 1979 and said, 'Hey, we've got a problem with Japan.' So we all got yanked upstairs and were told, 'Hey, we've got a problem with Japan.

Why didn't you guys tell us about it.' We reached into the drawer and pulled this thing out and said, 'We tried to tell you five years ago you had a problem with the Japanese.' It wasn't the same group that pulled us off then.

"We said to the assistant secretary, 'Here is the report. We've sort of kept it up-to-date; take a look at it.' SIA now is saying, 'Yeah, yeah, get this out; it tells our story pretty well.' So the assistant secretary says, 'Publish it; get it out.' But even then, when we came back to our own department, they still had concerns about it, because it was a little controversial. But anyway we finally got it out."[13]

Among the many criticisms of the semiconductor industry management and indeed of most American industry is its failure to see beyond the quarterly bottom line. But when most companies are dependent on stockholders for their capital, it is hard not to plan for quarterly profits.

"It's very popular to criticize management for that," declares Roger Altman, "but I think that's misguided. I don't know what else they are supposed to do. If you can't finance a business, you can't have a business. In this industry, which is very capital intensive and very cyclical, you just can't raise your capital for the next eight years. And your basic earnings are not big enough to enable you to get that much even if it was available. You couldn't rationalize it."

A case in point is MRC, a modest-sized equipment manufacturer that makes sputterers, etchers, and semiconductor materials. Its ever pressing need for capital was explained to me in the summer of 1987 in the living-color vocabulary of its president, Dr. Sheldon Weinig. "I run a $100 million company, and I'm in deep shit. I'm there for the very reason the Japanese scoff at us. I'm having trouble making quarters. I'm a public company and must respond to Wall Street. My company right now shouldn't be paying attention to quarters. We should be paying attention to the overall cycle of the electronics industry, which is out of phase with quarters.

"Right now I have a new product which has had tremendous acceptance but needs a few million dollars of additional engineering. It's a magnetically enhanced etching machine we call Aries. It's totally innovative, does remarkably interesting things. Nobody's buying a hundred of them, but IBM bought one here and one there, Siemens bought three, Matsushita has bought one, as has NEC and Mitsubishi. We are getting tremendous response on a onesy-twoosy basis. I'm the only American company making it; the only competitor in the world is Japanese, naturally.

"I understand now where I am with this device. I need to put a couple or 3 million into it right now in engineering and I will have something that's the wave of tomorrow. But right now my business is dog shit; everything you've heard about the electronics business is not happening today [summer 1987]. It's all future business. Customers are starting to make a little money, but they're not ordering new machines yet. So I'm saying to myself, I don't want to spend a few million because it will adversely affect my profits for this quarter. I need to hold the quarter positive because the banks will jump all over me if I don't.

"I'm thinking of making a bad engineering compromise based upon performance for those morons on Wall Street. Suddenly I see it clear as a bell, and it's frightening. Not because I'm going to go out of business but because I could hurt my technological position because of it. Whereas my competition, which is Takuda, which is part of the great Toshiba zaibatsu, haven't got these problems."[14]

So long as Wall Street is a major source of capital for American industry, it will be bound to the quarterly bottom line. It is but one of the major aspects in which the United States differs from the Japanese and most of the rest of the industrialized world in the manner in which its companies raise capital. "There's a big contrast in terms of access to and cost of capital," explains Roger Altman. "Now why is it different in Japan. One is the ownership structure of the semiconductor industry. Typically, each of the major Japanese semiconductor companies is part of the zaibatsu. NEC is part of the Sumitomo group, for example. Typically there are six, eight, twelve, even twenty major industrial enterprises linked together by ownership. Sumitomo Bank, for example, was both the largest shareholder in and the largest lender to NEC. And the various companies in the Sumitomo group were among the largest purchasers of NEC output.

"So you have cross-ownership promoting financing of NEC because the bank was the largest shareholder and also the largest lender. And some of the bigger entities in the group were also shareholders in the bank, and they were also big customers of NEC. So the structure of ownership and the ability of their banks to have a major position in their industry have contributed very favorably to the access and cost of capital to companies like NEC. During the late seventies many of the Japanese semiconductor companies ran huge losses. Did it terminate their ability to finance? Did it raise the cost of capital to them? No. Because there is an affiliate relationship. Our antitrust laws and banking laws prohibit such relationships. So our legal structure in terms of both banking law and antitrust law is much less favorable to semiconductor producers than in Japan.

"The second factor is the way the banking system works and can freely lend enormous amounts to companies that are losing money year after year. The answer is that if it's part of their national strategy, as so designated by MITI, which has a very subtle influence, they can do that. The banks know that the Bank of Japan, the central bank, stands behind those loans. When one of the banks poured money into Toyo Kogyo, Mazda, in the mid-seventies, it did so knowing that the Bank of Japan was behind it. If Mazda didn't turn around and those loans became worthless, they could in effect discount the paper at the Bank of Japan and get liquid. That's not the way it works in this country. Continental Illinois makes some bad loans, zip down the tubes. Now, Continental Illinois may remain as an entity, but the shareholders sure didn't do too well; they got nothing. So the way the banks are structured in Japan is the second big factor."

"The banking industry in Japan is still operating under a protected environment, but I think gradually that protection is being lifted," Roy Takata, former managing director of the Bank of Tokyo, explained to me in his New York office in the World Trade Center. Only just retired from his banking job, he is now the president of the American branch of Yamatane Securities, one of Japan's largest brokerage houses. The son of a banker, Takata spent much of his early life outside Japan. "My educational background was different from most Japanese of my age in that I had the overseas experience before the war. I was in Hawaii from 1934 to 1937, which helped me a great deal, of course, in speaking English. Then I went to China. But the competitive nature of the Japanese educational system even then was such that my father felt I had to stay in one place. And so, from the seventh grade on, I was brought up in Kyoto and went to Kyoto University, where I studied law. That was after the war. Then I had a Fulbright Scholarship to study in the U.S., where I went to NYU law school. I had already joined the Bank of Tokyo and took a leave of absence for that. I returned to the bank and after several years opened branches in Houston and Chicago. In Chicago I did everything from loans to running the computer, everything but act as the president of the bank. I spent eight years in the U.S. doing that. Then I returned to Japan as the senior loan officer and set up a computerized system."

With that background, Takata is in a unique position to compare the banking systems of the two countries. "The biggest difference is the regulated environment in interest rates," he points out. "The cost of capital depends on market conditions. Three or four years ago, when the interest rates in the U.S. were very high, it was probably half as expensive to borrow money in Japan. Now I don't think it is quite as cheap, but the Japanese interest rates are still probably among the lowest in the industrialized world. The Bank of Japan's discount rate was 3.5 percent. The prime rate was probably around 6 percent, but we also have a long-term prime rate which doesn't exist in the U.S. The short-term rate was artificially high, because the Bank of Japan wanted to keep it high. The long-term rate was allowed to float with the market, and it was below the short-term. Anything over one year is considered long-term."[15]

Because of their zabaitsu affiliations, and the remarkable savings habits of the Japanese people and the appreciation of the yen, Japanese banks have grown enormously in the last decade.

A 1988 survey by the *American Banker* showed that the world's ten largest banks, based on deposits, are all Japanese. Of the next fifteen largest banks, seven are Japanese, two are German, four are French, and two are British. Citibank, the largest American bank, dropped from its seventeenth position on the list last year to twenty-eighth.

"It's shocking how fast Japanese banks have come up to dominate the banking industry," said William M. Isaac, the former chairman of the Federal Deposit Insurance Corporation. The growth of the Japanese banks, he pointed out, reflected "a fundamental change in the competitiveness of the two systems."[16]

Even the nonbanking financial institutions, the "Wall Street" component of our system, is dwarfed by foreign competition. The largest of these American firms was the American Express Company, with $16.1 billion capitalization. That was good enough only for tenth place on the world list. Indeed, eight of the top ten companies were Japanese, headed by Nomuira Securities with $59.6 billion in capital.

Part of that enormous pool of capital comes from the aggressive, almost frantic savings habits of the Japanese people, an estimated 15 percent of their incomes. By contrast, U.S. savings have dipped from an average of 7 percent from 1953 to 1980 to 4.3 percent for the 1980s. The reasons for the disparity are both cultural and economic. With safety nets such as social security in place in America, there has been less of a perceived need to save for one's future.

"Savings is a Japanese habit," Roy Takata declares. "And despite that some economists say they shouldn't save as much as they do, I think it is a virtue of the Japanese people. The savings habit developed historically because consumer financing was not that readily available and pensions were not then very common. Now, of course, pensions are much more common, but the habit of saving is so ingrained that it has not changed."

Traditional economic theory states that the savings rate falls during a recession as families seek to maintain their standards of living by spending more of their incomes and savings. On the other hand, during periods of economic growth, savings was deemed more likely to rise. But during the past five years of expansion, savings have actually declined sharply in the United States. This makes less capital available to American banks that can be loaned to business and consumers. Moreover, the amount of their assets American banks can loan to commercial customers is sharply proscribed by the need to maintain fairly sizable reserves. Japanese banks by contrast are permitted by their regulators to maintain much lower capital reserves.

The strength of Japanese banks is definitely a factor in the ability of Japanese semiconductor companies to weather the boom and bust cycles of the marketplace and still invest in the latest technology. "They don't rely on the public markets the way we do. It's mostly bank financing in Japan," points out Roger Altman. "Most big American companies don't borrow very much from banks. The companies I work on are all big companies, and they don't borrow from banks. They have bank backup lines of credit, but they don't actually borrow from banks. They borrow from the public markets through their investment programs. Why don't our companies borrow from banks? Because the banks aren't prepared to hang in there. You've got limitations on all loans. Our banks are not as strong as the Japanese banks."

A web of regulations, most drawn early in the century, serve to inhibit the competitive position of American banks in the global marketplace. Says George D. Gould, undersecretary of the treasury for banking:

American Finance is a late 20th century business governed by early 20th century rules. . . . Responding to the problems of the past, the archaic laws still divide the financial business by geography and even by product, even as rapid advancements in technology, increased global economic competition and the integration of world markets have created a revolution.

The result, according to Gould, is "that financial institutions in the United States suddenly find themselves poorly equipped for the new world of global finance. Make no mistake, financial markets truly are global now and the competition is fierce."[17]

Gould and others in the Treasury Department have been calling for legislation that would allow U.S. banks to become more closely involved with their commercial customers, to assume many other financial services in the area of stocks and bonds that would enable them to better compete with foreign banks and make more money available through sources other than traditional ones to capital-intensive American industries.

Most American semiconductor companies are also not as strong as their Japanese counterparts. And that, too, is an increasingly important factor in an increasingly capital intensive industry. "Most of the major Japanese semiconductor makers," Takata admitted, "have the financial strength to finance new semiconductor manufacturing facilities on their own. In all likelihood, each of these major companies had their own funding programs, and the banks that were participating in it had no idea of what the money was specifically designated for."

Still another advantage is the investment philosophy of the Japanese. "The general investor in Japan has a much longer view and can wait longer for profit than American investors," notes Takata. "Most of the expansion by Japanese semiconductor companies during the slump, 1985–86, were financed by themselves. The difference in the aggressiveness and the confidence in the future is probably in the psychology of the long-range view held by the Japanese."

Research and Education

On the university level, the United States has perhaps its greatest advantage over the rest of the world. For unlike the Japanese,

European, and Asian systems, American universities are active and important participants in R & D in many fields, including semiconductors.

To see just how deeply involved some of our universities actually were in semiconductor research, I visited Columbia University. In 1984, Columbia set up within its School of Engineering the Microelectronics Sciences Laboratories. Funded by industry and the Office of Naval Research, the MSL seeks new materials and improved fabrication techniques for ICs. Its head is Professor Richard M. Osgood, a forty-four-year-old electrical engineer and physicist. "Our lab is oriented to coming up with innovations for microelectronics," he told me. "We do basic science which can lead to new ideas for the electronics industry. That's what our charter is. We have people here who do molecular-beam epitaxy [a process that can control the thickness and composition of a crystal film so precisely it can grow one atomic layer at a time]. I do laser processing and laser fabrication of electronic materials. We have one man here who works ion-beam fabrication techniques."

"Our emphasis here at the Microelectronics Science Laboratory has been very much on innovation." And that is where the mix of older professors, often with experience in industry, and young, eager graduate students can often result in fresh approaches that industrial research laboratories are less apt to take. "It turns out that university labs have a certain strength," declares Osgood with a grin. "They tend to have more people who are less experienced—the kids. And they're likely to come up with something special. . . . To some extent, to do inventions you've got to do things that are crazy. Universities are good at that."

One such idea, more serendipitous than crazy, led to a new technique for using optical fibers—the hair-thin, light-carrying filaments now used in phone lines—as communications links between chips. The innovation was discovered when one of MSL's young graduate students substituted an ultraviolet laser beam for visible light in an experiment. The laser light, normally used to etch features onto a chip, can be trapped at the point it strikes the surface. The trapped beam burrows into the chip's interior, creating a neat, narrow, light pipe into which an optical fiber can be inserted. Eventually, Osgood believes the technique will be applied to the manufacture of opto-electronic chips that will combine the capabilities of the photon for data transmission and switching—to link components that need to "talk" to each other—with the electron's redoubtable ability to perform data processing.

"We are not trying to do manufacturing research," claims Osgood. "People are pushing the resolution limits of silicon ever harder; they are constantly coming up with new materials. There is, for example, right now a lot of work in growing gallium arsenide on silicon. It's a hybrid technology that would combine the optical properties of gallium aresenide with the reliability and ease of fabrication of silicon. So one direction we are looking to some extent at is in combining different material technologies together, organizing a symphony of different materials, if you will. We work on that; AT&T has a very good program along the same lines, as does Cal Tech.

"Another area is that in making silicon integrated circuits one is constantly coming up with questions of how not to damage the substrate when laying down very fine features. We are trying to develop new ways of etching silicon so that it doesn't damage the surface as much. A new technique is to use ion beams or plasmas to impinge on the substrate and etch it. You put down a metal mask optically and the lines are etched where the mask is blank. We are developing new laser etching techniques which are simpler than the conventional methods and have less of a tendency to damage the substrate."[18]

I was given a tour of the MSL by one of Osgood's "kids," a twenty-four-year-old graduate student named Allen Willner. He is from Baltimore, a graduate of Yeshiva University in New York with a master's degree from Columbia, and is now working for his Ph.D. in electrical engineering. A knitted yarmulke is almost lost in his curly hair, and like so many of the younger engineers I have met in this field, he sports an equally curly beard. He leads me into a small room, about ten by fifteen feet, with cinder-block walls painted a light green. In the center and virtually filling the room is a large flat-topped table set on sturdy black metal legs. The legs, which are hollow, hold a gas that allows the tabletop to float, thereby eliminating the risk of disturbing the laser beam by vibration from a passing heavy truck on the street twelve stories below or from the subway that runs beneath the surface several blocks away.

Equipment sits on shelves mounted above the table and on benches that line the walls of the lab. Mounted on the table across from where we are standing is the laser, an eight-foot-long rectangular container with the same proportions as a florist's box for long-stemmed roses. Pulsing from the box is a shimmering, slender beam of green light about as thick as a pencil lead.

The beam strikes a mirror at the end of the table and is shifted ninety degrees to race along the short edge of the table, where it strikes another mirror and is again rotated ninety degrees and passed through a double crystal that changes it to a blue ultraviolet light. It then is directed to strike a piece of semiconductor material, which sits under an optical microscope directly in front of us. By using either a joystick, the kind used to play computer games but more expensive, or a preset computer program, the laser etches tiny lines, smaller than the eye can see, onto the semiconductor material. The lines, however, can be viewed under a microscope, which magnifies them about 150 times, or on a small television monitor mounted on the shelf about the table.

"This whole setup here is a laser processing lab." Willner explains. "This laser lases in a green light which we send through a doubling crystal, which changes it to ultraviolet light. The UV light has certain properties which are very useful for processing semiconductors. The energy per photon is high, and so we can do certain things we cannot do with ordinary light. We can scan across the surface of the material and deposit metal lines. Traditional methods require hundreds of lithographic processing steps to lay down circuit lines. What we are able to do here is in a single step use the laser to break apart a gas and leave a metal line in its wake.

"There are two ways to go about it. You have a gas that sits atop the surface. The laser photons have a certain amount of energy to break apart the gas, and the metal part of the gas migrates down to form whatever lines the laser traces. On the other hand, there is a thermal process called pyrolitic, which uses an intense laser beam. Instead of worrying how energetic the photons are, you just bombard the surface with high power. The gas molecules just sit on the surface. When the laser heats up the surface, they are just cracked apart. When they break apart, the lighter atoms are driven off, and only the metal molecules remain. If you're breaking apart the gas, it doesn't matter what surface you are on; you just draw a line. If you are heating the surface, which is a much faster process, then it depends on what surface you are on.

"The real advantage is that it's a one-step process. What industry is interested in is a very simple setup which is very cost effective. This whole setup here is every inexpensive. The standard processing equipment can run into hundreds of thousands of dollars. Here we have a laser, a doubling crystal, a microscope, some pumping equipment, and a gas. The whole thing is a couple of thousand dollars.

"We are also able to do chip repair, even if one connection is wrong with the chip, and there might be hundreds of thousands of connections. So what we are able to do is a spot correction. If you mass-produced a chip, it might be cheaper to just throw out the defective chip, but in the custom-logic fields or in high-power applications the chips can cost hundreds of dollars. This can be a very useful technique. This is especially true of ROMs, read only memory, where the memory is burned in permanently and if something goes wrong, you've lost your entire investment. So that is the in situ correction capability."[19]

At the other end of the country is Stanford University, another of the major university semiconductor research centers. Here on the sprawling California campus an ultramodern building to house a major effort in semiconductor research was recently completed at a cost of $14.5 million. Called the Center for Integrated Systems (CIS), it is a soaring glass-and-concrete structure of great arched skylights and steel trusses that form the exterior shell of what is in reality two buildings, one within the other. The inner building is actually a glass-walled 10,000-square-foot box. Within it is more than $15 million worth of state-of-the-art semiconductor-making equipment. Enfolding the glass box in a U shape is the exterior building housing offices and all of the mechanical blowers that constantly recirculate scrubbed air and dozens of stainless-steel gas-delivery columns that furnish the thirty-three different gases used in semiconductor production.

The laboratory and the U-shaped building that embraces it each rest on separate concrete foundations joined by a four-inch strip of rubber that damps all vibrations from the so-called mechanical attic and basement that houses the supporting machinery. Funded by twenty computer and semiconductor companies and with donations of equipment from another seventy companies, the CIS represents a $40 million commitment to semiconductor research in one of the most advanced research laboratories of its kind in the world. There are eighty faculty members from many of the Stanford physics, aeronautics, and chemical engineering departments affiliated with the CIS and some seventy-five postdoctoral students at work there. One unique aspect is the industry visitors, or industry residents, titles given to the twenty-five engineers and scientists from semiconductor and computer companies who do research and teach at the center. I was scheduled to meet with one of them, Brian Davies, of TI. Davies has been with TI for seven years and at the CIS for the past two years. I asked for him at the reception desk.

As I mentioned his name, a slender baby-faced young man whom I took to be an undergraduate, looked up. "I'm Brian Davies," he said, sticking out his hand.

Although he looked like a freshman, Davies was in fact thirty-three years old. At the age of eighteen, when most people enter college, he had already graduated number one in his class from Canada's University of Saskatchewan. "I grew up in Canada, went to the University there, and then went to graduate school in the U.S. at Cal Tech and MIT." At MIT, Davies began working in the field of AI. "I'd been at MIT on and off for seven years as a graduate student, but I didn't get my Ph.D. I just finally decided I was going to go out in the real world. I had gotten fed up with what I was doing. I was working in AI with medicine, and I was looking for the chance to apply AI to another kind of real-world problem, and there was this opportunity to apply AI to oil exploration. That's how I really decided to move to Dallas and join Texas Instruments. That really got me into the ground floor of AI at Texas Instruments. And things just blossomed from that. TI is now, some people think, the industry AI leader in both hardware and software.

"I work for TI at Stanford. That's my job, and that's what I tell people. When I first came here, we set up a division of my resources, the kinds of things I would concentrate on. I would be doing research that would presumably be of interest to TI—also relevant to Stanford. I mean, you can't come here and set up your own research project; it must be part of what's going on here. So part of what I would be doing would be strictly research, and many things would come out of that, including contacts with people, and so that would be valuable in itself. A second part of it was technology transfer, and that is just being here as a conduit for interesting things that would be going on at Stanford, making sure that the appropriate people at TI know what's going on at

Stanford in their areas. And a third part of it for TI was recruiting. I was here to talk to students and make sure they knew about TI, knew that TI was an exciting place to work and to make sure that they considered TI even though I can't force them to go to work for TI. And so TI profits in those ways from what I'm doing here.

"My salary is paid by TI, and they also pay an annual fee to the CIS to be a participant here. One of the benefits of that, for paying that, is that they're allowed to send somebody to Stanford to use the research facilities and have access to all the research being done here. Stanford is an open place. But one of the benefits of CIS is facilitated access. Companies get more of an opportunity to participate and utilize the benefits of the research by being a member of CIS than other companies might."

To Davies, CIS, with its emphasis on manufacturing research, seemed the perfect place to try and mate AI to manufacturing technology. "For little more than a year," he told me, "I've been the head of one of the projects here, the automation of semiconductor manufacturing. We want to understand how fabrication lines work and then automate them in various ways. Of course, here in the university we're not going to develop an automated fabrication line. We just don't have the resources to do that, so they have to be developed in various prototypes to illustrate to industry different ways of automating those lines.

"We have a lot of professional people around here, but most of the work, most of the real work, is done by graduate students, people who are in the learning phase. And so I hear Dr. Linville (director of the CIS) say this all the time: 'Our primary product is really the students. We have to get those students turned on to manufacturing.'

"We have twenty-five to thirty Ph.D. students. Their cycle time is four to five years, so you can figure that maybe a half a dozen a year, or something like that, will be going into industry. That's not a very large number, but there's gonna be a trickle-down effect. So one of the things that we're doing is to produce students who are aware of issues in manufacturing and know about problems in manufacturing. So they're not just knowledgeable about process technology; they know what it's really like to run the factory, know what the problems are in a factory.

"They're motivated; they're excited. In fact, one of our very brightest students out here—his primary desire is to go up there and run the fabrication line. But the companies have to make it worthwhile. The companies can't treat their fab managers as blue-collar workers or foremen or managers of blue-collar workers. They're not gonna be like that. The manager of tomorrow's fabrication lines is gonna be a Ph.D. They're gonna want to see that fabrication line as a sophisticated entity where they can basically sit in their office, run it through a computer terminal. And, of course, they have to deal with people as well. The Japanese give more status to manufacturing people, and in fact, in a Japanese company, from what I've heard, there's more prestige to being manufacturing manager than to be in research. It's a step down to go into research labs."

That has never been the case in the United States, but it is on the manufacturing lines that the Japanese have been beating the pants off of the American semiconductor companies. But until now no one, neither industry nor the universities, has addressed the need to make the manager of a semiconductor fabrication facility an important, prominent personage within the industry. "It's not just a matter of status," adds Davies. "It's more. 'We're going to raise the salaries. We're going to give them the corner offices.' And that's part of it. But the other part is to change the nature of semiconductor manufacturing, to change it into something more like design. We must give the manufacturing manager the power to design his factory, the power to design the operation of it, to give him the tools to turn that into a design task rather than a crisis-management task."[20]

The CIS is one of the best-funded university research facilities and programs in the nation. Its $15 million worth of equipment and $14.5 million physical plant is all state of the art, and its yearly operating and research budgets are guaranteed by twenty corporate sponsor-members and the federal government. Most of the major universities in the nation are not nearly as well off. Surveys by a number of government agencies and commissions show a steady deterioration in the laboratories and equipment in use in American universities. The National Science Foundation (NSF) has found that less than one-fifth of the instruments used by university laboratories is state of the art, while as much as one-half may not even be usable for research purposes. In engineering and the physical sciences one instrument system in four does not even work.

Such a state of disrepair not only hinders research; it forces leading researchers out of academia and adversely affects the quality of technical and scientific education in America. "In many universities the buildings are literally beginning to fall down around the ears of the people doing the work," said D. Allan Bromley, the vice-chairman of the 1986 White House Science Council Panel on the Health of U.S. Colleges and Universities and a professor of physics at Yale University. "The equipment is often older than the students and sometimes the faculty," he noted with ironic sadness. "The time has come when this will impact our international competitiveness."[21]

It is already beginning to affect the kinds of research being done in academia. Without adequate funding, sophisticated research is either not undertaken or progresses more slowly than it should. At Case Western Reserve University of Cleveland, which has invested heavily in new equipment and instrumentation for semiconductor research, experimentation is slowed by their failure to raise an additional million dollars for a clean room.

Even such well-funded programs as those at Columbia's Microelectronics Science Laboratory feel the pinch in certain areas. "I've been trying to build up this lab here at Columbia for the last several years, and so I know very well about the expense of this equipment," admits Professor Richard Osgood ruefully. "It costs a lot of dough to put it on the line. Even a well-equipped university like ours is still hurting when it comes to getting into the more exotic things. Even our MBE (multibeam epitaxy), which is a very good one in a university setting, is not going to be quite as elaborate as we'd like to have it. For example, the Japanese are doing a lot of work by adding an ion-beam generator to do focused ion-beam epitaxy, and we can't afford that here. That's another half a million dollars."

Just as each new generation of chips comes along faster and faster, and not coincidentally require new generations or major modifications of equipment and instrumentation, so, too, are the equipment needs of university laboratories in these areas subject to constant upgrading. What was state-of-the-art equipment when it is unpacked is antiquated four years later. Computers are outmoded at an even faster rate. And in microelectronics and many other fields, advances in instrumentation and computers are essential for technological progress.

"In so many fields the breakthroughs have been due to a new instrument with a different kind of approach, and tied to computers," declared Frank Press, president of the National Academy of Sciences. "So we have to make sure the universities have state-of-the-art equipment to back up the intellectual talent we know is there."[22]

Even when the money for equipment and facilities is available, the cost of maintaining them is often staggering. A 1987 study by the National Society of Professional Engineers noted that the added costs of maintaining and operating complicated computer systems were "devouring" college budgets. "The problems add up to students who will have less hands-on experience, and any experience they do have may be on outmoded equipment," declared Neil Schmitt, the dean of engineering at the University of Arkansas, who released the report to the press. "The ability of new graduates to contribute to U.S. technological competitiveness is directly related to the laboratory experience," he added.[23]

But fewer and fewer engineering students are getting that experience. "At the University of Wisconsin, writes *New York Times* reporter Steven Solomon, "a shortage of building space has made it impossible to set up enough properly equipped engineering teaching labs. Waiting lists for classes are so long that many engineering students are taking five years to get four-year undergraduate degrees. Some students spend all night in the hallways to be sure to get seats in class the next day."

For these and other reasons, the United States is not producing enough engineers to satisfy the needs of American industry or of the universities for graduate students and postdoctoral candidates to perform research and to instruct future generations of engineers. Much of that function is being turned over to the graduates of foreign engineering schools. A study by the NSF disclosed that 57 percent of all engineering doctorates awarded in the United States in 1987 went to foreign students. Moreover, the ratio of foreign students has been increasing at an annual rate of 2 percent for the last ten years. Should the trend continue, and there is unfortunately every reason to believe it will, then foreigners will outnumber Americans with advanced degrees in engineering by more than three to two in 1995.

These numbers represent a drastic decline in the number of Americans gaining advanced degrees in engineering. Indeed, the number of engineering doctorates gained by U.S. citizens was 50 percent lower in 1985 than it was in 1970. The most shocking aspect of those numbers is that they have been achieved despite the exploding demands for engineering skills in an increasingly high tech economy. Foreign-born engineers have rushed to fill that vacuum in American industry. The NSF study found that the proportion of the scientific/ engineering jobs held by foreign nationals increased from 10 percent in 1972 to 17 percent in 1982. And a greater proportion of those foreign workers held advanced degrees than did their American counterparts.

Jim Dykes, the American president of TSMC, points out that "the largest of all of the foreign student bodies in the U.S. come from Taiwan—32,500 young men and women. That didn't just happen over one year; that happened over quite a number of years, of course. The U.S. has some excellent engineers from Taiwan who have come out of Berkeley and Stanford, Penn State, Illinois, MIT."

According to Dykes, it's almost a genetic compulsion for the Taiwanese, who are for the most part the children of the Chinese who fled the mainland after the Civil War of 1948. "The Chinese are driven by the need for higher education. It's just been bred in them over the years. Because of the old bureaucratic system in mainland China, the only place you could succeed eventually was to become a bureaucrat, and the only way to get there was to be supereducated. So before businesses like this got started, people were overeducated, and they had to go work in the U.S. or Canada or Europe. We're starting to get some of them back, and some of them are coming back in our company. We hired several out of Intel, one out of Hewlett-Packard, and they are every bit as productive here as they were in that environment. They are good at computer programming, software, equipment design, process design. And as a matter of fact, if you peel the cover off of the U.S. companies and look at where the Chinese are, they're doing a hell of a lot of work right now."[26]

Another study published by the National Academy of Engineering in January 1988 found that foreign students comprised more than half the number of doctoral students and two-thirds of the number of postdoctoral appointments in the United States in 1983–85.

- Almost half of all assistant professors of engineering in U.S. colleges and universities under the age of 35 are foreign-born.
- Almost one in three doctorate engineers in industry is foreign-born and the number is increasing.
- Sixty percent of foreign-born doctorate engineers remain in the U.S. after graduation.

• More than half of the foreign engineering students in 1985 came from countries with cultures substantially different from that of the U.S.—Far East (31 percent), India (6 percent), and Middle East (20 percent).

"Foreign-born engineers are becoming an increasingly important component of our engineering work force," said the report. "Their presence is creating not only real opportunities, but also possible problems."

Among the major problems is the quality of instruction offered undergraduate engineers. "With decreasing numbers of American engineers in graduate programs," notes the report, "universities have increasingly relied on foreign students to fill their faculties. This has led to complaints from undergraduates that some foreign teaching assistants do not speak English well enough to communicate clearly."[27]

And the overwhelming number of undergraduate engineering students, as many as 90 percent, in American universities are English-speaking Americans. Upon graduation, the great majority of them go into industry, which pays far higher starting salaries than the stipends available for postgraduate work.

It is clear that foreign engineers will continue to play a major role in the American semiconductor industry and in the universities. Moreover, their's is, and will continue to be, a valuable contribution. But we are reaching a point where those contributions are critical to the continued well-being of our high-tech economy. Just as our industries have become dependent on foreign chips, so, too, have we become dependent on foreign engineering and technological talent. Not all of those high-tech mercenaries, especially those from developing countries, will want to continue working and teaching in the United States. As their own nations reach a level where they can begin to employ more and more of the engineers and scientists they had sent to the United States, they will reach out for them.

"Maybe less than half of our engineers are coming back, because they have good jobs in the United States. But we are very much trying to bring them back from the United States, because they have the good training there and we are now needing them here," Professor Kyu Tae Park of Korea's Yonsei University told me.[28]

Many Koreans in the United States are responding to the need. The heads of Daewoo's and Goldstar's semiconductor operations, for example, were both trained and worked in the United States. The two men have a combined total of thirty-three years of study and work in the States before answering the call to return home. Jung Uck Seo, the executive director of the Korean Telecommunications Project Development Center, which is actively engaged in semiconductor research, similarly gained his advanced degrees in the United States and worked on a number of research projects on a NASA fellowship at Texas A & M.

So much of our scientific and technological infrastructure is supported by foreigners that a sudden, or even gradual, reduction in their numbers will ultimately prove to be disastrous to the U.S. economy. We must therefore develop programs that will attract American engineers and scientists to pursue advanced degrees, to teach, and to perform research in their fields. Government and industry must develop programs that will make advanced eduction for American graduates as attractive as jobs in industry. Many more fellowships and scholarships must be made available, stipends must be increased, and facilities must be upgraded so that hands-on state-of-the-art training can be achieved.

But more than just equipment is needed. Somehow the idea that making things, creating a product, is something not only worth doing but something that is exciting to do, and something that will receive commensurate rewards. That is not the situation today.

"The best and the brightest in this country today are becoming financial analysts, arbitragers, and investment bankers," complains Norman Augustine, the president of Martin Marietta. "And with all due respect to those people, all they do is move money from here to there, but they do not contribute to the building of products that people can use. The average median graduate from the Harvard Business School will earn $72,000 a year. The guy who graduates with an industrial engineering degree and goes to work on the factory floor, will probably get $30,000. I think we have to reverse our priorities to where people who are focusing on manufacturing, on value added, are the ones who get more rewards."[29]

It is not an attitude encouraged or fostered in our educational system today. "There is a great deal of effort spent in the U.S. trying to figure out the proper allocation of the goodies that come out of this industrial state," declares Bob Noyce. "Whether it's how many lawyers we have or how many social workers or whatever it may be compared to the number of people who are producing the goods. And that shows up in the distribution of subject matter in our educational system. Much less emphasis is on the production of wealth and much greater emphasis is placed on the distribution of wealth."[30]

Even if we were to make graduate programs in engineering more attractive to young Americans, the problem will not go away. Part of it cannot be solved. The demographics of tomorrow are such that there will simply not be as many American youngsters entering college ten years from now as there are today. So a greater percentage of those college students must be encouraged to seek careers in the sciences and in engineering and to continue to pursue them on the postgraduate level.

Increasing the percentage of college-bound students, however, will be a virtual impossibility unless the present status of our elementary and secondary school education is improved dramatically. In the most recent, 1988, study of science achievement in the schools of seventeen countries, the United States fared very poorly. Three age levels, ten-year-olds, fourteen-year-olds, and seventeen-year-olds, were tested. At the ten-year-old level, the American youngsters scored best, ranking eighth, or right in the middle. Fourteen-year-olds did very poorly, ranking fourteenth, tied with Thailand and ahead of only the Philippines and Hong Kong, all considered developing countries. The results for this age group were particularly dismaying, since fourteen is the age when youngsters leave school in many of the countries studied. As a result, the achievement levels at that time are considered an indicator of the scientific literacy of the general public and the work force. The decline in the relative scores of the American fifth- and ninth-graders was called "one of the more disturbing things in the survey," by Richard N. Wolf of Teachers College, Columbia University. Wolfe, a coordinator for the study in the United States was also unhappy with the results of the eldest age group, the twelfth-graders. These students, all enrolled in second-year science courses, ranged thirteenth in biology, eleventh in chemistry, and ninth in physics. "Since these students are regarded as the cream of the crop, our kids don't look so good," noted Wolf with regret.

Addressing the American performance, the study concluded, "For a technologically advanced country, it would appear that a reexamination of how science is represented and studied is required."[31]

Erich Bloch, the director of the NSF, the agency funding the U.S. participation in the study, added, "These findings emphasize again the troubled state of science education in the United States."

Even more troubling is the state of our traditional elementary and secondary school education. So poor is the job being done at those levels that 80 percent of all colleges and universities now provide remedial reading and math classes. Competency in the three Rs, reading, writing, and arithmetic, once the staple of American public school education, is often not required. One result of our shamefully inadequate basic education is that 23 million people in the American labor force are functionally illiterate and lack the basic reading, writing, and calculating skills to compete in the job market. Another 47 million adult Americans are borderline illiterates.

Pride in education, the desire to learn, and the demand for excellence, once hallmarks of American primary and secondary education, have vanished from far too many of our schools. They must be restored not only in the wealthy and middle-class neighborhood schools of America but across the entire spectrum of society. For not only does our economy require highly trained engineers and scientists; it also requires highly skilled and equally well trained technical workers. As more and more jobs in the semiconductor and other industries become automated, the need for far greater technical skills on the part of the worker will grow.

American industry is already spending an estimated $210 billion each year on formal and informal training of its work force—a figure almost equal to the total monies spent by federal, state, and local governments on formal education in our schools. That is a stupendous sum, and yet it is not enough, nor is it spent as wisely as it might be. The problems of the American educational system are immense and are being addressed on many levels. Moreover, they are not properly the subject of this book. But they impinge mightily on all American business and industry and are a vital component in our ability to compete globally. And it is unquestionably a factor in the ability of the semiconductor industry to compete effectively.

Virtually every executive in the semiconductor industry with whom I spoke agreed on the importance of education. "Education is another major factor," Charlie Sporck told me at our first meeting at his offices in Santa Clara. "It terrifies me to read reports of large percentages of people who don't know how to read. The world that we are headed for can't tolerate that. Especially compared to our trading partners, who do not have those disadvantages. But if we are going to do something about that, it must begin with an active debate. The megamouths have got to start focusing on these issues."

The Ultimate Resource

We have grown complacent in and uncomprehending of a world that has changed before our very eyes. Once it was a world we ruled technologically, economically, politically. No longer. The technology of Japan, the Little Dragons, and Europe is equal to or superior to ours. And that is not all that has changed. While we were busily consuming the truly fine products of Japan and Europe, they were working harder than ever to create, develop, and manufacture even better goods. No one, we thought, could ever equal our energy, our creativity. Indeed, by virtually every measure, Americans have been among the most truly creative people in history. Nobel Prizes, new scientific discoveries, patents, new products, innovations—Americans far outstrip the rest of the world in those areas. But something strange has been happening; something we never believed was possible has taken place. And nowhere is it more apparent than in the semiconductor industry. From 1975 to 1982 the U.S. share of world IC patent activity declined from 43 percent to 27 percent. The Japanese share by contrast rose from 18 percent to almost half—48 percent. Since 1985 more than 40 percent of all the papers at the annual IEEE Solid State Circuits Conference, the largest and perhaps single most important technology meeting for the industry, were presented by Japanese engineers and scientists.[32]

In the final analysis, the Chip War will be decided by people. We can change our tax laws, alter our institutions to create megabanks and multibillion-dollar vertically integrated electronics companies, but in the final analysis our ultimate resource is our people. Once Americans were brash, confident, creative, and hard working. They were proud of what they did and believed in such homilies as an honest day's work for an honest day's pay. Such ideas are no longer in vogue. Pride and patriotism wax and wane like the fashions in women's clothing. We move money but create little. It is time to restore the American work ethic. Without it, no amount of access to foreign markets, no tariff barriers, will enable us to regain our competitiveness. For we must compete with people who display remarkable traits—loyalty, pride, and desire.

I think of the young women gathered about a flow chart in a quality control circle in Kyushu after working hours. I still have a vivid picture of Japanese garbagemen on a Sunday morning in Tokyo running full tilt, from pick-up point to pick-up point, because that was the way the job was supposed to be done.

I recall the answer to a question I asked that turned into a lecture on the difference in work ethics between Americans and Koreans. It came from Chang Soo Kim, the senior managing director of Goldstar's semiconductor group. "I worked in the U.S. many years, so I can tell you some of the feelings I have in that area. You know, person to person, there are much smarter persons in the U.S. labor force, both in the hourly workers as well as salaried, very smart engineers and very smart technicians, very capable. But I think the Koreans are working much better as a group than the U.S. labor force. The difference is the attitude and work ethics. The Koreans have a strong sense of duty and responsibility. And they work very hard. And, for instance, you know, we have a thirty-minute, say, lunch break. In the U.S., of course, we have set lunch-break hours and set coffee-break hours between the lunch breaks, before and after lunch breaks, before and after coffee breaks. People don't work that hard; they are not conscientious in the U.S.

"And going back to your question of where we get a competitive edge over other people, it is not simply the basic labor rate. But where we get the advantage is the efficiency of labor. For the same work hour, our people do work more and waste less. And that makes more of a difference, because the materials and the fixed cost are so high that the yield or the waste becomes significant. It's the cultural difference and work ethic. You can see some of this in the road-cleaning crews in the streets. There are no supervisors out there to supervise whether they're goofing off or they're working. But you can see, while you're driving by or walking on the road, they're always sweeping something, picking up something."

Along the way, they have also picked up our technology, our ideas, and our education. Technology and creativity are no longer exclusive. How they are utilized is the question. How we answer it will determine the outcome of the Chip War.

GLOSSARY

Algorithm A specific set of instructions for the solution of a problem.

Alignment The precise arrangement of the mask and the wafer in their proper positions with respect to each other.

Analog A continuous measurement of phenomena.

Assembly The final step in semiconductor manufacturing in which the device is encased in a plastic or ceramic package.

Binary A system of numbers that uses 2 as its base as opposed, for example, to the decimal system, which uses 10 as its base. The binary system has only two digits, 1 and 0.

Bipolar Transistors composed of semiconductor material containing negative and positive electrical poles.

Bit Either one of the two digits used in the binary system. A bit is the smallest unit of storage in a digital computer.

Byte A set of binary bits, usually eight, that are operated as a single unit.

Chip A single square or rectangular piece of semiconductor material containing transistors and other electrical devices and circuits.

CMOS Complementary metal oxide semiconductor, a manufacturing process that combines N-type and P-type transistors. Used most often to produce logic circuits.

CPU Central processor unit. That part of a computer that fetches, decodes, and executes the instructions stored in programs.

Data The basic elements of information that can be processed or generated by a computer.

Device Another word for a chip. Also a system or end product composed of chips and other components.

Die A single square or rectangular piece of semiconductor material containing transistors, and other electrical devices and circuits. Plural is dice.

Diffusion A means of doping or modifying the characteristics of semiconductor materials by baking wafers of the material in furnaces with controlled atmospheres or dopants.

Diode An electrical device with two terminals through which an electric current can pass only in one direction.

Discrete A semiconductor device containing only one active component, such as a transistor or a diode.

Dopant An element incorporated into semiconductor materials as an impurity to enhance their ability to conduct electricity. Dopants from the third column of the periodic table, such as aluminum boron, and gallium, produce P-type conductivity. Those from the fifth column, such as phosphorus, aresenic, and antimony, produce N-type conductivity.

Doping The process of introducing dopants into semiconductor materials.

DRAM Dynamic random access memory. A chip that stores information in the form of binary digits represented by the presence or absence of an electrical charge. The charge must be refreshed periodically.

EPROM Erasable programmable read only memory. A chip that permanently stores information that is used repeatedly, such as microcode or the electronic characters displayed on a television monitor. ROM chips are programmed after fabrication and prior to installation in devices. The erasable versions are wiped clean usually by exposure to ultraviolet light.

Gate A basic unit of measure of integrated circuit complexity. Gates are usually combinations of transistors, diodes, and their connecting circuits.

Insulator A poor conductor of electricity.

Integrated Circuit Also known as ICs, these semiconductor dice contain many transistors, diodes, wires, and other components that together form a complete device or chip and perform the function of an electronic circuit.

LED   Light-emitting diode. A semiconductor device that emits light whenever current passes through it.

Linear IC   An integrated circuit that stores information by analog rather than digital means.

LSI   Large-scale integration. The inclusion of 100 or more transistors and other components on a chip.

Mask   A screen, most often of glass, on which the circuit design of a chip is etched. Light passing through the mask photographically records the image of the circuit onto a wafer.

Micron   A unit of measure equivalent to one millionth of a meter or approximately .04 inch.

Microprocessor   The central processing unit of a computer on a single chip.

Monolithic device   A device containing all of its circuitry on a single chip.

MOS   Metal oxide semiconductor.

MSI   Medium-scale integration. Integrated circuits containing ten or more transistors and other components, but less than 100, on a single chip.

N-type   Semiconductor material containing a small amount of dopant atoms that have one extra electron. These electrons carry a negative charge and are free to wander, creating an electric current.

Parts   Another term for chips.

P-type   Semiconductor material containing a small amount of dopant atoms that have one less electron. Each dopant atom creates a "hole" or unoccupied space among the electrons. The holes carry a positive charge and move to create an electrical current.

RAM   Random access memory. The basic memory storage unit in a computer. Stores information temporarily in binary form and can be changed by the user. All information is lost when the power is turned off.

ROM   Read Only Memory. Permanently stores information that is used repeatedly. Unlike RAM memory, ROM cannot be changed.

Semiconductor A material that contains the properties of both an insulator and a conductor, thereby conducting electrons half as efficiently as the latter. The most common semiconductor materials are silicon, germanium, and gallium arsenide.

SRAM Static random access memory. A RAM that will hold information without the need to be periodically refreshed with an electrical charge, as do DRAMs.

SSI Small-scale integration. Integrated circuits containing fewer than ten logic gates.

Substrate The underlying material, usually silicon, on which a chip is built.

Transistor The basic device used to amplify or switch electrical current. Composed of semiconductor material, it replaced the vacuum tube in virtually all electronic uses.

ULSI Ultra-large-scale integration. Devices that contain 10,000 or more gates on a single chip.

VLSI Very large scale integration. Integrated circuits that contain 1,000 or more logic gates.

Wafer A thin, circular disk sliced from a silicon rod on which up to as many as 1,000 individual chips are fabricated and then cut into individual pieces for packaging.

NOTES

Chapter 1

1. Interview, Sandy Kane, Harrison, N.Y. (February 10, 1988).
2. Dataquest (a research organization specializing in the worldwide high-technology industries).
3. Michael Borrus, James Millstein, and John Zysman, *U.S.-Japanese Competition in the Semiconductor Industry*, (Berkeley: Institute of International Studies, University of California, 1982).
4. Charles Ferguson quoted in "The Japanese Strategy for Computer Supremacy," *Forbes*, February 9, 1987.
5. Interview, Jack Clifford, Washington, D.C. (June 4, 1987).
6. Andrew Grove, quoted in "Chip Dispute, Reading Between the Lines," *New York Times*, March 30, 1987.
7. Mark Potts, "The Chips Are Down in Semiconductor Industry," *Washington Post*, national weekly edition, June 1, 1987.
8. Peter H. Lewis, "The Executive Computer," *New York Times*, May 15, 1988.
9. Andrew Pollack, "Shortage of Memory Chips Hurts Computer Industry," *New York Times*, March 11, 1988.
10. "American Weapons, Alien Parts," *Science*, October 10, 1986.
11. Report of Defense Science Board Task Force on Defense Semiconductor Dependency, February 1987.
12. "Balancing the National Interest, U.S. National Security Export Controls and Global Economic Competition," National Academy of Sciences, 1987.
13. Dataquest.
14. Charles H. Ferguson. "The Competitive Decline of the U.S. Semiconductor Industry," testimony to the Subcommittee on Technology and the Law, Judiciary Committee, U.S. Senate, February 26, 1987.
15. "Electronic Components, U.S. Industrial Outlook," U.S. Department of Commerce.
16. Dataquest.
17. Interview, Martin Starr, New York City (May 6, 1987).

Chapter 2

1. Alex Osborne, *Running Wild, the Next Industrial Revolution* (Berkeley, Calif.:Osborne/McGraw-Hill, 1979).
2. I used a number of excellent books on the history of the computer includ-

ing Franklin M. Fisher, *IBM and the U.S. Data Processing Industry* (New York: Praeger, 1983), and Nancy B. Stern, *From ENIAC to UNIVAC* (Bedford, Mass.: Digital Press, 1981).

3. Much of the account of the development of the transistor was provided by Bell Laboratories archives. Among the most useful sources were "Three Men Who Changed Our World—25 Years Later," the Bell Laboratories Record, December 1972. A lengthy review of the discovery is also contained in "The Transistor: Two Decades of Progress," *Electronics,* February 19, 1968.

4. There are a number of excellent histories of Silicon Valley, including Everett Rogers and Judith Larson, *Silicon Valley Fever* (New York: Basic Books, 1984), and Gene Bylinsky, *Silicon Valley High Tech Window to the Future* (Hong Kong: Intercontinental Publishing Corp., 1988).

5. Interview, Robert Noyce, Santa Clara, Calif. (June 23, 1987).

6. An excellent profile of Jack Kilby and his role in the development of the integrated circuit is in T. R. Reid, *The Chip* (New York: Simon & Schuster, 1984).

7. Gordon Moore, quoted by Bob Noyce in "Microelectronics," *Scientific American,* September 1977.

8. Bob Byers, "Research Innovation and Society," Stanford University Annual Financial Report, 1983.

9. Interview, Charles Sporck, Santa Clara, Calif. (June 25, 1987).

10. "Fairchild Semiconductor the Lily of the Valley, 1957–1987," *Electronic News,* September 28, 1987.

11. Interview, Elliot Sopkin, Sunnyvale, Calif. (June 24, 1987).

12. Fred Warshofsky, *The 21st Century, the New Age of Exploration* (New York: Viking Press, 1969).

13. Charles Ferguson, "American Microelectronics in Decline: Evidence, Analysis, and Alternatives," Massachusetts Institute of Technology, VLSI Memo No. 85-284, November 1985.

14. "A Report on the U.S. Semiconductor Industry," U.S. Department of Commerce, September 1979.

15. An excellent book on computer history and workings is Stan Augarten, *Bit by Bit, an Illustrated History of Computers* (New York: Ticknor & Fields, 1984).

16. "A Revolution in Progress," Intel Company History, 1984.

17. Peter Lewis, "Not So Limited," *New York Times,* June 9, 1987.

18. Interview, Gene Hill, Santa Clara, Calif. (June 22, 1987).

19. "A Report."

20. Ibid.

21. Ferguson, "American Microelectronics."

22. Dataquest.

23. Interview, Bob Flowers, New York (April 3, 1987).

24. Quoted by Andrew Pollack, "A Look At Entrepreneurs: Doubt on American Ideal," *New York Times,* June 14, 1988.

25. Charles Ferguson, written testimony for the Subcommittee on Technology and the Law, Judiciary Committee, U.S. Senate, February 26, 1987.

26. Ferguson, "American Microelectronics."
27. Timothy Larimer, "Gold Collar Workers," San Jose *Mercury News,* May 24, 1987.
28. Reg Alvarez Torres, "Scully Draws Top Salary Again." San Jose *Mercury News,* May 11, 1987.
29. Don Hoefler, *Microelectronic News,* September 1, 1984.
30. Dataquest.
31. Ferguson, "Report on the U.S. Semiconductor."
32. Bernard Cole, "Here Comes the Billion-Transistor IC," *Electronics,* April 2, 1987, pp. 81–85.
33. Lenny Siegel, "Delicate Bonds: The Global Semiconductor Industry," Pacific Studies Center, 1980.
34. Ferguson, "American Microelectronics."
35. Michael Borrus, "Reversing Attrition: A Strategic Response to the Erosion of U.S. Leadership in Microelectronics," BRIE Working Paper no. 13, 1985.

Chapter 3

1. I am indebted for this and many other explanations of Japanese technology, institutions, culture, customs, and history to Akio Akagi, the science correspondent for NHK.
2. Interview, Tsuneo Mano, Atsugi, Japan (September 3, 1987).
3. For a detailed account of MITI from 1925 to 1975, see Chalmers Johnson, *MITI and the Japanese Miracle* (Palo Alto, Calif.: Stanford University Press, 1982).
4. Makoto Kikuchi, *Japanese Electronics* (Tokyo: Simul Press, 1983).
5. Akito Morita, *Made in Japan* (New York: E. P. Dutton, 1986).
6. Interview, Makoto Kikuchi, Tokyo (September 4, 1987).
7. There are a number of excellent histories of Japan. For me, the most useful was Edwin O. Reischauer, *The Japanese* (Cambridge, Mass.: Harvard University Press, 1977, 1981).
8. Hadley, quoted by Michael Schaller in *The American Occupation of Japan: The Origin of the Cold War in Asia* (New York: Oxford University Press, 1985).
9. Rodney Clark, *The Japanese Company* (New Haven: Yale University Press, 1979).
10. Schaller, *American Occupation.*
11. Michael Borrus, James Millstein, and John Zysman, *U.S.-Japanese Competition in the Semiconductor Industry* (Berkeley: Institute of International Studies, University of California, 1982).
12. Gene Gregory, *Japanese Electronics Technology: Enterprise and Innovation* (Tokyo: Japan Times Ltd., 1986).
13. "NEC Corporation: The First 80 Years," NEC Tokyo, 1984.
14. Interview, Atsuyoshi Ouchi, Tokyo (September 7, 1987).
15. Interview, Thomas Pugel, New York (May 29, 1987).
16. Interview, Hideo Yoshizaki, Tokyo (August 31, 1987).
17. Interview, Hajime Karatsu, Tokyo (September 12, 1987).

18. "QC Circle Koryo, Union of Japanese Scientists and Engineers, Tokyo, 1980.
19. For an excellent profile of W. Edwards Deming, see Mary Walton, *The Deming Management Method* (New York: Dodd, Mead, 1986).
20. *Kodansha Encyclopedia.*
21. Robert Cole, "What Was Deming's Real Influence?" *Across the Board,* February 1987.
22. Tomatsu Goto and Nobukutsu Mauabe, "How Japanese Manufacturers Achieve High IC Reliability," *Electronics,* March 13, 1980.
23. Much of the background information about Kyushu was provided by Hiroaki Mori, the director general of commerce, industry, tourism, and labor, and other officials of the Kumamoto prefectural government with whom I met.
24. Interview, Chiharu Okabe, Kumamoto City, Japan (September 11, 1987).
25. Interview, Toshimitsu Kohtaka, Kumamoto City, Japan (September 11, 1987).
26. "Can We Speak the Same Language?" *New York Times,* July 14, 1987.

Chapter 4

1. In Japan, where sex is accepted as a normal and healthy aspect of life and Victorian sexual attitudes never took root, the pornographic comic, I am advised, is often used as a device to ensure wakefulness late at night on the commute home. "It is a means of not falling asleep and thus missing the station where one should get off."
2. John Burgess, "It's a Warrior, It's a Drone, It's—It's Salaryman!" *Washington Post,* August 23, 1987.
3. Interview, William Taylor, New York (July 6, 1987).
4. Interviews, Koichi Shimbo, New York (1987–1988).
5. I first came across these lyrics in the shooting notes for a film, *The Genius of Japan,* I had done for the Hearst Corporation in 1983. In the course of researching this book, I contacted Mike Kitadeya of the Matsushita News Center, who informed me these were old lyrics that have since been replaced. The new ones are somewhat less obviously nationalistic and materialistic. The last verse reads:

> Finding happiness,
> Matsushita Denki
> (Electric).
> Animating joy
> everywhere,
> A world of dedication,
> Let us fulfill our hopes,
> Shining hopes
> Of a radiant dawning,
> With love, light, and a
> dream.

6. Interviews, Akio Akagi, New York, Tokyo (1986–1987).
7. *Kodansha Encyclopedia.*
8. Interview, Kunishiro Saito, Tokyo (September 1, 1987).
9. "Tokyo's Salarymen: Suffering a Lot," *Japan Times,* September 4, 1987.
10. "A Puzzling Toll at the Top," *Time,* August 3, 1987.
11. Robert Chapman Wood, "Japan's Economic Masochism," *Forbes,* September 21, 1987.
12. "The Japan They Don't Talk About, NBC White Paper," National Broadcasting Company, April 26, 1986.
13. Wood, "Masochism."
14. *The Genius of Japan* (notes for a film), Hearst Corporation, 1973.
15. Kathryn Graven, "The Home Front, Japanese Housewives Grow More Resentful of Executive Spouses," *Wall Street Journal,* September 10, 1987.
16. Leonard Silk, "A Lesson from Japan," *New York Times,* November 17, 1982.
17. Lynn Steen, "Mathematics Education: A Predictor of Scientific Competitiveness," *Science,* July 17, 1987.
18. "Education: Math and Aftermath," *Science News,* January 31, 1987.
19. Eliot Marshall, "School Reformers Aim at Creativity," *Science,* July 18, 1986.
20. Interview, Makoto Kikuchi, Tokyo (September 4, 1987).
21. Marshall, "School Reformers."
22. Carol Simons, "Secret of Japanese Schools," *Smithsonian* Magazine, March 1987.
23. Edwin O. Reischauer, *The Japanese* (Cambridge, Mass.: Harvard University Press, 1981).
24. Jared Taylor, *Shadows of the Rising Sun* (New York: Morrow, 1983).
25. Hideki Okeda, *An Introduction to Kanji* (Tokyo, 1975).
26. "Asian Languages Aid Mathematics Skills," *Science News,* September 19, 1987.
27. I was given a convincing demonstration of this ability of a Japanese PC to switch instantly from English characters to kanji and hiragana by Akio Akagi in his office at NHK Tokyo.
28. Marshall, "School Reformers."
29. The subjective rankings of the Japanese universities I owe to Akio Akagi, and most other Japanese whom I have questioned on the subject have agreed with his assessment.
30. Masayasu Kudo, "Educating the Establishment," *Business Tokyo,* October 1987.
31. Ibid.
32. "Next Year's Graduates Look for Work as Job Season Opens," *Japan Times,* September 6, 1987.
33. Marshall, "School Reformers."
34. "U.S.-Japan Study Aim Is Education Reform," *Science,* January 16, 1987.
35. Ibid.
36. "Report on the U.S. Semiconductor Industry," U.S. Department of Commerce, September 1979.

37. Clyde Haberman, "Japan Asks Why Scientists Go West to Thrive," *New York Times,* November 8, 1987.

38. Interview, Sheridan Tatsuno, San Jose, Calif. (June 22, 1987).

Chapter 5

1. "Vision of MITI Policies in the 1980s," quoted by Michael Borrus, James Millstein, and John Zysman in *U.S.-Japanese Competition in the Semiconductor Industry* (Berkeley: University of California, 1982).

2. Interview, Ryozo Hayashi, New York (June 30, 1987).

3. "IBM Is a Tiger Let Loose in a Field," *Bhungeishunju,* September 1982.

4. An excellent account of MITI's role in the Japanese VLSI project is contained in "The Effect of Government Targeting on World Semiconductor Competition, a Case History of Japanese Industrial Strategy and Its Costs for America," Semiconductor Industry Association, 1983.

5. Interview, Chuck Minihan, Hayward, Calif. (June 25, 1987).

6. "The Effect of Government Targeting."

7. Gene Gregory, *Japanese Electronics Technology: Enterprise and Innovation* (Tokyo: Japan Times, 1986).

8. "A Competitive Assessment of the U.S. Semiconductor Manufacturing Equipment Industry," Department of Commerce, March 1985.

9. Interview, William Finan, Washington, D.C. (June 3, 1987).

10. Interview, Michael Borrus, San Francisco (June 25, 1987).

11. Dataquest.

12. Interview, Shoichiro Yoshida, Tokyo (September 8, 1987).

13. Gregory, *Japanese Electronics Technology.*

14. Interview, John Suzuki, Tokyo (September 2, 1987).

15. Interview, Chang Soo Kim, Seoul (September 17, 1987).

16. "A Competitive Assessment."

17. "The Effect of Government Targeting."

18. Dataquest.

19. Interviews, Sheldon Weinig, Orangeburg, N.Y., and New York City, various times (1987–1988).

20. Interview, Hajime Karatsu, Tokyo (September 12, 1987).

21. Interview, Charles Dittrich, Washington, D.C. (June 4, 1987).

22. Interview Hideo Yoshizaki, Tokyo (August 31, 1987).

23. Interview, Michael Attardo, Harrison, N.Y. (February 5, 1987).

24. "The Effect of Government Targeting."

25. Dataquest.

26. Mark Mehler, "IC Equipment Makers Get Down to Business," *Electronic Business,* May 1987.

27. Gregory, *Japanese Electronics Technology.*

28. Interview, Hisao Oka, Tokyo (September 9, 1987).

29. Telephone interviews, Edward Bloch (1987–1988).
30. This description has come from personal observation in several semiconductor manufacturing plants, from interviews with engineers in these plants, and from a number of company booklets that describe the process. Among the most useful I have found are those put out by National Semiconductor, AMD, and IBM.
31. Robert Noyce, "Hardware Prospects and Limitations," in *The Computer Age: A Twenty-Year View* (Cambridge, Mass.: MIT Press, 1979).
32. Interview, Koichi Shimbo, New York (1987 and 1988).
33. Lenny Siegel, "Delicate Bonds: The Global Semiconductor Industry," Pacific Studies Center, January 1980.
34. Interview, Jack Clifford, Washington, D.C. (June 4, 1987).
35. "A Competitive Assessment."
36. "Something to Look Up To," *Electronic News,* September 28, 1987.
37. A number of excellent articles have appeared recently on X-ray lithography, including, "Quest for Fastest Computer Chip: International Contest Intensifies," *New York Times,* February 23, 1988; "Examining Competitive Submicron Lithography," *Semiconductor International,* February 1988; and "Chip Processing," *Electronics,* October, 16, 1987.

Chapter 6

1. David E. Rosenbaum, "Softening Tone, Jackson Delivers Dual Message," *New York Times,* November 1, 1987.
2. Telephone interview, John Streeter (November 17, 1987).
3. Stephen Phillips "Zenith Is Left in Spotlight," *New York Times,* July 24, 1987.
4. Interview, William Taylor, New York (July 6, 1987).
5. Interview, Jack Clifford, Washington, D.C. (June 4, 1987).
6. Michael Borrus, James Millstein, and John Zysman, "International Competition in Advanced Industrial Sectors: Trade and Development in the Semiconductor Industry," written report for the Joint Economic Committee, U.S. Congress, 1982.
7. "The Effect of Government Targeting on World Semiconductor Competition," Semiconductor Industry Association Report, 1983.
8. W. J. Sanders, "International Trade Policy," speech to the high technology meetings held by the Secretary of Commerce, February 1983.
9. "The Effect of Government Targeting."
10. Borrus et al., "International Competition."
11. Interview, Bob Flowers, New York (April 3, 1987).
12. Interview, William Egelhoff, New York (June 11, 1987).
13. Interview, anonymous magazine editor, Tokyo (September 2, 1987).
14. This now infamous memo has been quoted in numerous publications including

"Can This Be Silicon Valley," *Barron's,* March 30, 1987, and "Fallout from the Trade War in Chips," *Science,* November 22, 1985.

15. Clyde V. Prestowitz, Jr., *Trading Places: How We Allowed Japan to Take the Lead* (New York: Basic Books, 1988).

16. Thomas Pugel, "The Responses of Japanese and U.S. Firms to Semiconductor Trade Friction," IBEAR Research and Management Workshop, University of Southern California, April 1987.

17. Telephone interview, George Schneer (July 21, 1988).

18. "Fallout from the Trade War in Chips," *Science,* November 22, 1985.

19. Jerry Sanders, Speech to the Semiconductor Industry Association Workshop, Washington, D.C., February 4, 1983.

20. "The U.S.-Japan Semiconductor Trade Agreement," Semiconductor Industry Association, September 23, 1986.

21. Ibid.

22. "Europe Parts Group Files Japan Anti-Dumping Case," *Electronic News,* December 8, 1986.

23. Telephone interview, Clyde V. Prestowitz, Jr. (May 3, 1988).

24. Susan Chira, "Japan Denies Dumping, Says Chip Pact Is Intact," *New York Times,* March 18, 1987.

25. James Fallon, "Sub-FMV Japan DRAMS, EPROMS Thrive In European Gray Market," *Electronic News,* March 9, 1987.

26. Jack Robertson, "Sanctioning Dependence," *Electronic News,* April 27, 1987.

27. Interview, Charlie Sporck, Santa Clara, Calif. (June 25, 1987).

28. Robertson, "Sanctioning Dependence."

29. Gene Norrett and AEA Survey, both quoted by Thomas G. Donlan, "Can This Be Silicon Valley?" *Barron's,* March 30, 1987.

30. Bruce Smart, Jr., quoted by David E. Sanger, "Japanese Chip Dumping Has Ended, U.S. Finds," *New York Times,* October 7, 1987.

31. Frank Press, "Technological Competence and the Western Alliance," in Andrew Pierre, ed., *A High-Tech Gap, Europe-America* (New York: Council on Foreign Affairs, 1987).

32. Philip Abelson, "The Trade Deficit," *Science,* May 8, 1987.

33. "The Effect of Government Targeting."

34. "One Year of Experience Under the U.S.-Japan Semiconductor Agreement," Semiconductor Industry Association, September 1987.

35. Interview, Tetsuo Matsui, Tokyo (September 2, 1987).

36. Interview, Hiseo Oka, Tokyo (September 9, 1987).

37. Interview, Bob Noyce, Santa Clara, Calif. (June 23, 1987).

38. "The Impact of Japanese Market Barriers in Microelectronics," Semiconductor Industry Association, 1985.

39. Interview, William Finan, Washington, D.C. (June 3, 1987).

40. Borrus et al., "International Competition."

41. Interviews, Sheldon Weinig, New York City and Orangeburg, N.Y., various times (1987–1988).

Chapter 7

1. Michael Berger and Don Shapiro, "A New Wave Rises on the Pacific Rim," *Electronics,* April 2, 1987.

2. Tommy Koh, "Asia's Four Tigers Aren't Bad Tigers," *Japan Times,* September 7, 1987.

3. Clyde Farnsworth, "U.S. Pressing Koreans on Trade," *New York Times,* March 7, 1988.

4. "Low Cost, High Growth," *Time,* October 19, 1987.

5. "U.S. Property Rights Push Achieves Results in Korea," *Korea Business World,* September 1987.

6. Telephone interview, Sam Korin (May 11, 1987).

7. Interview, Hsin Chu, Taiwan (September 14, 1987).

8. Excerpt from a National Institute of Technology introductory videotape.

9. Interview, Chang Husan Lin, Taipei (September 14, 1987).

10. Interview Ding-Hua Hu, Hsin Chu, Taipei (September 15, 1987).

11. Interview, Bob Tsao, Hsin Chu, Taiwan (September 15, 1987).

12. Interview, James Dykes, Hsin Chu, Taiwan (September 15, 1987). Just about a year after this interview, Dykes resigned from Taiwan Semiconductor, citing personal reasons. He was replaced by Morris Chang.

13. Phillip Abelson, "Research and Development in South Korea," *Science,* April 1, 1988.

14. Susan Chira, "Boom Time in South Korea: An Era of Dizzying Change," *New York Times,* April 7, 1987.

15. Interview, Jung Uck Seo, Seoul, Korea (September 17, 1988).

16. "An Historical Overview," Lucky Goldstar Annual Review, 1986

17. Interview, Kyu Tae Park, Seoul, Korea (September 18, 1987).

18. There are a number of excellent historical overviews of Korea, including William E. Henthorn, *A History of Korea* (New York: Free Press, 1971, and Yi Ki-Baek, "A New History of Korea (Cambridge, Mass.: Harvard University Press, 1984).

19. *Science Achievement in Seventeen Countries: A Preliminary Report* (New York: Pergamon, 1988).

20. I am indebted for much of this information to Mr. H. S. Lee, general manager of international trade and relations, the Electronics Industries Association of Korea, in Seoul.

21. Interview, Chang Soo Kim, Seoul, Korea (September 18, 1987).

22. "Korea, Takeoff or Explosion," "Adam Smith's Money World," Public Broadcasting System, June 1, 1987.

23. Andrew Tanzer, "Korea Chips Away," *Forbes,* February 25, 1985; and Dataquest.

24. Interview, In Sang Lim, Seoul, Korea (September 19, 1987).

25. Pascal Zachary, "Intel to Get DRAM Chips from Samsung," San Jose *Mercury News,* June 25, 1987.

26. Hyeon Gon Kim, quoted in "A New Wave Rises."

27. Berger and Shapiro, "A New Wave Rises."
28. Interview, Chang Hong Jo, Seoul, Korea (September 18, 1987).

Chapter 8

1. Interview, Franco Malerba, Milan (March 2, 1988). Malerba is also the author of what is perhaps the most comprehensive book on the European semiconductor industry, *The Semiconductor Business* (Madison: University of Wisconsin Press, 1986).
2. Interview, Sheldon Weinig, New York City and Orangeburg, N.Y., various times (1987–1988).
3. Dataquest.
4. Interview, Malcolm Penn, Zurich (March 1, 1988).
5. "Philips Net Falls 51% in Qtr." *Electronic News,* February 29, 1988.
6. Philips Annual Report, 1987.
7. Interview, Cees Krijgsman, Eindhoven, Netherlands (February 29, 1988).
8. Interview, David Heard, Nijmegen, Netherlands (February 29, 1988).
9. Interview, Paul De Ruwe, Nijmegen, Netherlands (February, 29, 1988).
10. Siemans Annual Report, 1987, and additional information provided by company spokesman Martin Weitzner.
11. Interview, Helmuth Murrmann, Munich (February 29, 1988).
12. Interview, Carlo Ottaviani, Agrate Brianza, Italy (March 2, 1988).
13. Pasquale Pistorio is one of the most peripetatic of the world's semiconductor executives, and he was unfortunately in Asia when I was in Europe. Much of his background, therefore, was provided by Carlo Ottaviani and other SGS executives. Additional material, including many of the Pistorio quotes, are from "Manager's Return to Italian Roots Salvages SGS," *International Management,* May 1984; "How Europe Can Win in the Vital Hi-Tech Market, *The Manchester Guardian,* July 5, 1984; and "Europe's Takeover Kings," *Fortune,* July 20, 1987.
14. "Cooperative Spirit Guides European Chipmakers," Semiconductor International, February 1988; "EEC: Uniting to Meet High-Tech's Challenge, *Science* September 4, 1987
15. Interview, Enrico Villa, Agrate Brianza, Italy (March 2, 1988).
16. "ASM Seeks to Cut Stake in N. V. Philips Venture," *Electronic News,* June 27, 1988.

Chapter 9

1. Duncan Hunter, quoted by Clyde Farnsworth in "Toshiba, Norway Concern Assailed in Soviet Sale," *New York Times,* May 1, 1987.
2. "Soviet Acquisition of Militarily Significant Western Technology: An Update," September 1985.

3. Casper Weinberger, quoted in "Moscow's Prying Eyes," *Newsweek,* September 30, 1985.

4. Richard N. Perle, "Like Putting the KGB into the Pentagon," *New York Times,* June 30, 1987.

5. "Soviet Acquisition."

6. Interview, John Konfala, Alexandria, Va. (June 3, 1987).

7. Interview, George Menas, Alexandria, Va. (June 3, 1987).

8. "Soviet Acquisition."

9. Ibid.

10. *Meeting the Espionage Challenge: A Review of United States Counterintelligence and Security Programs* (Washington, D.C.: U.S. Government Printing Office, 1986).

11. "Silicon Valley: No. 1 Soviet Spying Target," *U.S. News & World Report,* August 12, 1985.

12. The relevant sections of the Farewell file were provided by the Defense Department and are extracted from Thierry Wolton, *Le KGB en France,* (Paris: PG Editions Grasset & Fasquelle, 1986).

13. "Soviet Acquisition."

14. "Moles Who Burrow for Microchips," *Time* (June 17, 1985).

15. Interview, L. Stephen Walton, Washington, D.C. (June 4, 1987).

16. *Balancing the National Interest: U.S. National Export Controls and Global Economic Competition* (Washington D.C.: National Academy Press, 1987).

17. Ibid.

18. George Menas, telephone interview (July 28, 1988).

19. "Estimate of Direct Economic Costs Associated with U.S. National Security Controls," William Finan, Quick, Finan & Associates, 1986.

20. Ibid.

21. This story was told to me by a Japanese journalist who did not wish to be identified.

22. "Soviet Acquisition."

23. Perle, "The KGB."

24. Bill Gertz, "Fast-Lane Lifestyle in Silicon Valley Breeds Spies, Thieves, Drug Dealers," *Washington Times,* October 1, 1985.

25. "Information Thieves Are Now Corporate Enemy No. 1," *Business Week,* May 5, 1986.

26. Gertz, "Spies, Thieves, Drug Dealers."

27. Telephone interview, Ken Rosenblatt (November 12, 1988).

28. Malone, quoted in Gertz, "Fast-Lane Lifestyle."

29. Ibid.

30. Calvin Sims, "Wounded by Patent Piracy," *New York Times,* May 13, 1987.

31. Interview, Chikara Hayashi, Tokyo (September 5, 1987).

32. Kyun Won Kim, quoted in "U.S. Property Rights Push Achieves Results in Korea," *Korea Business World,* September 1987.

33. Tom Dunlap, quoted in John Thompson, "Intel Presses Customs: Hold NEC V Series," *Electronic News,* August 24, 1987.
34. "Texas Instruments to Report a Gain," *New York Times,* March 4, 1988.
35. John Eckhouse, "Toshiba Accused of Copying U.S. Chips," *San Francisco Chronicle,* June 23, 1987.
36. "National Semiconductor to Pay IBM $3 million," *Chicago Tribune,* January 11, 1984.
37. Among the many news stories that covered the case, one of the most comprehensive was David Tinnin, "How IBM Stung Hitachi," *Fortune,* March 7, 1983.
38. Interview, Sheridan Tatsuno, San Jose, Calif. (June 22, 1987).

Chapter 10

1. Michael Malone, "Fear and Xenophobia in Silicon Valley," *Wall Street Journal,* February 23, 1987.
2. William Safire, "Goodby, Mr. Chips," *New York Times,* January 26, 1987.
3. Jack Robertson, "Say Near-Buy of Fairchild Periled Cray R & D," *Electronic News,* March 30, 1987.
4. "2 In Cabinet Fight Sale to Japanese," *New York Times,* March 12, 1987.
5. Interview, Tom Pugel, New York (May 29, 1987).
6. Gene Gregory, "Negative Solutions," *Japan Times,* April 13, 1987.
7. Interview, Sadao Inoue, Tokyo (September 9, 1987).
8. An excellent account of this and many other events in the Fairchild saga are contained in "Fairchild Semiconductor, the Lily of the Valley, 1957–1987," *Electronic News,* September 28, 1987.
9. I am indebted to Jay Cooper, a semiconductor analyst with the Brokerage firm of Eberstadt Fleming, for many of the numbers and financial background of the Fairchild story.
10. Malone, "Fear and Xenophobia."
11. Ibid.
12. "Why National Came to the Fairchild Fire Sale," *Business Week,* September 14, 1987.
13. Interview, Jay Cooper, New York (December 11, 1987).
14. Interview, Charles Sporck, New York (January 26, 1988).
15. "Hitachi to Open Dallas Plant," *Electronic News,* February 8, 1988.
16. "Duties, Strong Yen May Spur Japan IC Production Here," *Electronic Engineering Times,* July 14, 1986.
17. "Nippon Steel Funds Semicon Co.," *Electronic News,* May 18, 1987.
18. Kenichi Ohmae, quoted in "Japanese Takeovers a Trickle," *New York Times,* July 4, 1987.
19. The Mostek story is covered in an excellent two-part series "Mostek, '70s Memory Mogul, Passes into Oblivion in Merger," *Electronic News,* January 18 and 25, 1988.

20. David Sanger, "Bar by Pentagon Is Seen on Plessey Bid to Harris," *New York Times,* September 23, 1987.
21. Clyde V. Prestowitz, quoted by Steven Prokesch, "Stopping the High Tech Giveaway," *New York Times,* March 22, 1987.
22. Interview, Bernie Vonderschmitt, San Jose, Calif. (June 22, 1987).
23. Dataquest.
24. Again, Jay Cooper of Eberstadt Fleming is the source for much of my information on joint ventures in the semiconductor industry.
25. "Motorola to Use Toshiba Tech for 1-Megabit DRAM Facility," *Electronic News,* December 8, 1986.
26. Telephone interview, Jack Carsten (February 10, 1988).
27. Interview, William Egelhoff, New York (June 11, 1987).
28. David E. Sanger, "NEC Wants Part in U.S. Chip Project," *New York Times,* August 15, 1988.
29. Telephone interview, George H. Kuper (February 17, 1988).
30. Jerry Sanders, Speech at an AMD Sales Conference, August 1976.
31. Interview, Sheridan Tatsuno, San Jose, Calif. (June 22, 1987).

Chapter 11

1. Interview, Norman Augustine, Bethesda, Md. (July 13, 1987).
2. Charles Fowler, action memorandum to the secretary of defense, February 9, 1987.
3. Timothy Stone, quoted by Andrew Pollack, "Japan's Growing Role in Chips Worrying U.S.," *New York Times,* January 5, 1987.
4. *Foreign Production of Electronic Components and Army Systems Vulnerabilities* (Washington D.C.: National Academy Press, 1986).
5. Ibid.
6. Interview, Bill Finan, Washington, D.C. (June 3, 1987).
7. *Foreign Production.*
8. John Cushman, "Navy Sale to Japan Meets Opposition," *New York Times,* February 5, 1988.
9. Eliot Marshall, "American Weapons, Alien Parts," *Science,* October 10, 1986.
10. "Report of the Defense Science Board Task Force on Defense Semiconductor Dependency," Office of the Under Secretary of Defense for Acquisition, Washington, D.C., February 1987.
11. Interview, Sheldon Weinig, New York City and Orangeburg, N.Y. (1987–1988).
12. Telephone interview, George Krug (August 3, 1988).
13. August Witt quoted in Marshall, "American Weapons."
14. Telephone interview, Sven Roosild (June 30, 1987).
15. Interview, Bob Noyce, Santa Clara, Calif. (June 23, 1987).
16. Interview, Charlie Sporck, New York (January 26, 1988).

17. "AT&T Wins Gallium Arsenide Development Contract from Pentagon," AT&T press release, April, 20, 1987.
18. Marshall, "American Weapons."
19. Interview, Allen Willner, New York (May 15, 1987).
20. Interview, Richard Osgood, New York (May 15, 1987).
21. Sanger, David E., "Big Worries Over Small GCA," *New York Times,* January 19, 1987.
22. Interview, James Dykes, Hsin Chu, Taiwan (September 15, 1987).
23. Interview, Michael Dertouzos, Cambridge, Mass.: (November 18, 1983).
24. Interview, Ryozo Hayashi, New York (June 30, 1987).
25. Edward Feigenbaum, *The Fifth Generation, Artificial Intelligence and Japan's Challenge to the World* (Reading, Mass.: Addison-Wesley, 1983).
26. David Lammers, "Having It Both Ways," *Electronic Engineering Times.*
27. Interview, Sadao Inoue, Tokyo (September 9, 1987).
28. Pollack, "Japan's Growing Role."
29. Robert Price, quoted in "See Japan Threat to U.S. Super CPU Lead," *Electronic News,* February 29, 1988.
30. "Strategic Computing," Third Annual Report, DARPA, February 1987.
31. "VHSIC: The Very High Speed Integrated Circuits Program," Office of the Undersecretary of Defense for Research and Engineering, Department of Defense, 1985
32. Dan Hutcheson, quoted in "A Supercomputer on a Single Chip," *Fortune,* September 29, 1986.
33. *Foreign Production.*

Chapter 12

1. Interview, Sandy Kane, Harrison, N.Y. (February 10, 1988).
2. SEMATECH Press Conference, New York, January 26, 1988.
3. Ibid.
4. Ibid.
5. Telephone interview, Jack Clifford (February 3, 1988).
6. Bob Noyce, quoted by Jack Robertson, "Kabuki DRAM-a," *Electron News,* August 15, 1988.
7. Tsuyoshi Kawanishi, quoted in Robertson, "Kabuki DRAM-a,"
8. Interview, Bob Noyce, Santa Clara, Calif. (June 23, 1987).
9. Ibid.
10. Interview, Michael Attardo, Harrison, N.Y. (February 5, 1987).
11. Interview, Bernie Vonderschmitt, San Jose, Calif. (June 22, 1987).
12. Interview, Roger Altman, New York (June 1, 1987).
13. Interview, Jack Clifford, Washington, D.C. (June 4, 1987).
14. Interview, Sheldon Weinig, New York and Orangeburg, N.Y. (1987–1988).
15. Interview, Roy Takata, New York (August 7, 1987).

16. Nathaniel C. Nash, "Japan's Banks: Top 10 in Deposits," *New York Times,* July 20, 1988.
17. George Gould, "Arithritic Laws Are Crippling United States Banks," *New York Times,* August 12, 1987.
18. Interview, Richard Osgood, New York (May 15, 1987).
19. Interview, Allen Willner, New York (May 15, 1987).
20. Interview, Brian Davies, Palo Alto, Calif. (June 23, 1987).
21. Steven Solomon, "Will Old Machines Kill New Ideas?," *New York Times,* May 24, 1987.
22. Ibid.
23. "Teaching Engineering: Budget Woes," *New York Times,* June 8, 1988.
24. Solomon, "Old Machines."
25. "Foreign Citizens in U.S. Science and Engineering," National Science Foundation, Washington, D.C., 1987.
26. Interview, James Dykes, Hsin Chu, Taiwan (September 15, 1987).
27. *Foreign and Foreign Born Engineers in the United States* (Washington, D.C.: National Academy Press, 1988).
28. Interview, Kyu Tae Park, Seoul (September 18, 1987).
29. Interview, Norman Augustine, Bethesda, Md. (July 13, 1987).
30. Interview, Bob Noyce, Santa Clara, Calif. (June 23, 1987).
31. "U.S. Science Students Near Foot of Class," *Science,* March 11, 1988
32. Dataquest.